TRANSPORT PHENOMENA
IN MEDICINE AND BIOLOGY

TRANSPORT PHENOMENA IN MEDICINE AND BIOLOGY

MARSHALL MIN-SHING LIH

Division of Engineering
National Science Foundation
Washington, D.C.

and

National Biomedical Research Foundation
Georgetown University Medical Center
Washington, D. C.

A WILEY-INTERSCIENCE PUBLICATION

JOHN WILEY & SONS
New York · London · Sydney · Toronto

Library of Congress Cataloging in Publication Data:

Lih, Marshall Min-Shing, 1936–
   Transport phenomena in medicine and biology.

   (Biomedical engineering and health systems: a Wiley-Interscience series)
   "A Wiley-Interscience publication."
   Includes bibliographical references.
   1. Biological transport—Mathematical models.
I. Title. [DNLM: 1. Biological transport. 2. Biomedical engineering. QT34 L727t 1974]

QH509.L53   574.8′75   74-6059
ISBN 0-471-53532-X

Printed in the United States of America

10 9 8 7 6 5 4 3 2 1

This book is dedicated to my natural parents

Kun-Hou* and Marion Liang Lih

and my academic parents

Professor and Mrs. O. A. Hougen

*To his loving memory

Series Preface

> "To love
> is not to gaze steadfast
> at one another:
> it is to look together
> in the same direction"
>
> St. Exupery, *Terre des Hommes*

The provision of good universal health care has only just become a social imperative. Like such other interlocking systems as transportation, water resources, and metropolises, the recognition of the need to adopt a systems approach has been forced upon us by the evident peril of mutual ruin resulting if we do not. The required systems approach is necessarily interdisciplinary rather than merely multidisciplinary, but this always makes evident the very real gulf between disciplines. Indeed, bridging the gulf requires much more time and patience than necessary just to learn the other discipline's "language." These different languages simultaneously represent and hide the whole gestalt associated with any discipline or profession. Nevertheless sufficient bridging must be achieved so that interdisciplinary teams can tackle our complex systems problems, for such teams constitute the only form of "intelligence amplification" that we can presently conceive.

Fortunately the existence of urgent problems always seems to provide the necessary impetus to work together in the same direction, using St. Exupery's thought, and, indeed, as a result of this to understand and value each other more deeply.

This series started with its emphasis on biomedical engineering. It illustrates the application of engineering in the service of medicine and biology. Books are required both to educate persons from one discipline in what they need to know of others, and to catalyse a synthesis in the core subject generally known as biomedical engineering. Thus one set of titles will aim at introducing the biologist and medical scientist to the quantitatively based analytical theories and techniques of the engineer and physical scientist. This set covers instrumentation, mathematical modeling, signal and system analysis, communication and control theory, and computer simulation techniques. A second set for engineers and physical scientists will cover basic material on biological and medical systems with as quantitative and compact a presentation as is possible. The omissions and simplifications necessitated by this approach should be justified by the increased ease of transferring the information, and more subtly by the increased pressure this can bring to bear upon the search for unifying quantitative principles.

A second emphasis has become appropriate as society increasingly demands universal health care, because the emphasis of medicine is now rapidly shifting from individual practice to health care systems. The huge financial burden of our health care systems (currently some 6% of GNP and increasing at about 14% per year) alone will ensure that engineers will be called upon, for the technological content is already large and still increasing. Engineers will need to apply their full spectrum of methodologies and techniques, and, moreover, to work in close collaboration with management professionals, as well as with many different health professionals. This series, then, will develop a group of books for mutually educating these various professionals so that they may better achieve their common task.

This present volume presents a broad treatment of the problems involved in the transport phenomena of medicine and biology. It is becoming clear that a basic knowledge in this area is important to an increasing variety of problems. In treatment the book is partly complementary to an earlier volume in this series, namely, *Transport Phenomena in the Cardiovascular System* by Stanley Middleman.

JOHN H. MILSUM

Vancouver, 1973

Preface

The basic objective of this book is to introduce to the biomedical researcher or practitioner the physical principles underlying the various known or observed physiological phenomena, especially those involving the transport of mass, momentum, and energy, with the hope that this will make these phenomena more interesting and easily understood, so that rote memory will be reduced to a minimum.

At the risk of generalizing my personal experience, the biology class was not one of those I looked forward to during my high school days. Whether it was because of the inherent nature of the subject matter or because of the method of my teacher, I am not sure. I suspect a little of both, although it could well be that I was never intended to be a biologist or physician. In any case we were committed to memorize all body parts, bones, muscles, and so on, so that we could reproduce them in beautiful detail from memory. Consequently, we saw very little science in it.

However, my entire outlook changed 15 years later when I was presented the opportunity to work in the biomedical field after I had had my formal education plus research and teaching experience in the chemical engineering field. Now I could see the physical cause behind a biological effect. I had been freed from most of the rote-memory work because I could now derive most of the physiological phenomena from the very few laws of nature and fundamental principles through simple reasoning or mathematical logic. For example, I could now describe the cardiac cycle in a way similar to that I would use in describing to my students the working cycle of the automobile engine based on thermodynamic principles.

Furthermore, I can now view the entire human body as a miniature chemical plant for which I can simply draw a schematic block diagram. In analyzing the body function in compartments, I can treat each separate organ as a unit operation in a chemical plant. For instance, one can apply the same basic principles and methodology in analyzing heat exchangers to the analysis of placental and renal transports. In other words, by working with models, one can solve a basic problem and apply the result across discipline lines.

This interdisciplinary approach is especially appealing in processes and operations involving the transport of mass, momentum, and energy. Approximately two decades ago, based on the mathematical analogies between these three types of transport in engineering applications, these three subjects were gradually organized into a single unified subject called transport phenomena with the first publication of the monumental treatise by Bird, Stewart, and Lightfoot. It was greeted by sweeping enthusiasm.

The application of the same approach in the analysis of biological phenomena and in the design of medical apparatus such as oxygenators and artificial kidneys came as a natural and logical development because almost all physical phenomena involved in physiological systems are nothing but the transport of mass, momentum, and/or energy.* An increasing number of biomedical researchers have had the subject of transport phenomena as part of their basic training. This should not be surprising, however. After all, some of the pioneering and most important theoretical work and experimental observations on this subject have been made by physicians or medical researchers in their study of the human body, especially the circulatory system.

Since my involvement in biomedical work, I have had various opportunities, ranging from informal sessions to formal talks at national meetings, to discuss with my medical colleagues transport phenomena and its physiological applications—sometimes in exchange for some medical education for myself. I have found these occasions most challenging and rewarding. Through these cross-disciplinary contacts, a set of notes gradually evolved. Unfortunately, a series of personal and professional occurrences have sidetracked my attempt to organize it into a book until now.

In developing this volume, I have been guided by the thought that in order for it to be practical and serve a useful purpose, I must limit its length, and hence its scope, rather than be overly ambitious and try to be everything to everybody. Drawing on my own experience as well as that of some of my colleagues, I have chosen to concentrate on the description of the phenomena by using mathematical language and explaining them in physical terms based on the mathematical result. Thus, I have arranged the topics according to their physical, rather than physiological, classification. I regard this as a bold experiment on the biomedically oriented reader.

In addition to being a sourcebook for biomedical researchers and

*Another transport phenomenon, that of the transport of electrons, also falls within the same set of analogies with mass, momentum, and energy transfer. However, that is not within the scope of this book.

practitioners on problem solving, this should be an ideal textbook for a three-credit course to be offered between the third and fifth* years in one's professional training. Reasonable knowledge in anatomy and physiology is assumed, but only little in chemistry, physics, and mathematics beyond the freshman level is required to start. The review of mathematics (Chapter 2) should make this book reasonably self-contained. Considerable effort was spent in collecting and selecting appropriate examples and exercise problems from a wide range of applied fields but the instructor and students are encouraged to go beyond them.

In reviewing the development of this book, I must first express my deep appreciation to Prof. R. B. Bird for introducing me to the subject of transport phenomena. His excellent teaching and writing have greatly enriched my academic career. Thanks are also due to Dr. R. S. Ledley for initiating me into biomedical work; to Dr. L. E. Baker for providing the first opportunity to expose the subject to a national audience; to Dr. H. F. Foncannon and the American Association for the Advancement of Science for the opportunity of test-using the manuscript in the Chautauqua-type short course program; and to Dr. J. H. Milsum, Dr. Mary Conway, Mr. A. W. Frankenfield, Ms. Catherine Pace and the Wiley production staff for their continual interest and extraordinary patience. Many professional colleagues provided encouragement, assistance, and advice, especially Dr. H. K. Huang and Dr. Robert H. M. Kwok, F.A.A.P., Dr. L. W. Ross, and those others who have provided me with their publications or background discussions; unfortunately I cannot name them one by one. I would also like to thank Mr. James Bailor for final checking and Miss Bernadette Laniak for typing the manuscript, much of it under remote control. Lastly, but not the least, I deeply appreciate the understanding of my family— June, Matthew, Andrew, and Angela—who have been kind enough to let me ruin some of their vacations.

Much of the material has been collected as a result of my research sponsored by the National Institutes of Health and the National Science Foundation. The U.S. Department of Commerce under the State Technical Service Act also sponsored my Transport Phenomena for Biomedical Scientists Workshop. It is my sincere hope that this book is commensurate with all these moral and material supports. I welcome suggestions and criticisms.

M. M. Lih

Washington, D. C.
August 1974

---

*That is, first year in medical or graduate school.

Contents

TRANSPORT PHENOMENA
IN MEDICINE AND BIOLOGY

CHAPTER ONE

Introduction

1.1 TRANSPORT PHENOMENA: WHAT AND WHY

The concept of transport phenomena provides a concise and organized way of studying three different subjects concerning the transport of three distinctive physical entities—mass, momentum, and energy—based on their *mathematical* analogies. Although vastly different in origin and nature, processes involving the movements of these entities under similar circumstances can be described by the same mathematical formulation and, hence, can be solved using the same mathematical procedures and techniques. All these subjects—mass transfer (such as diffusion), momentum transfer (more commonly known as *fluid dynamics*), and energy transport (heat transfer)—are important in biomedical work because the human body is essentially a miniature biochemical plant in which all these processes and their interactions take place. Unfortunately, because of necessary specialization and the limitation on the length of education, most biomedical scientists and engineers master only one or two of these separate subjects. For instance, a physiologist might be an expert in diffusion, being interested in intercellular transport, while a mechanical engineer might specialize in fluid dynamics, heat transfer, or both, making him competent in hemodynamic work, such as design of artificial cardiac-assist devices.

However, in such a highly interdisciplinary field as biomedicine a researcher or practitioner can no longer perform adequately without crossing the boundaries of his own little domain. For instance, a physiologist studying the rate of reabsorption of solute and fluids by the walls of renal tubule would find acquaintance with the equation of motion in fluid dynamics, in addition to the laws of diffusion, essential to his work. Another studying thermoregulation of the body would find that both the convection due to fluid flow and the metabolic energy generation must be considered along with thermal conduction. On the other hand, an engineer designing a hemodialyzer must have a detailed knowledge of mass transport through membranes and of fluid mechanics, in addition to physiology and chemistry.

1

In this day and age of rapid technological development and the knowledge explosion, one can no longer afford the time to learn these three subjects one by one. Presenting them as a unified set and teaching them via analogies represent a great increase in efficiency, comparable to killing three birds with one stone. One needs only to learn the first subject well in detail; the other two should fall in place automatically.

It becomes quite obvious, then, that we, as biomedical researchers and practitioners, whether scientists or engineers, have much to gain by acquainting ourselves with the fundamental principles and problem-solving techniques in transport phenomena. Perhaps even more important, they also provide a common "language" for all those who are engaged in this worthwhile interdisciplinary endeavor—biomedical research and practice.

1.2 SCOPE

In this book we shall limit ourselves to presenting the fundamental principles and mathematical description of and solution to problems in transport processes as observed, especially those of physiological interests. Actual examples from the biomedical research literature will be used to illustrate these principles and to demonstrate the problem-solving techniques and procedures. With the exception of Chapters 9 and 10, little attempt will be made to discuss the advanced theories and mechanisms usually associated with certain physiological processes.

It must also be emphasized that we shall approach the subject from the continuum point of view, considering the fluid or tissue to be a uniform and continuous material. This is, of course, not exactly valid. In the first place, for instance, blood is a suspension of cells in plasma, a two-phase flow proposition. As another example, the tissue is dispersed with numerous fine capillary vessels from which it obtains its oxygen and nutritional supplies and to which it discharges its metabolic wastes, including carbon dioxide. However, we can simplify this complex situation by using such quantities as *apparent* viscosity, in the former instance, and *effective* thermal conductivity and *effective* diffusivity, in the latter, as will be discussed in Chapter 9 (Sections 9.2 and 9.3) and Sections 6.2 and 6.3, respectively. In short, by using an overall or average quantity, this approach treats a multiphase system as if it were a single phase.

Even without considering these "physically visible" nonuniformities, the continuum assumption is not exactly true at the molecular level. However, the dimensions of most systems encountered in biomedical work are so much above those of the molecules involved that the continuum approach becomes quite valid. For those who are interested in the molecular theory of transport phenomena, the monumental treatises of Hirschfelder, Curtiss, and Bird [1] and of Chapman and Cowling [2] are highly recommended.

1.3 TYPES OF BIOMEDICAL TRANSPORT PROBLEMS

As already pointed out in the Preface, in the spatial aspect, transport problems mainly fall into three categories:

1.   Differential (distributed parameter, which varies with position)
2.   Interphase
3.   Macroscopic (lumped parameter, which considers the system as a whole)

These will be discussed in Chapters 6, 7, and 8, respectively. However, in actual application we find it not only useful but sometimes necessary to look at a distributed-lumped parameter system [3] (which is a series of varying lumped-parameter subsystems).

As far as time-dependency is concerned, transport processes are classified into the *steady-state* (time-independent) and the *unsteady-state* (time-dependent); in the latter category two types are most prominent, the *start-up* and the *periodic*.

Of particular relevance in biomedical work is another dimension of classification based on the level of detail. We shall first list them:

1.   The living body as a single unit
2.   The body as a combination of functional systems
3.   The system as interconnected organs
4.   The individual organ or body part
5.   The compartments of organ or body part
6.   The parts of organ
7.   The individual cell
8.   The components of cell

By the mere consideration of all possible combinations of these classifications, we can precisely identify each problem as being of a special type represented by one of the boxes in the three-dimensional matrix in Figure 1.3-1.

In this book, particularly in Chapters 6–10, we shall discuss many examples from the research literature. Although it is impractical and would serve no useful purpose to cover all of the possible problem types in these examples, we will do many of them. As a preview, let us mention a few, so as to provide some perspectives.

When we think of the body as a single unit, one thing that immediately

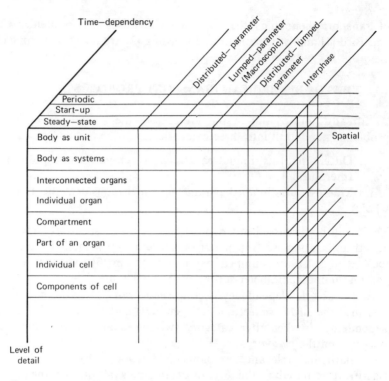

Figure 1.3-1. Matrix of types of transport problems.

comes to mind is the way a physician keeps track of his patients' daily food and water intake and output, amount of exercise, and the like. In engineering this is often referred to as the *black-box approach* and the type of analysis used is mainly the macroscopic one, although interphase transport also occurs every day in the form of body-heat dissipation and perspiration. Under normal conditions the input and response are periodic, following the biological rhythm. However, under pathological conditions they go through a transient stage and reach a new level of magnitude. At times they could be irregular. One analogy to this approach is that of a chemical plant of which the board of directors is only interested in the annual (or quarterly) sales and expenditure, raw material consumption, and production rate. It cares little, if at all, about either the technical or administrative details of this plant.

At the next level, the body is viewed as a combination of the circulatory, respiratory, digestive, nervous, and other systems, between which there are intimate relationships and interactions. The analogies between the circulatory system and the network of pipelines and between the neural system and the automatic control system in a chemical plant are so obvious that they hardly require any elaboration. Again, the analysis is usually macroscopic but encompasses both steady-state and time-dependent aspects.

In considering the individual organs or parts, we can view them spatially in one of many ways. Temperature profiles in limbs [4] [see Figure 6.2-10(a) and (b)] and oxygen distribution in the lung are examples of distributed-parameter systems usually described by differential equations. On the other hand, each half of the lung has also been modeled as a large, single alveolus [5] (Figure 1.3-2) similar to the way the stomach [6] (see Example 8.8-2) and the heart are treated as lumped-parameter systems.

If we treat each organ as a black box, then when we connect several of them into a system, we have a distributed-lumped parameter system such as the circulatory system schematically shown in Figure 1.3-3. This is similar to looking at a flow diagram of a section of a chemical plant in which different reactors, distillation columns, and filtration units are drawn as boxes connected by arrows representing stream flows. In this approach we are merely interested in the rates of input and output of each box, their characteristics, and how they affect each other. We do not care about the internal distribution of material or temperature.

When we study the transport of nutritional material and metabolic waste (including oxygen and carbon dioxide) between the blood and the tissue, we are reminded of the fact that each capillary is surrounded by an

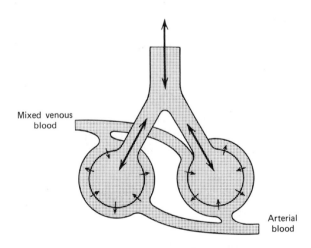

Figure 1.3-2. Two-alveolus model of the lung [5].

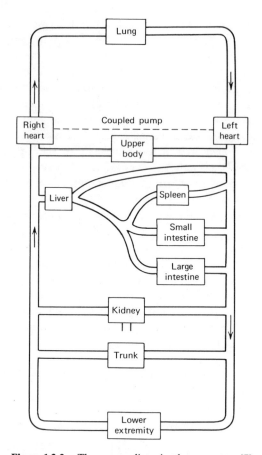

Figure 1.3-3. The mammalian circulatory system [7].

annular cylinder of tissue with interstitial fluid between them. Analysis of such a system, normally regarded as a distributed-parameter one, is relatively simple. But what about the millions of such units in each organ? How can we perform, and combine the results of, millions of such analyses in a meaningful way?

This is where the compartment approach comes in. In it we divide each organ into three "overall" compartments—capillary, interstitial, and cellular compartments, as shown in Figure 1.3-4—each of which is considered a lumped-parameter system. Quite often the longitudinal change of properties in the capillary compartment is quite large.* Thus, it is desirable

*This is because the capillaries, while in parallel, are approximately 1 mm in length. The oxygen tension, for example, drops from 100 to 40 mmHg in this short span.

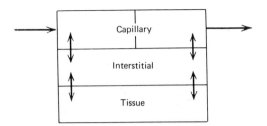

Figure 1.3-4. Schematic diagram of compartments of an organ.

sometimes to subdivide the capillary compartment into the arterial-end and venous-end sections, making a total of four for each organ.

There is no definite rule as to how compartments are to be divided. Based on a particular application or analysis, the same organ can be compartmentalized any number of ways. For instance, in studying the respiratory system, Warner and Seagrave [3] divide the simple macroscopic model [Figure 1.3-5(a)] into seven compartments as shown in Figure 1.3-5(b). Each of these so-called pools is considered as a stirred tank reactor. As mentioned earlier in this section, the entire system is of the distributed-lumped-parameter type.

Organs are not the only things that can be compartmentalized. Figure 1.3-6 shows a 54-compartment model for studying drug distribution in the mammalian body [8]; as shown, longer sections of the blood vessels are also divided into compartments. Of interest here is the difference in the volume distribution of the compartments. In the large vessels, longitudinal pressure drop is linear and there is no concentration decrease. Thus, the compartments are divided almost equally in volume. However, in the capillary compartments there is "lateral" mass transfer, so that the concentration tends to drop rapidly at first and then level off, perhaps somewhat in the exponential-decay fashion. Therefore, in order to divide equally the concentration drop, the first (arterial-end) compartment is made smaller than the second (venous-end) compartment, approximately 40% of the total combined volume. An implication of this statement is that if we were to study some other properties or characteristics, such as temperature distribution, whose behavior may be different from the above, we might have still another way of division. Again, in compartments we also have steady-state, as well as time-dependent, analysis.

More detailed still, but before reaching the cell level, we can consider fundamental constituent units of the organ, such as nephrons of the kidney (Figure 1.3-7) and alveolar sacs of the lung (see Example 10.4-1). In the former the problem is often similar to the macroscopic analysis of a heat exchanger, while in the latter, oxygen distribution in the primary lobule [10] is a distributed-parameter problem. Interphase transport is also involved in the former.

The cell is the ultimate building block of organs and tissues. At this level we tend to think of it as being so small that it can only be analyzed as a lumped-parameter system. However, bird eggs are often sufficiently large to warrant their being considered as distributed parameter systems [11]. At times even parts of it, such as the membrane, can be "enlarged" to be considered similar to the walls of an organ or blood vessel.

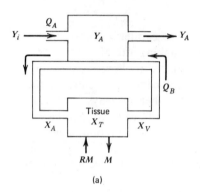

(a)

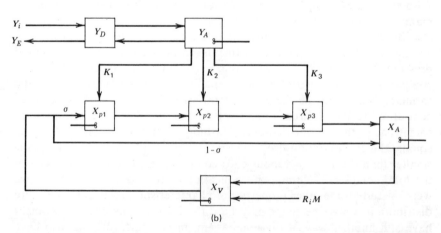

(b)

Figure 1.3-5. Macroscopic and compartment models of the respiratory system [3]. (a) Simplest macroscopic model: X_A in equilibrium with Y_A; X_V in equilibrium with X_T; $Q_B = Q_B(M)$; $Q_A = Q_A(M)$; R, respiratory quotient, ml. CO_2 produced/ml. O_2 consumed; M, metabolic oxygen consumption rate. (b) Distributed-lumped parameter respiratory model: Y_D, dead space pool; Y_A alveolar space, pool; X_{p1}, X_{p2}, X_{p3}, pulmonary capillary pools; X_A arterial pool; X_V venous-tissue pool; σ, fraction not shunted.

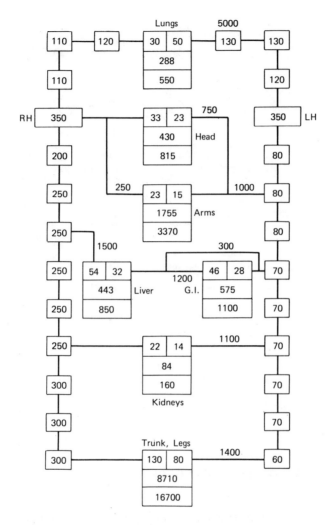

Figure 1.3-6. Fifty-four compartment model of the human body [8]. Numbers in boxes indicate volumes in cm³; those outside indicate flow rates in cm³/min.

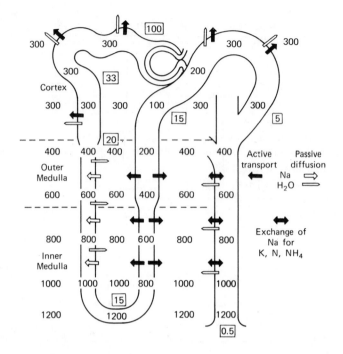

Figure 1.3-7. Schematic diagram of water and ion exchange in the nephron during elaboration of hypertonic urine [9]. Numbers in boxes indicate estimated percentages of glomerular filtrate remaining within tubule of various levels. Others indicate concentrations of tubular urine and peritubular fluid in milliosmoles per liter.

1.4 PROBLEM-SOLVING PROCEDURE

If one looks into the research literature of any field, including the biomedical areas, he will find that in a large number of projects, considerable funds and effort were expended, but the results obtained and conclusions reached are rather insignificant in both quantity and quality. This is because in many cases the author-researcher does not know how to perform the necessary theoretical analysis to fit his data. And worse yet, this lack of analytical ability makes it impossible for him even to identify the important variables and parameters he should have examined before the experimental work began. Thus, to improve one's theoretical problem-solving capability should be the first order of business for everyone, including the experimentalist, already in or preparing to enter the biomedical field.

Generally speaking, all problems in the physical sciences are solved by the combination of the equations of change and the equations of state, as will be explained later in Chapters 3 and 4. This principle is simple enough, and the mathematics involved is usually quite straightforward, albeit tedious at times. But, hardly a day passes without many students, and even scientists, encountering seemingly insurmountable difficulties in analyzing these problems in a logical manner. Many even have a psychological barrier against such problem solving. In most cases the cause can be traced to a lack of capability in translating the physical problem into mathematical language and setting up the working equations in reasonably simple form via the use of assumptions and insights. Such difficulties can be alleviated if one realizes that normally there is a logical sequence of steps to follow. One version of such a sequence is described below:

1.   Read the problem carefully and get the facts straight. Draw a diagram or sketch of the system as you go along. Also label on the diagram all pertinent quantities (in mathematical symbols).

2.   Invoke the governing physical laws and principles. This is where the equations of change and state come in. Examples are Newton's second law of motion, Fick's law of diffusion, and the conduction equation.

3.   If necessary, impose reasonable physical assumptions so as to simplify the problem. Make clear what the "model" is. For example, in the diffusion of γ-globulin in cylindrical muscle, if we neglect longitudinal concentration gradient, the model is that of an infinite cylinder.

4.   Translate the problem into working equations via mathematical symbolism. The use of dimensionless quantities to tidy up the equations is highly desirable.

5.   Again, impose mathematical assumptions if they are needed in order to render the problem mathematically solvable. For example, if the magnitudes of two additive terms differ by a factor of more than 10^2, the smaller one can be neglected except under extraordinary circumstances.

6.   Solve the problem mathematically. This is usually straightforward, and compared with the other steps, this is not the most important one, because if one cannot do it, he can always ask the help of a mathematician. But few mathematicians can take the responsibility of translating from biomedical to mathematical languages. The biomedical scientist has to do the setting-up himself.

7.   Check the solution by substituting it back into the problem (such as differential equation and boundary and initial conditions) to make sure that the result is mathematically correct (such as satisfying the differential equation and the boundary and initial conditions).

8.   Present the result in usable form such as graphs and dimensional equations.

9.   Interpret the result, such as investigating the limiting cases: If time reaches infinity, would it give reasonable results? And so on. Also, examine the error caused by the assumptions and the effect of their removal.

In almost all of the examples in this book, the reader will see how these steps are at work. Then he should find with even greater satisfaction how easily most of the exercise problems can be handled in a similar manner.

1.5 APPLICATIONS IN DESIGN

In case the above discussions have misled the reader to believe that transport-phenomena principles are useful only in analyzing physiological events in the living body, a few words are in order about their application in the engineering design of medical apparatus. Many of them, especially artificial kidneys and oxygenators, require detailed analysis of the heat and mass transfer along with fluid dynamics to achieve practicality and reliability. Even the seemingly theoretical analysis of axial migration of blood cells has important implications for the oxygen transport rate in oxygenators of certain designs. We shall not forget these applications. In fact, we will use some of them as examples in Chapter 8.

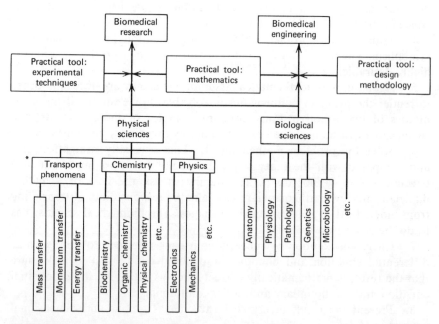

Figure 1.6-1.   Role of transport phenomena in biomedicine.

1.6 TRANSPORT PHENOMENA: WHERE

After discussing what the field of transport phenomena itself is and before we enter into detailed presentation of the subject matter, it is appropriate for us to have an overview of where it fits in the vast field of biomedicine. There may be a number of views on this, but that represented by Figure 1.6-1 seems to be a reasonable approach. In short, it is one of the physical foundations of the biomedical field. Physical and biological scientists and engineers would greatly benefit by cooperating and learning from one another. It is to be hoped that this book will contribute somewhat to this end, at least in breaking down the "language" barriers between these groups.

PROBLEMS

1.A Modeling of Body Material and Energy Balance

Draw a schematic diagram representing the material and energy intake and output of the human body as a whole.

1.B Body Water Balance

Draw separate diagrams on the water balance of the human body with numerical labels representing some extreme cases that may be of physiological interest as computed from the following typical daily data:

Intake

Metabolism	300∼ 400 ml
Food	1000∼1500 ml
Drinking	As needed
Total	2500 ml

Output

Respiration	400∼ 500 ml
Insensible loss	500∼ 800 ml
Urine	1000∼1500 ml
Feces	90∼ 200 ml
Perspiration	As required
Total	2500 ml

1.C Modeling of Fluid Transport in Plants

Draw a schematic diagram representing the transport and storage of fluids, including water, salts, amino acids, and photosynthetic products, in a typical plant.

1.D Idealized Model of a Leaf

Bugliarello [12] offers the diagram here of an idealized cross-section of a leaf. Label the arrows with the materials they represent and identify the important components in this model.

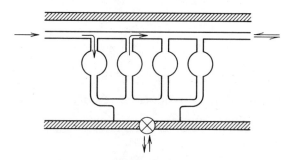

Figure 1.D.   Idealized model of a leaf [12].

1.E Modeling of Fluid and Solute Transport in the Nephron

Draw a schematic diagram of the nephron with peritubular capillaries showing the flow rates and concentrations of the substances in both the renal blood stream and urine involved in the filtration, secretion, and reabsorption processes. List all conditions assumed.

1.F Comparison of the Human Body with an Automobile

Draw two schematic diagrams, one representing the human body and the other an automobile, and compare them in the following aspects, among others you may think of:

1. Raw feed
2. Refine fuel
3. Energy generating process
4. Waste disposal

5. Functional systems (propulsion, control, respiratory, etc.)
6. Organs or body parts
7. Compartments
8. Cells.

REFERENCES

1. Hirschfelder, J. O.; Curtiss, C. F.; and Bird, R. B. *Molecular Theory of Gases and Liquids*, rev. ed. Wiley: New York, 1964.

2. Chapman, S., and Cowling, T. G. *Mathematical Theory of Non-Uniform Gases*, 2nd ed. Cambridge Univ. Press: Cambridge, 1951.

3. Warner, H. R., and Seagrave, R. C. In A. L. Shrier and T. G. Kaufmann, Eds., *Mass Transfer in Biological Systems*, Chem. Eng. Prog. Symp. Series, No. 99, Vol. LXVI. American Institute of Chemical Engineers: New York, 1970, p. 12.

4. Nevins, R. G., and Darwish, M. A. In J. D. Hardy, A. P. Gagge, and J. A. J. Stolwijk, Eds., *Physiological and Behavioral Temperature Regulation*. Thomas: Springfield, Ill., 1970, Chapter 21.

5. Comroe, J. H., Jr.; Forster, R. E., II; DuBois, A. B.; Briscoe, W. A.; and Carlsen, E. *The Lung: Clinical Physiology and Pulmonary Function Tests*, 2nd ed. Year Book: Chicago, 1962.

6. Öbrink, K. J., and Waller, M. *Acta Physiol. Scand.*, **63**, 175 (1965).

7. Bischoff, K. B., and Brown, R. G. In E. F. Leonard, Ed., *Chemical Engineering in Medicine*. Chem. Eng. Prog. Symp. Series. No. 66, Vol. LXII, 1966, p. 33.

8. Bischoff, K. B. In D. Hershey, Ed., *Chemical Engineering in Medicine and Biology*. Plenum: New York, 1967, p. 417.

9. Pitts, R. F. *Physiology of the Kidney and Body Fluids*. Year Book: Chicago, 1963.

10. Kylstra, J. A.; Paganelli, C. V.; and Lanphier, E. H. *J. Appl. Physiol.*, **21**(1), 177–184 (1966).

11. Kashkin, V. V. *Biophysics*, **60**, 97 (1961).

12. Bugiarello, G. In E. F., Leonard, Ed., *Chemical Engineering in Medicine*, Chem. Eng. Prog. Symp. Series, No. 66, Vol. LXII. American Institute of Chemical Engineers: New York, 1966, p. 49.

CHAPTER TWO
Mathematical Background

In this chapter we shall provide the reader with a uniform mathematical foundation for the material to be covered in this book. For some this chapter is unnecessary if they have already had adequate mathematical knowledge through their training as physical scientists or engineers. Even for those others who do not have such preparation it may not be necessary to study thoroughly the entire chapter if they need only a modest working knowledge to understand the main principles and theories. But, for those who wish to be able to follow all of the examples and do all of the problems, this chapter in effect makes this book self-contained. It must be emphasized, however, that because of limitations of space, we can only present the theorems, laws, equations, and the like as they are (the "recipe" approach), with a minimum amount of proof. Moreover, only the aspects most pertinent to the material in this book will be explained and in "layman's language" rather than formal mathematical jargon. The reader will be referrred to specific texts if he wishes to follow the rigorous approach or know the intricacies of the subject matters.

2.1. FUNCTIONS AND VARIABLES

We say that y is a function of x if for every value of x, there are one or more corresponding values of y.

For example, in dealing with the variation of the apparent viscosity, η_r, of blood relative to that of plasma with hematocrit (h, volume fraction of red blood cells in the whole blood) for a dilute blood suspension, we often use the simple relationship developed by Einstein [1] for spherical particles in a neutrally buoyant fluid:

$$\eta_r = 1 + Bh, \qquad (2.1\text{-}1)$$

where $B \approx 2.5$ as determined by Einstein and other investigators [2,3].* The relative viscosity η_r is then a function of the hematocrit h. In this particular case η_r is said to be a *single-valued function* of h because there is only one value of η_r for each value of h.

*The symbol $\approx$ means "approximately equal to."

16

On the other hand, functions can be *multiple-valued*. A simple example of this is the parabola $y^2 = x$. As graphically shown in Figure 2.1-1, except for $x = 0$, there are two values of $y(\pm\sqrt{x})$ for each value of x, since the squares of both $+\sqrt{x}$ and $-\sqrt{x}$ are x or y^2 i.e., whether or not the negative sign exists.

In the blood viscosity example above, since we can vary the blood cell concentration h at will and the viscosity η_r changes with it, we call h the *independent variable* and η_r the *dependent variable*.

Very frequently the dependent variable is a function of more than one independent variable. Well-known examples of this are the P–V–T (pressure–volume–temperature) relationship of gases and the dependence of the rate of dissipation of heat from the human body on the ambient temperature and humidity, wetness and temperature of the skin, wind velocity, and other factors. Such functions are called *multivariate functions*.

Variables and functions can be *continuous* much the same way we can (theoretically at least) obtain the exact degree of lighting desired by properly adjusting the dimmer. They can also be *discrete*, similar to the situation with a kitchen blender where we are limited to four or eight speeds obtainable through the selection buttons.

A function can be expressed mathematically (analytically), graphically, or in tabulated form. Equation 2.1-1 is a mathematical expression. A more exotic one is the following equation describing the axial velocity component, v_z, of the flow in the renal tubule when there is a linearly decreasing radial component [4]:*

$$v_z = \left[1 - \left(\frac{r}{R}\right)^2\right]\left\{\frac{2Q_0}{\pi R^2} - \frac{2}{R}(2a_0 z + a_1 z^2) - \frac{a_1 R}{2}\left[\frac{1}{3} - \left(\frac{r}{R}\right)^2\right]\right\}, \quad (2.1\text{-}2)$$

where r and z are, respectively, the radial and axial coordinates; R is the radius; Q_0 the volumetric flow rate at the entrance of the tubule; and a_0 and a_1 are, respectively, the zero-order and linear coefficients of the radial component function $a_0 + a_1 z$.

Graphical representations of functions are frequently most useful in providing insight to a problem or its solution. For example, in Michaelis–

---

*The author regrets having to introduce here many physical terms somewhat prematurely. We have to write them down to make the explanation complete. But a thorough understanding of these terms is not necessary at this point as the main objective here is to show the *appearance* of a function.

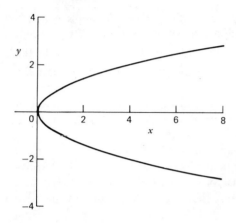

Figure 2.1-1. The parabola $y^2 = x$.

Menten kinetics involving competitive inhibition [5],

$$
\text{(1)} \qquad A_1 \qquad + \qquad A_2 \qquad \overset{k_1}{\underset{k_1'}{\rightleftarrows}} \qquad A_3
$$

Substrate   Free enzyme   Complex

site

$$
\text{(2)} \qquad\qquad\qquad A_3 \qquad \overset{k_2}{\underset{k_2'}{\rightleftarrows}} \qquad A_2 \qquad + \qquad A_4
$$

Complex   Free enzyme   Product

site

$$
\text{(3)} \qquad A_2 \qquad + \qquad A_6 \qquad \overset{k_3}{\underset{k_3'}{\rightleftarrows}} \qquad A_7
$$

Free enzyme   Poison   Dead-end

site   product

the production rate can be mathematically derived to be

$$
r_4 \left(\begin{array}{c} \text{rate of product} \\ \text{formation} \end{array} \right) = \frac{k_2 c_1 c_e}{K_I(1 + K_{III}c_6) + c_1}, \qquad (2.1\text{-}3)
$$

where c_i is the concentration of species i $(i = 1, 2, \ldots, 7)$, $c_e = c_2 + c_3 + c_7$, the

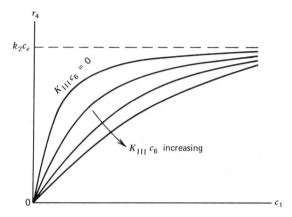

Figure 2.1-2.   Rate of product formation as a function of substrate concentration, according to Michaelis–Menten kinetics involving competitive inhibition.

total enzyme concentration; t is the time;

$$K_I = \frac{k_1' + k_2}{k_1}, \quad \text{and} \quad K_{III} = \frac{k_3}{k_3'}.$$

Although Eq. 2.1-3 may be difficult for some to perceive, its graphical representation, as shown in Figure 2.1-2 is very enlightening. It shows the variation of r_4 with respect to c_1. One can see from Eq. 2.1-3 that $K_{III}c_6$ also affects r_4. That is, for each $K_{III}c_6$ value we can plot an r_4 versus c_1 curve. The result is that we have a family of curves showing r_4 as a function of c_1 at various $K_{III}c_6$ values.

Figure 2.1-2 shows that at $K_{III}c_6 = 0$, the curve approaches an asymptotic value of $k_2 c_e$ for c_1 approaching infinity, while for small values of c_1, the curve is nearly a straight line (with a slope of $k_2 c_1 c_e / K_I$). These are the essential features of the "regular" Michaelis–Menten kinetics (i.e., without reaction 3, the reversible temporary poisoning reaction). This can easily be verified by setting $K_{III}c_6 = 0$ in Eq. 2.1-3. As $K_{III}c_6$ increases above zero, the other curves show the effect of the inhibition by different degrees of poisoning.

A family of curves such as those in Figure 2.1-2 enables us to relate three quantities without going into three-dimensional modeling, an example of which is the $P-V-T$ model available for teaching thermodynamics. However, if there are more than three quantities to be related, many more graphs would be needed were coverage to be complete. That is why quantities are grouped into dimensionless quantities whenever and wherever possible, as will be seen later throughout this book.

Tabulated representations are usually reserved for reporting experimental data, discrete variables and functions, and functions that do not have a theoretical and predictable behavior. These data can also be represented graphically except that they will not be in the form of smooth curves. Instead, they will appear as zig-zag lines and histograms. A well-known example is the stock-market index charts.

There are a number of frequently used functions we should be familiar with; they are listed in Table 2.1-1. Some are referred to in other sections, in which case we shall simply cite the sections. For others, we shall sketch their behaviors as well as give their definitions:

Table 2.1-1 Some Common Functions

Trigonometric	$\sin x$ (sine)*
	$\cos x$ (cosine)*
	$\tan x$ (tangent)
	$\cot x$ (cotangent)
Exponential	e^x (exponential rise)*
	e^{-x} (exponential decay)*
Logarithmic	$\ln x$ (natural log, base e)
	$\log x$ (common log, base 10)
Hyperbolic	$\sinh x$ (hyperbolic sine)*†
	$\cosh x$ (hyperbolic cosine)*†
	$\tanh x$ (hyperbolic tangent)†
	$\coth x$ (hyperbolic cotangent)†
Bessel functions	$J_0(x)$ (0 order, first kind)*
	$Y_0(x)$ (0 order, second kind)*
	$I_0(x)$ (0 order, modified, first kind)*
	$K_0(x)$ (0 order, modified, second kind)*

*For graphs, see Table 2.10-1; $e = 2.7182818\ldots = \lim_{n\to\infty}(1 + \frac{1}{n})^n$.
†$\sinh x = (e^x - e^{-x})/2$, $\cosh x = (e^x + e^{-x})/2$, $\tanh x = 1/\coth x = \sinh x/\cosh x$.

The trigonometric and hyperbolic tangent and cotangent functions are sketched in Figure 2.1-3, and the logarithmic functions, in Figure 2.1-4.

Lastly, we have to say a few words about the frequent confusion between the terms *constant, parameter,* and *variable.* Some seem to consider parameters as constants, and others consider them variables. No one should be blamed for this confusion, because one mathematical dictionary clearly (or, rather, unclearly) states that a parameter is "an arbitrary

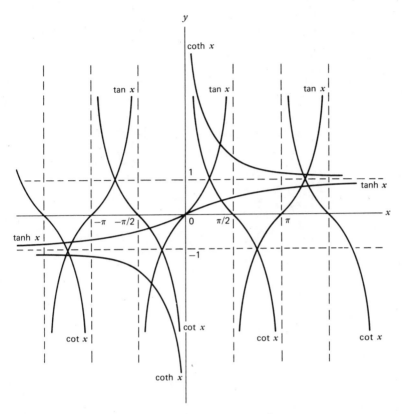

Figure 2.1-3. Some trigonometric and hyperbolic functions.

constant or a variable in a mathematical expression which distinguishes various specific cases."* It is obvious, then, that whether it is a constant or variable depends on circumstance and usage. For example in Eq. 2.1-3 when we are talking about the dependence of r_4 on c_1, we concentrate on the variation of r_4 with that of c_1, while maintaining $K_{III}c_6$ constant. However, that is not to say that $K_{III}c_6$ is a universal constant, because K_{III} can change with temperature and c_6 changes as the reaction progresses. We merely say that for each value of $K_{III}c_6$, we have an r_4 versus c_1 curve.

As further examples, in Eq. 2.1-2 we have the coordinates r and z as independent variables, a_0, a_1, Q_0, and R as parameters; in Eq. 2.1-1 B is not

*James, G., and James, R. C. *Mathematics Dictionary*, 3rd ed. Van Nostrand: Princeton, N.J., 1968.

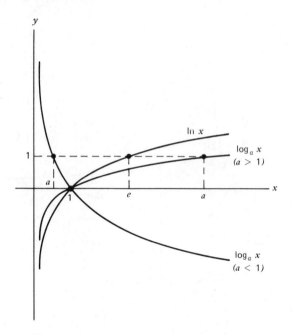

Figure 2.1-4. Logarithmic functions. $a = 10$ for common log: $\log x = \log_{10} x$; $a = e = 2.718$ for natural log: $\ln x = \log_e x$

a parameter but a universal constant—although it is subject to minor revisions as better experimental techniques are developed to measure it—and h is the independent variable.

2.2 DERIVATIVES AND DIFFERENTIATION

Let us begin by considering a simple single-valued continuous function

$$y = f(x).$$

For a certain value, x_0, of x we have a corresponding value of $y_0 = f(x_0)$. Now if we have a small deviation Δx away from x_0, we also have a corresponding deviation Δy away from y_0. Then we can write

$$\Delta y = f(x_0 + \Delta x) - f(x_0)$$

$$\Delta x = (x_0 + \Delta x) - x_0$$

so that

$$\frac{\Delta y}{\Delta x} = \frac{f(x_0 + \Delta x) - f(x_0)}{(x_0 + \Delta x) - x_0},\tag{2.2-1}$$

where $f(x_0 + \Delta x)$ means "$f(x)$ evaluated at $x = x_0 + \Delta x$."

An example of this can be found in the parabola $y = x^2$, as plotted in Figure 2.2-1. If we consider the point $H(1, 1)$ and exaggerate by taking a deviation in x of 2, we find point $A[(x_0 + \Delta x), y_0 + \Delta y = f(x_0 + \Delta x)]$ to be $(3, 9)$ and the slope of straight line a joining A and H is $\Delta y / \Delta x$, or $(9 - 1)/(3 - 1) = 4$, at this point. Now if we diminish the size of Δx from 2 to 1, we have point $B(2, 4)$ and the straight line b joining B and H has a slightly smaller slope; that is, $\Delta y / \Delta x = (4 - 1)/(2 - 1) = 3$. Now if we continue diminishing the size of Δx, we will find the succession of points A, $B, \ldots$, gradually approaching H and the slope of the joining line gradually decreasing—that is, until this point merges with H. At this time we write

$$\frac{dy}{dx} = \lim_{\Delta x \to 0} \frac{\Delta y}{\Delta x} = \lim_{\Delta x \to 0} \frac{f(x_0 + \Delta x) - f(x_0)}{\Delta x},\tag{2.2-2}$$

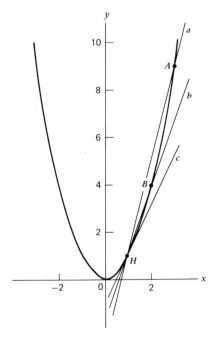

Figure 2.2-1. Derivative as tangent to the parabola $y = x^2$.

where dy/dx is called the *derivative* of y with respect to x at $x = x_0$. More specifically, we call it the *first* derivative. The expression $\lim_{\Delta x \to 0} \Delta y / \Delta x$ is said to represent the "limit of $\Delta y / \Delta x$ as Δx approaches zero." From Figure 2.2-1, clearly dy/dx is the slope of the tangent (line c) to the parabola at $(1,1)$. This is the mathematical interpretation of the first derivative.

Physically, if we define the following:

$c(t)$ = total concentration (as a function of time)
$s(t)$ = distance (as a function of time)
$v(t)$ = velocity (as a function of time)
$p(x)$ = pressure (as a function of position)
$T(x)$ = (steady-state) temperature (as a function of position),

then the following first derivatives have these meanings:

$\dfrac{dc}{dt}$ = time rate of change of total concentration

$\dfrac{ds}{dt}$ = velocity v

$\dfrac{dv}{dt}$ = acceleration

$\dfrac{dp}{dx}$ = pressure gradient

$\dfrac{dT}{dx}$ = temperature gradient,

where by "gradient" we mean "change per unit distance."

These are but a few examples of the derivative. We must note here that in reality quantities such as temperature and velocity can be functions of *both* position and time. This we shall discuss in Section 2.6 and Chapter 4 but not at the present time.

Next, we must ask ourselves how we obtain the derivative, given a specific function, such as

$$y = x^3.$$

According to Eq. 2.2-1, we must first compute

$$\frac{\Delta y}{\Delta x} = \frac{(x_0 + \Delta x)^3 - x_0^3}{\Delta x},$$

which after expansion and cancellation gives

$$\frac{\Delta y}{\Delta x} = 3x_0^2 + 3x_0 \Delta x + \Delta x^2.$$

Now, if we let Δx approach zero, as suggested by Eq. 2.2-2, we can obtain

$$\frac{dy}{dx} = \lim_{\Delta x \to 0} \frac{\Delta y}{\Delta x} = 3x_0^2. \qquad (2.2\text{-}3)$$

This means that at a specific point x_0, the derivative dy/dx is equal to $3x_0^2$. Then, at an arbitrary point x, the derivative of $y = x^3$ is

$$\frac{dy}{dx} = f'(x) = 3x^2, \qquad (2.2\text{-}4)$$

in which $f'(x)$ is another way of expressing the "derivative of $f(x)$ with respect to x."

From this and similar functions, we can see that the derivative of a power function is equal to the exponent multiplied by the function with subsequent lowering of the power by 1. Mathematically, we write

$$\frac{d}{dx}(x^n) = nx^{n-1}. \qquad (2.2\text{-}5)$$

This process we call *differentiation*.

Now if we have a longer function such as

$$y = x^3 - 6x^2 + 9x + 1, \qquad (2.2\text{-}6)$$

following a similar procedure to the above, we can find

$$\frac{dy}{dx} = 3x^2 - 12x + 9. \qquad (2.2\text{-}7)$$

Comparing Eqs. 2.2-6 and 2.2-7 we can find the following additional rules:

1. The derivative of a constant is zero.
2. The constant coefficient in front of a term does not interfere with the differentiation process. We simply multiply it with the exponent when the latter "comes down."
3. The derivative of the sum of several terms is equal to the sum of the derivatives of the individual terms.

All of these rules can be covered by a more general mathematical formula

$$\frac{d}{dx}(ax^n + bx^{n-1} + \cdots + px^2 + qx + r)$$

$$= nax^{n-1} + (n-1)bx^{n-2} + \cdots + 2px + q, \qquad (2.2\text{-}8)$$

where $a, b, \ldots, n, \ldots, p, q$, and r are constants.

Table 2.2-1 Important Formulas for Differentiation*
($a, b =$ constants)

General

(a) $\dfrac{da}{dx} = 0$

(b) $\dfrac{d}{dx}[af(x) + bg(x)] = af'(x) + bg'(x)$

(c) $\dfrac{d}{dx}[f(x)g(x)] = f(x)g'(x) + f'(x)g(x)$

(d) $\dfrac{d}{dx}\left[\dfrac{f(x)}{g(x)}\right] = \dfrac{g(x)f'(x) - f(x)g'(x)}{[g(x)]^2}$

(e) $\dfrac{d}{dx}\sqrt{f(x)} = \dfrac{1}{2\sqrt{f(x)}} \cdot \dfrac{df}{dx}$

Specific

(1) $\dfrac{d}{dx}(x^n) = nx^{n-1}$

(2) $\dfrac{d}{dx}(\sin x) = \cos x$

(3) $\dfrac{d}{dx}(\cos x) = -\sin x$

---

*The usefulness of this table can be greatly expanded if we
consider $h = h(f)$ and $f = f(x)$; then

$$\frac{d}{dx}\{h[f(x)]\} = \frac{dh}{df} \cdot \frac{df}{dx}.$$

Therefore, for example, $(d/dx)\sin u = \cos u(du/dx)$ if $u = u(x)$.

There are more rules, some of which govern general relationships such
as the product or quotient of functions, while others deal with specific
functions such as e^x, $\sin x$, $\cos x$ and $\ln x$. We do not have the space to
derive them one by one. Instead, we summarize the important ones in
Table 2.2-1 so that the reader will have a handy reference. More complete
tables are available elsewhere [6, 7, 8].

Table 2.2-1 (*Continued*)

---

(4) $\dfrac{d}{dx}(\tan x) = \sec^2 x$

(5) $\dfrac{d}{dx}(\sinh x) = \cosh x$

(6) $\dfrac{d}{dx}(\cosh x) = \sinh x$

(7) $\dfrac{d}{dx}(\tanh x) = \text{sech}^2 x$

(8) $\dfrac{d}{dx}(\sin^{-1} x) = \dfrac{1}{\sqrt{1-x^2}}$ †

(9) $\dfrac{d}{dx}(\cos^{-1} x) = -\dfrac{1}{\sqrt{1-x^2}}$ †

(10) $\dfrac{d}{dx}(\ln x) = \dfrac{1}{x}$ $(x > 0)$

(11) $\dfrac{d}{dx}(e^x) = e^x$

(12) $\dfrac{d}{dx}(a^x) = a^x \ln a$

---

† The reader should remember that $\sin^{-1} x$ does *not* mean $1/\sin x$, nor does $\cos^{-1} x$ mean $1/\cos x$. It simply means $\arcsin x$ and $\arccos x$, or the inverse of $\sin x$ and $\cos x$, respectively; that is, if $\sin^{-1} x = y$, $\cos^{-1} x = z$, then $x = \sin y = \cos z$. The same applies to $\tan^{-1} x$, $\sinh^{-1} x$, $\cosh^{-1} x$, and so on.

2.3 THE SECOND DERIVATIVE: MAXIMUM AND MINIMUM

If we further differentiate the first derivative, we obtain the second derivative. For example, the second derivative of the function in Eq. 2.2-6

$$y = x^3 - 6x^2 + 9x + 1$$

is

$$\frac{d^2 y}{dx^2} = \frac{d}{dx}\left(\frac{dy}{dx}\right) = \frac{d}{dx}(3x^2 - 12x + 9) = 6x - 12.$$

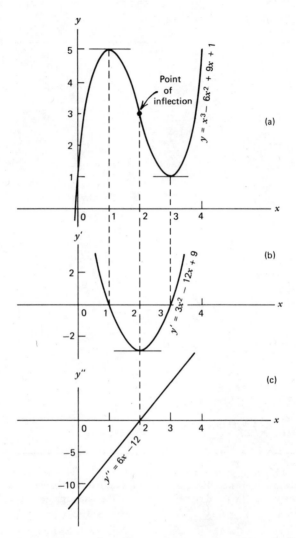

Figure 2.3-1. The function $y = x^3 - 6x^2 + 9x + 1$ and its derivatives. Note the differences in the vertical scales.

28

It is appropriate to examine further the meanings of the derivatives. For this, we have plotted this function and both its first and second derivatives against the same x scale (Figure 2.3-1). First of all we can see that whenever the function $f(x)$ has a maximum (on top of a hump) or minimum (at the bottom of the valley) its first derivative $f'(x)$ is always zero.* This is naturally true because we can see, either from the $f'(x)$ chart (b) or by drawing tangents all along the $f(x)$ curve, that in the neighborhood of the maximum the slope of $f(x)$ decreases from large to small positive values and finally to negative values, with increasing x. In between, it goes through $f'(x) = 0$ as shown on Part (b) of the charts and the horizontal tangent on top of the maximum of the $f(x)$ curve. The opposite is true for the minimum about which the slope increases from negative, passing through zero and into the positive region. Again we can see the horizontal tangent line, or $f'(x) = 0$, at the point where the minimum occurs.

Then one question may arise: Since both the maximum and the minimum occur when the slope, or the first derivative, is equal to zero, how can we tell one from the other without the graphical representation? This can be answered by observing the second derivative. We can see that in the neighborhood of the maximum, $f'(x)$ is always decreasing and its slope, the second derivative $f''(x)$, is always negative, as shown in both charts (b) and (c). No doubt $f''(x)$ is constantly increasing, but before it reaches $x = 2$, it is negative. Again the opposite is true for the minimum. Therefore, we can conclude that when $dy/dx = 0$ and if $d^2y/dx^2 = (-)$ then $y = f(x)$ has a local maximum; but if $d^2y/dx^2 = (+)$, then $y = f(x)$ has a local minimum. In short, when the slope is zero the extremum is a maximum if the curve is concave downward (i.e., when the curve lies beneath the horizontal tangent); the extremum is a minimum if the curve is concave upward (i.e., when it lies above the horizontal tangent).

One further point can be most conveniently illustrated by imagining ourselves sitting in a car climbing and descending a hill as represented by this curve $y = x^3 - 6x^2 + 9x + 1$. Suppose we can carefully control the car such that its horizontal velocity component (i.e., on the x scale) is constant. We can feel that as we ascend the hill starting from $x = 0$, the rate of ascent will decrease until we reach the top when we can no longer climb further (rate of ascent = 0) and start downhill. As we do this, we pick up speed because of the increasingly negative slope until we reach $x = 2$, at which point the rate of descent reaches a maximum. Beyond this point, we are

---

*More specifically, we call it *local* maximum or minimum, because it is only relative to points within its immediate vicinity. For higher-order functions, there may be several maxima and minima and only one of them will be the largest and one smallest, the true maximum and minimum.

still descending but at a decreasing rate of descent until it reaches 0 (at $x = 3$), where we again start climbing. The point at which the increase of rate of descent (or ascent, in an opposite case) is 0 is called the *point of inflection*, because the curve changes from being concave downward to concave upward (or vice versa).

This illustration also points up a common usage of the derivatives. If they are derivatives of position, s, with respect to time, t, then d^2s/dt^2 is the *acceleration*. When d^2s/dt^2 is negative, we are decelerating.

A useful theorem involving derivatives is L'Hopital's rule for evaluating an indeterminate expression. This rule states that in evaluating a quotient involving a parameter t, such as $f(t)/g(t)$, at a certain value of t, if this value of t (say, t_1) causes the entire expression to be indeterminate, such as $0/0$ or ∞/∞, it can be calculated by taking the quotient of the derivatives; that is,

$$\lim_{t \to t_1} \frac{f(t)}{g(t)} = \lim_{t \to t_1} \frac{df/dt}{dg/dt}. \qquad (2.3\text{-}1)$$

If the limit of the quotient of derivatives is still indeterminate, we can continue taking higher derivatives until a definite value is reached.

2.4 INTEGRATION

In terms of mathematical procedure, integration is the inverse of differentiation. Thus, if we want to find the integral of, say, x^3, we are really asking ourselves: If we considered x^3 as being the derivative $f'(x)$, what would be $f(x)$? Of course, this particular function is not too difficult to "reconstruct"; obviously it should be $\frac{1}{4}x^4$ because it gives x^3 back upon differentiation.

However, an apparent uncertainty occurs here. That is, we will be equally correct if we say that the integral of x^3 is $\frac{1}{4}x^4 + 5$—or, in fact, $\frac{1}{4}x^4$ plus any constant, because the derivative of a constant is zero anyway.* Therefore, it is customary to add the arbitrary constant C at the end of all integration formulas. Thus, in this case we formally write

$$\int x^3 \, dx = \tfrac{1}{4}x^4 + C.$$

This constant is to be evaluated by using the boundary or initial condition. One integration constant is generated for each integration step taken.

*This is also apparent in Section 2.3 where we can see that changing the value of the constant term in $x^3 - 6x^2 + 9x + 1$ will only displace the curve in Figure 2.3-1 (a) vertically, while changing absolutely nothing in Figure 2.3-1 (b) and (c).

Thus, if we start integrating from a second derivative, two boundary conditions are needed to evaluate the two integration constants produced.

As in the case of differentiation, there are various standard formulas for integration, the most important of which we list in Table 2.4-1. Again, the reader is referred to standard tables [6, 7, 8] for more complete listings.

Formula (c) in Table 2.4-1, generally known as the *integral-by-parts* formula, is one of the most useful because it can combine with other formulas to perform integrations not listed, thereby greatly expanding the utility of the table. For example, if we want to find the integration $\int x \cosh x \, dx$, we can regard x as u and $\cosh x \, dx$ as dv, so that $du = dx$ and $v = \sinh x$. To complete it, we write

$$\int \underbrace{x}_{u} \underbrace{\cosh x \, dx}_{dv} = \underbrace{x}_{u} \underbrace{\sinh x}_{v} - \int \underbrace{\sinh x}_{v} \underbrace{dx}_{du}$$

$$= x \sinh x - \cosh x + C. \qquad (2.4\text{-}1)$$

From this, we can also show that in general, we can write

$$\int \underbrace{x^n}_{u} \underbrace{\cosh x \, dx}_{dv} = \underbrace{x^n}_{u} \underbrace{\sinh x}_{v} - \int \underbrace{(\sinh x)}_{v} \underbrace{n x^{n-1} dx}_{du}$$

$$= x^n \sinh x - n \int x^{n-1} \sinh x \, dx. \qquad (2.4\text{-}2)$$

We can see that we have now lowered the power on x by 1. Thus, as long as n is an integer, we can perform this operation n times in succession until we reach $\int x^{n-n} \sinh x \, dx = \int \sinh x \, dx$, whereupon we can use formula (15) in Table 2.4-1 directly. This is what makes the integration-by-parts formula so useful. An important application of it in hemorheology is in the derivation of the Rabinowitsch equation (see Eq. 9.1-16).

The integral of a function $f(x)$ between two—upper and lower—limits is called a *definite integral* and written as $\int_a^b f(x)dx$, which in mathematical procedure means "evaluate the integrated function at upper-limit value b and subtract from it that evaluated at lower-limit value a." Geometrically, it gives the area under the curve represented by the function between the two limits. To illustrate this, we divide the entire area into n small vertical strips each Δx in width* as shown in Figure 2.4-1. For each of these strips, we choose a medium value, say, x_i, corresponding to which we have a value of the function $f(x_i)$. The area of the strip is approximately equal to

---

*Actually these strips do not have to be of equal width, because they will be shrunk to infinitesimal width anyway. But we do this to make the explanation simpler.

Table 2.4-1 Important Formulas for Integration

(a, b = constants; u, v = functions of x; C = integration constant)

General

(a) $\quad \int a\,dx = ax + C$

(b) $\quad \int (au + bv)\,dx = a\int u\,dx + b\int v\,dx$

(c) $\quad \int u\,dv = uv - \int v\,du$

Specific

(1) $\quad \int x^n\,dx = \dfrac{x^{n+1}}{n+1} + C$

(2) $\quad \int \dfrac{dx}{x} = \ln x + C$

(3) $\quad \int e^{ax}\,dx = \dfrac{1}{a}e^{ax} + C$

(4) $\quad \int \ln x\,dx = x\ln x - x + C$

(5) $\quad \int \dfrac{dx}{a^2 + x^2} = \dfrac{1}{a}\tan^{-1}\dfrac{x}{a} + C$

(6) (7) $\quad \displaystyle\int \dfrac{dx}{a^2 - x^2} = \begin{cases} \dfrac{1}{a}\tanh^{-1}\dfrac{x}{a} + C \\[2mm] \dfrac{1}{2a}\ln\dfrac{a+x}{a-x} + C \quad (a^2 > x^2) \end{cases}$

(8) (9) $\quad \displaystyle\int \dfrac{dx}{x^2 - a^2} = \begin{cases} -\dfrac{1}{q}\coth^{-1}\dfrac{x}{a} + C \\[2mm] \dfrac{1}{2a}\ln\dfrac{x-a}{x+a} + C \quad (x^2 > a^2) \end{cases}$

(10) $\quad \int \sin x\,dx = -\cos x + C$

(11) $\quad \int \cos x\,dx = \sin x + C$

(12) $\quad \int \tan x\,dx = \ln \sec x + C$

(13) $\quad \int \sin ax \sin bx\,dx = \dfrac{\sin(a-b)x}{2(a-b)} - \dfrac{\sin(a+b)x}{2(a+b)} + C$

(14) $\quad \int \cos ax \cos bx\,dx = \dfrac{\sin(a-b)x}{2(a-b)} + \dfrac{\sin(a+b)x}{2(a+b)} + C$

(15) $\quad \int \sinh x\,dx = \cosh x + C$

Table 2.4-1 (*Continued*)

(16) $\int \cosh x \, dx = \sinh x + C$

(17) $\int e^{ax} \sin nx \, dx = \dfrac{e^{ax}(a \sin nx - n \cos nx)}{a^2 + n^2} + C$

(18) $\int e^{ax} \cos nx \, dx = \dfrac{e^{ax}(n \sin nx + a \cos nx)}{a^2 + n^2} + C$

(19) $\int x^n \sin ax \, dx = -\dfrac{x^n}{a} \cos ax + \dfrac{n}{a} \int x^{n-1} \cos ax \, dx$

(20) $\int x^n \cos ax \, dx = \dfrac{x^n \sin ax}{a} - \dfrac{n}{a} \int x^{n-1} \sin ax \, dx$

the rectangular area $f(x_i) \cdot \Delta x$ because the strip is very narrow, so that the areas "gained" and "lost" by this approximation are approximately equal to each other. In other words, we can write:

$$\left\{ \begin{array}{l} \text{Area under curve} \\ \text{between } a \text{ and } b \end{array} \right\} \approx f(x_1) \cdot \Delta x + f(x_2) \cdot \Delta x + \cdots$$

$$+ f(x_i) \cdot \Delta x + \cdots + f(x_n) \cdot \Delta x$$

$$= \sum_{i=1}^{n} f(x_i) \cdot \Delta x, \qquad (2.4\text{-}3)$$

where the $\sum_{i=1}^{n}$ (summation) sign means "sum over all terms as i varies from 1 to n." This is called the *summation* process.

If we refine the division of this area to make more and narrower strips, our approximation becomes increasingly better until these strips become infinitesimally narrow or infinite in number, at which point Eq. (2.4-3) becomes exact. Thus, we write

$$\left\{ \begin{array}{l} \text{Area under curve} \\ \text{between } a \text{ and } b \end{array} \right\} = \int_a^b f(x) \, dx = \lim_{\substack{\Delta x \to 0 \\ n \to \infty}} \sum_{i=1}^{n} f(x_i) \cdot \Delta x. \qquad (2.4\text{-}4)$$

We can see that this "limit" process is similar to that in the definition of differentiation (Eq. 2.2-3), except that now it is used in the reverse operation.

Physically, the integral has various meanings depending on the application. For example, in Figure 2.4-2 if v is velocity and it varies with time t as shown, then the area under this curve, which does not have to be

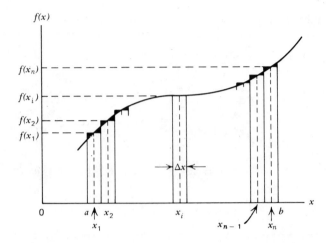

Figure 2.4-1. Summation as an approximation of integration.

regular, represents the distance traveled between times $t = t_0$ and $t = t_1$. We can thus write

$$s = \int_{t_0}^{t_1} v\, dt$$

for this case. Again we can see that this is just the opposite of the meaning of differentiation regarding velocity and distance (Section 2.2).

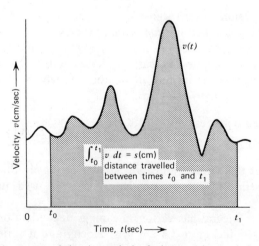

Figure 2.4-2. Distance traveled as integral of velocity over time period t_0 to t_1.

2.5 INFINITE SERIES

A powerful method of expressing functions is by *infinite series*. Quite often a function can be analytically integrated only if it is expanded in infinite series, as will be shown later in Example 2.5-3.

Normally an infinite series is expressed as

$$u_1 + u_2 + u_3 + \cdots + u_n + \cdots \qquad (2.5\text{-}1)$$

or, in shorthand form,

$$\sum_{n=1}^{\infty} u_n, \qquad (2.5\text{-}2)$$

where the summation operation is from $n = 1$ to $n = \infty$.

A series is *convergent* if there is a finite value for the sum of the series. According to Cauchy's principle, a necessary and sufficient condition for this to be true is that for each positive number ϵ, no matter how small, there exists a positive integer N such that $|S_n - S_m| < \epsilon$ whenever $m \geqslant N$ and $n \geqslant N$, where S_k is the *partial sum* of the first k terms of the series.

There are a number of ways of testing convergence and divergence. The simplest is by comparing a series with another that is known to be convergent or divergent. If each term of the series to be tested is less than, or equal to, the corresponding term of a known convergent series, then it is also a convergent one. On the other hand, if each term of it is greater than, or equal to, the corresponding term of a known divergent series, it is divergent.

However, this applies only to series with positive terms. For *alternating series* such as

$$u_1 - u_2 + u_3 + \cdots + (-1)^{n+1} u_n + \cdots, \qquad (2.5\text{-}3)$$

where $u_1, u_2, \ldots$ are all positive, it is convergent if $u_{n+1} \leqslant u_n$ for all n and if $\lim_{n \to \infty} u_n = 0$. It is also true that an alternating series is convergent if the corresponding series of absolute values of its terms is convergent.

But the most powerful test of convergence is the *ratio test*, which states that the infinite series

$$u_1 + u_2 + u_3 + \cdots + u_n + \cdots$$

converges if

$$\lim_{n \to \infty} \left| \frac{u_{n+1}}{u_n} \right| < 1$$

and diverges if

$$\lim_{n\to\infty}\left|\frac{u_{n+1}}{u_n}\right|>1.$$

If

$$\lim_{n\to\infty}\left|\frac{u_{n+1}}{u_n}\right|=1,$$

it yields no information.

Example 2.5-1. In the *geometric* series

$$1+x+x^2+\cdots+x^n+\cdots,$$

if $x<1$, it converges and has $1/(1-x)$ as its sum, a finite value. In this case, we can show, algebraically,

$$\frac{1}{1-x}=1+x+x^2+\cdots. \tag{2.5-4}$$

At other times, a function cannot be expanded into series so simply. In these cases, we resort to the Taylor's theorem, which can be written*

$$f(x)=f(a)+f'(a)(x-a)+\frac{f''(a)}{2!}(x-a)^2+\cdots$$

$$+\frac{f^{(n)}(a)}{n!}(x-a)^n+\cdots, \tag{2.5-5}$$

a special case of which is Maclaurin's series, when $a=0$:

$$f(x)=f(0)+f'(0)\cdot x+\frac{f''(0)}{2!}x^2+\cdots+\frac{f^{(n)}(0)}{n!}x^n+\cdots. \tag{2.5-6}$$

*$n!$ is called "n factorial" and is equal to $n(n-1)(n-2)\cdots3\cdot2\cdot1$ and $f^{(n)}(x)$ is the nth derivative of $f(x)$.

Example 2.5-2. Following Eq. 2.5-6, we can, for example, expand $\sin x$ into series form. If we write

$$f(x) = \sin x, \qquad f(0) = 0$$
$$f'(x) = \cos x, \qquad f'(0) = 1$$
$$f''(x) = -\sin x, \qquad f''(0) = 0$$
$$f'''(x) = -\cos x, \qquad f'''(0) = -1$$
$$f''''(x) = \sin x, \qquad f''''(0) = 0,$$

$$\vdots \qquad\qquad \vdots$$

then

$$\sin x = x - \frac{x^3}{3!} + \frac{x^5}{5!} - \frac{x^7}{7!} + \cdots. \qquad (2.5\text{-}7)$$

Following a similar procedure, many more functions can be expanded into series, as shown in Table 2.5-1. It may be of interest to note in the table that, for example, formulas (7) and (8) can actually be obtained from (1) and that (2) is only a special case of (1) (for $a = e = 2.718\ldots$). These series can also be used to verify identities between trigonometric and hyperbolic trigonometric functions, as shown in the following example.

Example 2.5-3. In Table 2.5-1, formulas (1), (5), and (6) are considered fundamental because they come from the Maclaurin series and nothing else. Substituting (1) into the fundamental definitions for hyperbolic sine and cosine functions (see footnote of Table 2.1-1), (7) and (8) can be obtained. If we further substitute ix for x in (7) and (8), we have *

$$\sinh ix = ix - \frac{ix^3}{3!} + \frac{ix^5}{5!} - \frac{ix^7}{7!} + \cdots$$

$$\cosh ix = 1 - \frac{x^2}{2!} + \frac{x^4}{4!} - \frac{x^6}{6!} + \cdots.$$

Comparing them with Eqs. (5) and (6) in Table 2.5-1, we can conclude that

$$\sinh ix = i \sin x \qquad (2.5\text{-}8)$$

$$\cosh ix = \cos x. \qquad (2.5\text{-}9)$$

*Remembering that $i^2 = -1$, $i^4 = (-1)^2 = 1$, $i^6 = i^2 = -1,\ldots$.

This is the same basis on which we can interchange the $e^{imx} \sim e^{-imx}$ type solution to a differential equation with the $\sin mx \sim \cos mx$ type and the $e^{mx} \sim e^{-mx}$ type with the $\sinh mx \sim \cosh mx$ type in Section 2.10 (with different constants C_1 and C_2, C_1' and C_2', of course).

Example 2.5-4. As an example of the practical utility of the infinite series, let us consider the velocity distribution of the flow of suspension in

Table 2.5-1 Useful Infinite Series

(1) $e^x = 1 + x + \dfrac{x^2}{2!} + \dfrac{x^3}{3!} + \dfrac{x^4}{4!} + \cdots$

(2) $a^x = 1 + x \ln a + \dfrac{(x \ln a)^2}{2!} + \dfrac{(x \ln a)^3}{3!} + \cdots$

(3) $\ln x = 2\left[\dfrac{x-1}{x+1} + \dfrac{1}{3}\left(\dfrac{x-1}{x+1}\right)^3 + \dfrac{1}{5}\left(\dfrac{x-1}{x+1}\right)^5 + \cdots \right]$ $(x > 0)$

(4) $\ln(1+x) = x - \dfrac{x^2}{2} + \dfrac{x^3}{3} - \dfrac{x^4}{4} + \cdots$ $(1 > x - 1)$

(5) $\sin x = x - \dfrac{x^3}{3!} + \dfrac{x^5}{5!} - \dfrac{x^7}{7!} + \cdots$

(6) $\cos x = 1 - \dfrac{x^2}{2!} + \dfrac{x^4}{4!} - \dfrac{x^6}{6!} + \cdots$

(7) $\sinh x = \dfrac{1}{2}(e^x - e^{-x}) = x + \dfrac{x^3}{3!} + \dfrac{x^5}{5!} + \dfrac{x^7}{7!} + \cdots$

(8) $\cosh x = \dfrac{1}{2}(e^x + e^{-x}) = 1 + \dfrac{x^2}{2!} + \dfrac{x^4}{4!} + \dfrac{x^6}{6!} + \cdots$

(9)* $(x+y)^n = x^n + nx^{n-1}y + \dfrac{n(n-1)}{2}x^{n-2}y^2 + \dfrac{n(n-1)(n-2)}{3!}x^{n-3}y^3 + \cdots$
   $(|y| < |x|)$

(10) $(1 \pm x)^{-n} = 1 \mp nx + \dfrac{n(n+1)}{2!}x^2 \mp \dfrac{n(n+1)(n+2)}{3!}x^3 + \cdots$ $(|x| < 1)$

*This is actually not an infinite series, for finite values of n. It is often referred to as the *binomial expansion* and is quite useful also.

circular tube, which was found to be [9]

$$v_z = \frac{\Delta p R^2}{4\mu_0 L} \cdot \frac{2}{k} \int_1^{\xi} \frac{\xi d\xi}{\xi^n - K},$$ (2.5-10)

which, although seemingly a simple integration to perform, has no analytic solution as yet. However, a series solution can be found [10] if it is expanded uniformly in infinite series:

$$\frac{1}{\xi^n - K} = -\frac{1/K}{1 - \xi^n/K}$$

$$= -\frac{1}{K}\left[1 + \frac{\xi^n}{K} + \left(\frac{\xi^n}{K}\right)^2 + \cdots\right]$$

$$= -\frac{1}{K}\sum_{j=0}^{\infty}\left(\frac{\xi^n}{K}\right)^j$$ (2.5-11)

since in that particular problem $K > 1$ and $0 \leqslant \xi \leqslant 1$ (see Example 9.6-2), thus making $\xi^n/K < 1$. Substituting Eq. 2.5-11 into Eq. 2.5-10, we can then perform the integration in straightforward fashion* and the result is

$$v_z = \frac{\Delta p R^2}{4\mu_0 L} \cdot \frac{2}{k} \sum_{j=0}^{\infty} \frac{1 - \xi^{jn+2}}{(jn+2)K^{j+1}},$$ (2.5-12)

which can be plotted as shown in Figure 2.5-1.

2.6 PARTIAL DIFFERENTIATION

If we have a function dependent upon more than one variable—for example, $f(x,y,z)$—we can first define the partial derivatives, each of which is the derivative with respect to one of the independent variables, while the other independent variables are being held or treated as constants. Symbolically, we write

$$\left(\frac{\partial f}{\partial x}\right)_{y,z}, \qquad \left(\frac{\partial f}{\partial y}\right)_{x,z}, \qquad \text{and} \qquad \left(\frac{\partial f}{\partial z}\right)_{x,y}.$$

---

*This involves interchanging the order of summation and integration. For an infinite series, this can be done if and only if the series converges uniformly. Our case meets this requirement.

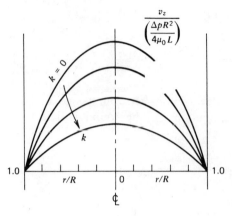

Figure 2.5-1. Velocity distribution of flow of suspension [9, 10].

Then we can write the chain rule in differential form, relating the total differential to the partial ones:

$$df = \left(\frac{\partial f}{\partial x} \right)_{y,z} dx + \left(\frac{\partial f}{\partial y} \right)_{x,z} dy + \left(\frac{\partial f}{\partial z} \right)_{x,y} dz. \qquad (2.6\text{-}1)$$

The truth of this can be seen from Figure 2.6-1 in finite difference form for the slightly simplified case of $f = f(x, y)$, where the total change Δf from A to B is taken in two steps. First, we go from A to the intermediate point I along the x-direction (keeping y constant). Then, we complete the change by going from I to B (keeping x constant). By simple trigonometry, we can see that

$$\Delta f = II' + BI''$$

$$= \left(\begin{array}{c} \text{slope} \\ \text{of } AI \end{array} \right) \cdot AI' + \left(\begin{array}{c} \text{Slope} \\ \text{of } IB \end{array} \right) \cdot II''$$

$$= \left(\frac{\Delta f}{\Delta x} \right)_{\text{Keeping } y \text{ constant at } y = y_1} \Delta x + \left(\frac{\Delta f}{\Delta y} \right)_{\text{Keeping } x \text{ constant at } x = x_2} \Delta y. \qquad (2.6\text{-}2)$$

As previously stated, if we let all of the increments approach zero, we can write the above in differential form:

$$df = \left(\frac{\partial f}{\partial x} \right)_{y} dx + \left(\frac{\partial f}{\partial y} \right)_{x} dy,$$

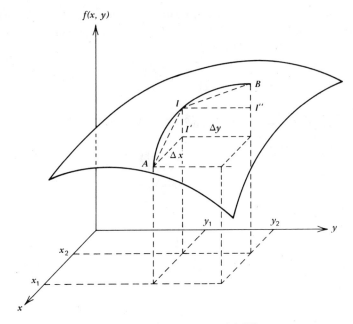

Figure 2.6-1. Total difference consists of partial differences.

which can be extended to Eq. 2.6-1 for a function $f(x,y,z)$ or, even more generally, for $f(x_1,x_2,\ldots,x_n)$, to

$$df = \sum_{i=1}^{n} \left(\frac{\partial f}{\partial x_i} \right) dx_i. \qquad (2.6\text{-}3)$$

Here we have neglected the quantities to be held constant for each of the various partial differentiations. This is permissible as long as there is no ambiguity as to the quantities to be held constant. Usually they are all quantities other than the one with respect to which the partial differentiation is performed.

The following case illustrates a common error and shows that we must keep in mind what is being held constant. Although it is correct to write, say,

$$\frac{df}{dx} = \frac{1}{\left(\dfrac{dx}{df} \right)}, \qquad (2.6\text{-}4)$$

it is generally incorrect to write a similar expression for partial derivatives. As an example, let us take the rectangular-cylindrical coordinate transformation (see Section 2.17); we have

$$
\begin{array}{llll}
\text{(a)} & x = r\cos\theta & \text{(d)} & r = \sqrt{x^2 + y^2} \\
\\
\text{(b)} & y = r\sin\theta & \text{(e)} & \theta = \tan^{-1}\dfrac{y}{x} & \text{(2.6-5)} \\
\\
\text{(c)} & z = z & \text{(f)} & z = z
\end{array}
$$

and from Eq. 2.6-5(a)

$$\frac{\partial x}{\partial r} = \cos\theta, \tag{2.6-6}$$

but it would be disastrous to take a short-cut and write

$$\frac{\partial r}{\partial x} = \frac{1}{\partial x/\partial r} = \frac{1}{\cos\theta}. \tag{2.6-7}$$

In fact, if we use Eq. 2.6-5(d), we can show that

$$\frac{\partial r}{\partial x} = \frac{1}{2} \cdot \frac{1}{\sqrt{x^2 + y^2}} \cdot 2x = \frac{x}{r} = \cos\theta. \tag{2.6-8}$$

Then which one is correct, Eq. 2.6-7 or 2.6-8? The answer will become clear if we consider that in writing $\partial r/\partial x$, we think of r as a function of x, y, and z; and so we are really saying $(\partial r/\partial x)_{y,z}$ (i.e., the partial derivative keeping y and z constant. But, Eq. 2.6-7 is really only an erroneous "adaptation" of Eq. 2.6-6, in which θ and z are kept constant, because in formula (a) of Eq. 2.6-5, x is expressed as a function of r, θ, and z.* Therefore, the conclusion is that Eq. 2.6-7 is wrong and Eq. 2.6-8 is correct. This illustrates the importance of writing down the quantities to be kept constant if there is any ambiguity about them.

In Eq. 2.6-1, $df, dx, \dots$ are shown by themselves and are called differentials. Going one step further, we can change this equation into the chain rule in its usual derivative forms of which there are two. First, if all of x, y, and z are functions of still another variable, say, t, f is also a function, by substitution, of t for which we write $f[x(t), y(t), z(t)]$. Then we

---

*Generally speaking; but here z drops out.

can, so to speak, "divide the entire equation by dt" and form the total derivative df/dt:

$$\frac{df}{dt} = \left(\frac{\partial f}{\partial x}\right)_{y,z}\frac{dx}{dt} + \left(\frac{\partial f}{\partial y}\right)_{x,z}\frac{dy}{dt} + \left(\frac{\partial f}{\partial z}\right)_{x,y}\frac{dz}{dt}. \qquad (2.6\text{-}9)$$

This way we express a direct relationship between f and t in terms of the indirect one between f and x,y,z and those between x,y,z and t. The practical application of this equation will be introduced in Chapter 4.

In the second form, if x,y, and z are functions of more than one variable, say, r, θ, and z, formally we write $f[x(r,\theta,z),y(r,\theta,z),z(r,\theta,z)]$. Again, we can "divide Eq. 2.6-1 by dr" but now holding θ and z constant. Thus, we form the following partial derivative:

$$\left(\frac{\partial f}{\partial r}\right)_{\theta,z} = \left(\frac{\partial f}{\partial x}\right)_{y,z}\left(\frac{\partial x}{\partial r}\right)_{\theta,z} + \left(\frac{\partial f}{\partial y}\right)_{x,z}\left(\frac{\partial y}{\partial r}\right)_{\theta,z}$$

$$+ \left(\frac{\partial f}{\partial z}\right)_{x,y}\left(\frac{\partial z}{\partial r}\right)_{\theta,z} \qquad (2.6\text{-}10)$$

and similarly $(\partial f/\partial\theta)_{r,z}$ and $(\partial f/\partial z)_{r,\theta}$. An example of this is when x,y,z and r,θ,z are, respectively, rectangular and cylindrical coordinates. Then, for formulas (a), (b), and (c) in Eq. 2.6-5 we have

$$\left(\frac{\partial x}{\partial r}\right)_{\theta,z} = \cos\theta, \qquad \left(\frac{\partial y}{\partial r}\right)_{\theta,z} = \sin\theta, \qquad \left(\frac{\partial z}{\partial r}\right)_{\theta,z} = 0,$$

so that Eq. 2.5-10 becomes

$$\left(\frac{\partial f}{\partial r}\right)_{\theta,z} = \left(\frac{\partial f}{\partial x}\right)_{y,z}\cos\theta + \left(\frac{\partial f}{\partial y}\right)_{x,z}\sin\theta, \qquad (2.6\text{-}11)$$

which relates the partial derivatives with respect to the cylindrical coordinates to those with respect to the rectangular coordinates. This is quite useful in vector and tensor transformations, as will be shown in Section 2.17. Other examples of the application of the chain rule are the relationships between thermodynamic quantities such as V,S,H,G,T, and P.

Similar to our discussion in Section 2.3, we can take the second partial derivatives such as

$$\frac{\partial^2 f}{\partial x^2}, \frac{\partial^2 f}{\partial y^2}, \ldots$$

and, in addition, what we might call the *mixed derivatives,*

$$\frac{\partial^2 f}{\partial x\,\partial y} = \frac{\partial}{\partial x}\left(\frac{\partial f}{\partial y}\right), \qquad \frac{\partial^2 f}{\partial y\,\partial x} = \frac{\partial}{\partial y}\left(\frac{\partial f}{\partial x}\right), \cdots$$

and higher-order ones. In general $\partial^2 f/\partial x\,\partial y = \partial^2 f/\partial y\,\partial x$; that is, the order of differentiation is immaterial if all partial derivatives involved are continuous functions.

Also, parallel to what we discussed previously regarding the univariate function $f(x)$, we have the following for $f(x,y)$: If at a certain point

$$\frac{\partial f}{\partial x} = 0, \qquad \frac{\partial f}{\partial y} = 0,$$

and

$$\left(\frac{\partial^2 f}{\partial x\,\partial y}\right)^2 - \frac{\partial^2 f}{\partial x^2}\cdot\frac{\partial^2 f}{\partial y^2} < 0,$$

we have a local maximum if $\partial^2 f/\partial x^2 < 0$ but a local minimum if $\partial^2 f/\partial x^2 > 0$.

Sometimes a maximum or minimum is taken with qualifying conditions called *constraints,* in which case $\partial f/\partial x, \partial f/\partial y,\ldots$ are not individually 0. Instead, if the function to be maximized or minimized is written $f(x,y,z)$ and there are two constraints $g_1(x,y,z) = 0$ and $g_2(x,y,z) = 0$, we define a functional

$$F = f + \lambda_1 g_1 + \lambda_2 g_2, \tag{2.6-12}$$

where λ_1 and λ_2 are called the Lagrangian multipliers. Taking $\partial F/\partial x, \partial F/\partial y$, and $\partial F/\partial z = 0$ yields the constrained extremum (maximum or minimum).

In general, if we have a function of n variables $f_1(x_1,x_2,\ldots,x_n)$ to be extremized with m constraints $g_1(x_1,x_2,\ldots,x_n) = 0$, $g_2(x_1,x_2,\ldots,x_n) = 0,\ldots,$ $g_m(x_1,x_2,\ldots,x_n) = 0$, then the criteria are

$$\frac{\partial}{\partial x_i}\left(f + \sum_{j=1}^{m}\lambda_j g_j\right) = 0 \qquad (i = 1,2,\ldots,n). \tag{2.6-13}$$

We note here that the m λ_j's are unknown. But they can be obtained along with the n x_i's. In other words, we want to solve for $(m+n)$ unknowns for which we have the n criteria represented by Eq. 2.6-13 and the m constraints, a total of $(m+n)$ simultaneous equations.

Here we must take note of the Leibnitz formula for the differentiation of integrals. If $I(t)$ is such a function of t that

$$I(t) = \int_{a_1(t)}^{a_2(t)} f(x,t)\, dx, \qquad (2.6\text{-}14)$$

where the limits of integration $a_1(t)$ and $a_2(t)$ are both functions of t instead of constants, the derivative of $I(t)$ with respect to t can be shown to be

$$\frac{dI}{dt} = \frac{d}{dt} \int_{a_1(t)}^{a_2(t)} f(x,t)\, dx$$

$$= \int_{a_1(t)}^{a_2(t)} \left[\frac{\partial}{\partial t} f(x,t) \right] dx + f[a_2(t),t]\frac{da_2}{dt} - f[a_1(t),t]\frac{da_1}{dt}. \qquad (2.6\text{-}15)$$

We can see here that it is not enough simply to move the derivative sign inside the integral and differentiate the integrand because the boundaries are variables as well. In the graphical representation in Figure 2.6-2, $I(t)$ is the area under the curve $f(x,t)$ and between the boundaries $a_1(t)$ and $a_2(t)$, all of which are varying. The first term on the right-hand side of Eq. 2.6-15 gives the rate of change of this area due to the movement of the curve $f(x,t)$ with t, while the other two represent those due to the movement of the boundaries $a_1(t)$ and $a_2(t)$.

Sometimes only one of the boundaries is variable, in which case either the second or third term on the right-hand side of Eq. 2.6-15 will drop out.

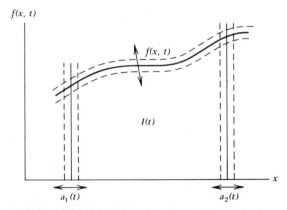

Figure 2.6-2. Integral with variable boundaries. (area under curve)

If both boundaries are invariant, then this equation reverts to the simple operation of directly moving the differentiation inside the integral, since both da_1/dt and da_2/dt are zero for this special case.

2.7 MULTIPLE INTEGRATION

A function of more than one variable can be integrated in succession with respect to the several variables. For example, referring to Figure 2.7-1, we can write

$$\int_{y_1}^{y_2}\int_{x_1}^{x_2} f(x,y)\,dx\,dy \qquad (2.7\text{-}1)$$

and extend the discussion in Section 2.4 to show that if $f(x,y)$ represents a curved surface, Eq. 2.7-1 represents the volume under the surface bounded by the limits x_1, x_2 and y_1, y_2.

As a practical example, if $\rho(x,y,z)$ is the nonuniform density distribution of a system, then

$$\int\int\int_V \rho(x,y,z)\,dx\,dy\,dz$$

is the mass of the system. In application to transport phenomena, for example, if $v(x,y)$ represents the velocity distribution of a fluid in a pipe as a function of its cross-sectional x,y position, then the following integral represents the volumetric flow rate of the flow, as evidenced by the units:

$$\int\int_A \underbrace{v(x,y)}_{}\ \underbrace{dx}_{}\ \underbrace{dy}_{}[=]\text{cm}^3/\text{sec.}^*$$

$$\text{cm/sec}\ \ \text{cm}\ \text{cm}$$

The multiple integral divided by the extent of the region over which it is taken is what we usually refer to as the average quantity and is denoted in this book by the symbol $\langle\ \rangle$. In the above example, for instance,

$$\langle v\rangle\equiv\frac{\int\int_A v(x,y)\,dx\,dy}{\int\int_A dx\,dy}=\frac{1}{A}\int\int_A v(x,y)\,dx\,dy\ ^\dagger \qquad (2.7\text{-}2)$$

is the area-averaged velocity because A, the cross-sectional area, is the

---

*The symbol $[=]$ means "has the units of."

†The symbol $\equiv$ means "identically equal to" or "equal to, by definition."

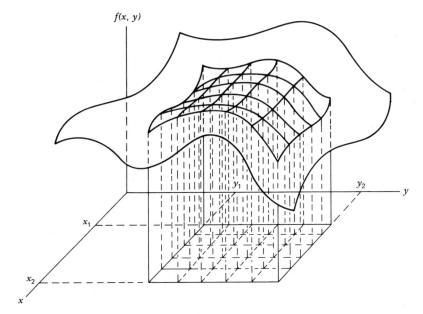

$f(x, y)$

Figure 2.7-1. Integration of $f(x,y)$ gives volume under surface $f(x,y)$.

region over which the integral is taken. On the other hand, if $c(x,y,z)$ and $T(x,y,z)$ are, respectively, the nonuniform concentration and temperature distributions of a material in a container with volume V, then

$$\langle c \rangle \equiv \frac{\int \int \int_V c(x,y,z)\,dx\,dy\,dz}{\int \int \int_V dx\,dy\,dz} = \frac{1}{V} \int \int \int_V c(x,y,z)\,dx\,dy\,dz \quad (2.7\text{-}3)$$

and

$$\langle T \rangle \equiv \frac{\int \int \int_V T(x,y,z)\,dx\,dy\,dz}{\int \int \int_V dx\,dy\,dz} = \frac{1}{V} \int \int \int_V T(x,y,z)\,dx\,dy\,dz \quad (2.7\text{-}4)$$

are the volume-average concentration and temperature (see Example 6.2-6 and Section 9.6).

It should be noted here that while the elementary volume $dx\,dy\,dz$ is most convenient for systems whose geometries are rectangular in nature, the cylindrical and spherical coordinates are easier to work with in others. As can be seen in Figure 2.7-2 (a) and (b), the elementary volume in

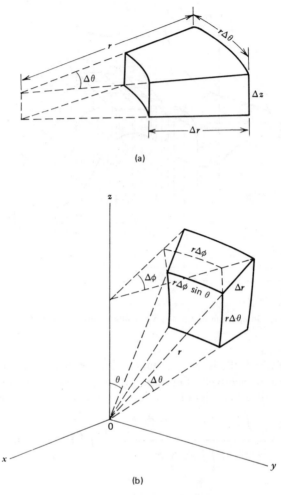

(a)

(b)

Figure 2.7-2. Elementary volume in (a) cylindrical coordinates and (b) spherical coordinates.

cylindrical coordinates is $r\,dr\,d\theta\,dz$, while in spherical coordinates it is $r^2\sin\theta\,dr\,d\theta\,d\phi$. Additional details on this can be found in Section 2.17, where coordinate transformation is discussed.

The limits of integration in the three types of coordinate systems are usually as those given in Figure 2.7-3 (a–c) if the right system is used and if it is properly placed.

Sometimes the integration becomes very simple because the integrand (the function to be integrated) turns out to be a function of only one

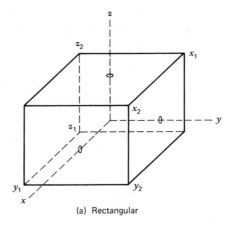

(a) Rectangular

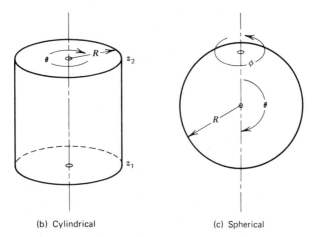

(b) Cylindrical   (c) Spherical

Figure 2.7-3. Limits of integration for full rectangle, cylinder, and sphere in their respective geometries. (a) $\int_{z_1}^{z_2} \int_{y_1}^{y_2} \int_{x_1}^{x_2} f(x,y,z)\,dx\,dy\,dz$; (b) $\int_{z_1}^{z_2} \int_{0}^{2\pi} \int_{0}^{R} f(r,\theta,z)\,r\,dr\,d\theta\,dz$; (c) $\int_{0}^{2\pi} \int_{0}^{\pi} \int_{0}^{R} f(r,\theta,\phi)\,r^2 \sin\theta\,dr\,d\theta\,d\phi$.

variable. An example of this is the parabolic velocity distribution of the laminar flow of a liquid in a circular tube:

$$Q\left(\begin{array}{c}\text{Volumetric}\\\text{flow rate}\end{array}\right) = \frac{\Delta p R^2}{4\mu L} \int_0^{2\pi} \int_0^R \left[1 - \left(\frac{r}{R}\right)^2\right] r\,dr\,d\theta, \quad (2.7\text{-}5)$$

where Δp is the pressure drop over the tube of length L and radius R, and

μ is the viscosity of the liquid. We can see that the integration with respect to θ merely yields a constant multiplication factor 2π because θ is not involved in the integrand. Thus, Eq. 2.7-5 becomes

$$Q = \frac{\pi \Delta p R^2}{2\mu L} \int_0^R \left[1 - \left(\frac{r}{R} \right)^2 \right] r \, dr. \qquad (2.7\text{-}6)$$

Usually such an integration can be made much easier to perceive if we make the variables dimensionless. In the present case we let $\xi = r/R$ and consequently, $dr = R d\xi$ so that we have

$$Q = \frac{\pi \Delta p R^4}{2\mu L} \int_0^1 (1 - \xi^2) \xi \, d\xi$$

$$= \frac{\pi \Delta p R^4}{2\mu L} \left[\tfrac{1}{2}\xi^2 - \tfrac{1}{4}\xi^4 \right]_0^1$$

$$= \frac{\pi \Delta p R^4}{2\mu L} \left[\tfrac{1}{2} - \tfrac{1}{4} \right]$$

$$= \frac{\pi \Delta p R^4}{8\mu L}. \qquad (2.7\text{-}7)$$

2.8 ORDINARY DIFFERENTIAL EQUATIONS

In solving most problems in the physical sciences, the pertinent physical laws, such as Newton's second law of motion and the first law of thermodynamics, are applied to the situation at hand and transformed into equations via mathematical symbolism. For simple problems, this results in algebraic equations that are relatively easy to solve. The results are then translated and interpreted in physical terms.

However, the equations resulting most frequently involve derivatives. For example, if Fick's law of diffusion governs, we will probably have a first derivative because of the concentration gradient term. On the other hand, if the second law of motion is involved, we will have a second derivative because it deals with acceleration, which is the second derivative of position. These equations have to be solved to yield such quantities as concentration distribution and distance of travel. They are called *differential equations*.

When a differential equation involves an unknown quantity that is a function of only another variable, it is called an *ordinary differential equation.* However, many times this dependent variable is a function of more than one variable (such as time and position) and partial derivatives are involved; we then have *partial differential equations.* It has been said that the entire world operates on partial differential equations. We will discuss them in a later section (Section 2.13) but will confine ourselves to ordinary differential equations here.

The *order* of a differential equation is that of its highest derivative. A *linear* differential equation is one in which the dependent variable and its derivatives do not form products or appear in *nonlinear* (i.e., other than zero- and first-order) form. A *homogeneous* differential equation is one in which there is no term that does not involve the dependent variable or its derivatives. For example, a differential equation of the general form

$$a_0(x)\frac{d^n y}{dx^n} + a_1(x)\frac{d^{n-1}y}{dx^{n-1}} + \cdots + a_{n-1}(x)\frac{dy}{dx} + a_n(x)y = f(x) \quad (2.8\text{-}1)$$

is a nonhomogeneous *n*th-order linear ordinary differential equation, while simpler-looking ones such as

$$\frac{d^2 y}{dx^2} + \left(\frac{dy}{dx}\right)^2 = \sin x \quad (2.8\text{-}2)$$

and

$$\frac{d^2 y}{dx^2} + e^y = 0 \quad (2.8\text{-}3)$$

are actually nonlinear. The *degree* of a nonlinear differential equation is equal to the power to which the highest derivative is raised. Thus, Eq. 2.8-2 is of the second order but only of the first degree.

Some differential equations are easy to solve, requiring only simple integration. For example, if we have

$$\frac{d^2 y}{dx^2} = \sin x, \quad (2.8\text{-}4)$$

by realizing that it can be written as

$$\frac{d}{dx}\left(\frac{dy}{dx}\right) = \sin x$$

we can integrate once

$$\frac{dy}{dx} = -\cos x + C_1$$

Table 2.8-1 Solutions to Ordinary Differential Equations [7, 8]

A. First Order, First Degree

Type	Differential Equation	Solution
1. Separable	$y' = \dfrac{F(x)}{G(y)}$ $y' = \dfrac{G(y)}{H(x)}$	$\displaystyle \int G\,dy = \int F\,dx + C$ $\displaystyle \int \frac{dy}{G} = \int \frac{dx}{H} + C$
2. Exact	$M(x,y) + N(x,y)y' = 0$ with $\dfrac{\partial M}{\partial y} = \dfrac{\partial N}{\partial x}$	$\displaystyle \int M\,dx + \int \left(N - \frac{\partial}{\partial y} \int M\,dx \right) dy = C$ or $\displaystyle \int N\,dy + \int \left(M - \frac{\partial}{\partial x} \int N\,dy \right) dx = C$
3. Homogeneous	$y' = H\!\left(\dfrac{y}{x} \right)$	$\displaystyle \ln x + \int \frac{dv}{v - H(v)} = C$ in which $\quad v = \dfrac{y}{x}$
4. Linear	$y' + P(x)y = Q(x)$	$\displaystyle y = e^{-\int P\,dx}\left[\int Qe^{+\int P\,dx}dx + C \right]$ Mnemonic: the minus and plus signs occur in the exponents in alphabetical order.

B. First Order, Higher Degree

Type	Differential Equation	Solution
Solvable for y	$y = F(x,p)$ where $p = y'$	(1) Differentiate w.r.t. x to get $p = G(x,p,dp/dx)$ (2) Solve (1) to get $H(x,p,C) = 0$ (3) Eliminate p between (2) and d.e. to get $J(x,y,C) = 0$
Solvable for x	$x = F(y,p)$ where $p = y'$	(1) Differentiate w.r.t. y to get $1/p = G(y,p,dp/dy)$ (2) Solve (1) to get $H(y,p,C) = 0$ (3) Eliminate p between (2) and d.e. to get $J(x,y,C) = 0$
Containing no y	$F(x,p) = 0$ where $p = y'$	$\langle$or$\rangle$ $\left\{\begin{array}{l}\text{(1) Solve for } p \text{ to get } y' = G(x)\\ \text{(2) Integrate (1) to get } y = \int G\,dx + C\\ \text{(1) Solve for } x \text{ to get } x = H(p)\\ \text{(2) Differentiate (1) w.r.t. } y \text{ to get } 1/p = F'(p)(dp/dy)\\ \text{(3) Solve (2) to get } y = \int pF'(p)\,dp + C\\ \text{(4) Eliminate } p \text{ between (3) and d.e. to get } J(x,y,C) = 0\end{array}\right.$
Containing no x	$F(y,p) = 0$ where $p = y'$	Analogous to previous case

Table 2.8-1 (*Continued*)

C. Special Solutions of $y'' + Ay' + B = 0$ $\qquad m_\pm = \dfrac{-A \pm \sqrt{A^2 - 4B}}{2}$

Nature of $m_\pm$	Differential Equation	Solution
1. Real, unequal $m_+ \neq m_-$	$y'' - (m_+ + m_-)y' + m_+ m_- y = 0$	$y = C_+ e^{m_+ x} + C_- e^{m_- x}$
2. Real, equal $m_+ = m_- = m$	$y'' - 2my' + m^2 y = 0$	$y = (C_1 + C_2 x)\,e^{mx}$
3. Real, sum equals zero $\;m_+ = -m_- = m$	$y'' - m^2 y = 0$	$y = C_+ e^{mx} + C_- e^{-mx}$ $= C_1 \cosh mx + C_2 \sinh mx$
4. Imaginary, sum equals zero $\;m_+ = -m_- = in$	$y'' + n^2 y = 0$	$y = C_+ e^{inx} + C_- e^{-inx}$ $= C_1 \cos nx + C_2 \sin nx$ $= C \sin (nx + C')$
5. Complex conjugates $\;m_\pm = m \pm in$	$y'' - 2my' + (m^2 + n^2)y = 0$	$y = C_+ e^{(m+in)x} + C_- e^{(m-in)x}$ $= e^{mx}(C_1 \cos nx + C_2 \sin nx)$ $= Ce^{mx} \sin (nx + C')$

D. Second Order, Special Types

Type	Solution
1. $y'' = f(y)$	(1) Multiply by $2y'$ to get $(d/dx)(y')^2 = 2fy'$ (2) Integrate to get $(y')^2 = 2\int f(y)\,dy + C_1$ (3) Take square root of (2) and integrate again to get $\displaystyle\int \dfrac{dy}{\pm\sqrt{2\int f\,dy + C_1}} = x + C_2$

2. $y'' = f(y')$

(1) Let $y' = p(x)$ so that equation becomes $p' = f(p)$

(2) Integrate to get $x = \int dp/f(p) + C_1$

(3) Then $y = \int p\,dx = \int \dfrac{p\,dp}{f(p)} + C_2$

(4) Eliminate p between (2) and (3) to get $F(x,y,C_1,C_2) = 0$

3. $f(y'',y',x) = 0$

$\left\{\begin{array}{l} \text{(1) Let } p=y' \text{ and get } f(p',p,x)=0 \text{ where } p'=f(p) \\ \text{(2) If possible get solution of (1) as } p = F(x,C_1) \\ \text{(3) Integrate (2) and get } y = \int F(x,C_1)dx + C_2 \end{array}\right.$

$\langle \text{or} \rangle$

$\left\{\begin{array}{l} \text{(1) Let } p=y' \text{ and get } f(p',p,x)=0 \\ \text{(2) If possible get } x = G(p,C_1) \text{ from (1)} \\ \text{(3) Then } y = \int p\,dx = \int pG'(p,C_1)\,dp + C_2 \;\langle \text{or} \rangle\; y = pG(p,C_1) - \int G(p,C_1)\,dp + C_2\text{*} \\ \text{(4) Eliminate } p \text{ from (2) and (3)} \end{array}\right.$

4. $f(y'',y',y) = 0$

$\left\{\begin{array}{l} \text{(1) Let } p=y' \text{ and get } f(pp',p,y)=0 \\ \text{(2) Solve (1) to get } p = F(y,C_1) \\ \text{(3) Integrate (2) to get } x = \int (dy/F) + C_2 \end{array}\right.$

$\langle \text{or} \rangle$

$\left\{\begin{array}{l} \text{(1) Let } p=y' \text{ and get } f(pp',p,y)=0 \\ \text{(2) Solve to get (1) in form } y = G(p,C_1) \\ \text{(3) Then } x = \int (1/p)dy = \int (G'(p,C_1)/p)\,dp + C_1 \;\langle \text{or} \rangle\; x = \dfrac{y}{p} + \int \dfrac{G(p,C_1)}{p^2}\,dp + C_2 \\ \text{(4) Eliminate } p \text{ from (2) and (3)} \end{array}\right.$

*Where $G' = dG/dp$.

and twice to obtain

$$y = -\sin x + C_1 x + C_2.$$

We note here that the number of integration constants is equal to the order of the differential equation.

However, most physical problems result in differential equations more complex than the above. Thanks to generations of distinguished mathematicians, various techniques have been developed to solve these equations. Instead of describing them one by one, we simply list them in Table 2.8-1 in recipe fashion. They are actually special cases of the general nonhomogeneous differential equation

$$y'' + P(x)y' + Q(x)y = R(x), \qquad (2.8\text{-}5)$$

of which the corresponding homogeneous form is

$$y'' + P(x)y' + Q(x)y = 0 \qquad (2.8\text{-}6)$$

by simply dropping the nonhomogeneous part $R(x)$.

Mathematical theories tell us that if $y_1(x)$ and $y_2(x)$ are two separate functions that both satisfy Eq. 2.8-6, then their linear combination is also a solution.* Thus, we can write the *homogeneous* (or *complementary*) *solution* y_h as

$$y_h = c_1 y_1(x) + c_2 y_2(x), \qquad (2.8\text{-}7)$$

where c_1 and c_2 are arbitrary constants.

The theorem further states that if any *particular solution*, y_p, satisfies Eq. 2.8-5 as Eq. 2.8-7 does Eq. 2.8-6, the *complete solution* is the sum of the complementary and particular solutions, or

$$y = y_h + y_p$$
$$= c_1 y_1(x) + c_2 y_2(x) + y_p \qquad (2.8\text{-}8)$$

Although Parts C and D of Table 2.8-1 have given some standard forms for the homogeneous solution, the particular integrals sometimes involve guesswork. However, if the nonhomogeneous differential equation is of the constant-coefficient type, that is,

$$ay'' + by' + cy = f(x), \qquad (2.8\text{-}9)$$

and if $f(x)$ is a function for which repeated differentiation yields only a finite number of independent derivatives, the particular integral can be

---

*The reader can prove this by simply substituting Eq. 2.8-7 into Eq. 2.8-6, while noting that $y_1'' + P(x)y_1' + Q(x)y_1 = 0$ and $y_2'' + P(x)y_2' + Q(x)y_2 = 0$. This is known as the principle of superimposition.

Table 2.8-2 Particular Solution for $ay'' + by' + cy = f(x)$*

$f(x)$	$y_p(x)$
k	A
kx^m	$A_m x^m + A_{m-1} x^{m-1} + \cdots + A_1 x + A_0$
ke^{mx}	$A e^{mx}$
$k \cos mx$ and/or $k' \sin mx$	$A \cos mx + B \sin mx$
$kx^m e^{nx} \cos px$ and/or $k'x^m e^{nx} \sin px$	$(A_m x^m + A_{m-1} x^{m-1} + \cdots + A_1 x + A_0) e^{nx} \cos px$ $+ (B_m x^m + B_{m-1} x^{m-1} + \cdots + B_1 x + B_0) e^{nx} \sin px$

*Where k, k', m, n, p, A, B, A_m, $A_{m-1},\ldots,A_1$, A_0, B_m, $B_{m-1},\ldots,B_1$, B_0, a, b, and c are constants. The homogeneous solution for this can be obtained from Part C of Table 2.8-1.

obtained by a well-defined procedure. This class of functions and their corresponding particular integrals are listed in Table 2.8-2.

The *undetermined coefficients** A, B, $A_m,\ldots,A_0$, $B_m,\ldots,B_0$ can then be evaluated by substituting into the nonhomogeneous differential equation Eq. 2.8-9 and equating coefficients of like-power terms.

For the more general case of Eq. 2.8-5 with variable coefficients *and* with $R(x)$ not being of one of the forms in Table 2.8-2, a method known as *variation of parameters* is used in which the particular solution is a combination of the two functions, y_1 and y_2,† satisfying the homogeneous differential equation, Eq. 2.8-6, but via two variable coefficients $u_1(x)$ and $u_2(x)$; that is,

$$y_p = u_1(x)y_1(x) + u_2(x)y_2(x). \tag{2.8-10}$$

After proper substitutions and manipulations, u_1 and u_2 can be found from the following integrations:

$$u_1(x) = -\int \frac{y_2}{y_1 y_2' - y_2 y_1'} R(x)\, dx \tag{2.8-11}$$

$$u_2(x) = \int \frac{y_1}{y_1 y_2' - y_2 y_1'} R(x)\, dx. \tag{2.8-12}$$

*Hence, the "method of undetermined coefficients."
†Which can be obtained from Part D of Table 2.8-1 or from Sections 2.9 and 2.12.

2.9 SERIES SOLUTIONS AND BESSEL FUNCTIONS

One type of linear ordinary differential equation not included in Table 2.8-1 is the following:*

$$\frac{1}{r}\frac{d}{dr}\left(r\frac{dy}{dr}\right) \pm b^2 y = 0,\qquad(2.9\text{-}1)$$

where b is a constant. It cannot be solved by any of the methods heretofore described. Therefore, we resort to the series solution in which we assume that the solution appears in the following general infinite-series form:

$$y = a_0 + a_1 r + a_2 r^2 + a_3 r^3 + \cdots$$

$$= \sum_{i=0}^{\infty} a_i r^i.\qquad(2.9\text{-}2)$$

Differentiating this step by step gives

$$\frac{dy}{dr} = a_1 + 2a_2 r + 3a_3 r^2 + \cdots$$

$$r\frac{dy}{dr} = a_1 r + 2a_2 r^2 + 3a_3 r^3 + \cdots$$

$$\frac{d}{dr}\left(r\frac{dy}{dr}\right) = a_1 + 4a_2 r + 9a_3 r^2 + \cdots$$

$$\frac{1}{r}\frac{d}{dr}\left(r\frac{dy}{dr}\right) = \frac{a_1}{r} + 4a_2 + 9a_3 r + 16a_4 r^2 + \cdots,\qquad(2.9\text{-}3)$$

which, if substituted into Eq. 2.9-1 with the plus sign, along with Eq. 2.9-2, yields

$$\frac{a_1}{r} + 4a_2 + 9a_3 r + \cdots + b^2 (a_0 + a_1 r + a_2 r^2 + \cdots) = 0$$

*An alternate form of this is

$$\frac{d^2 y}{dr^2} + \frac{1}{r}\frac{dy}{dr} \pm b^2 y = 0.$$

The plus-or-minus sign here actually means that there are two cases to be considered, one with the plus sign and the other with the minus sign, as will be seen later.

or

$$a_1 r^{-1} + (4a_2 + b^2 a_0) r^0 + (9a_3 + b^2 a_1) r^1 + (16a_4 + b^2 a_2) r^2 + \cdots = 0.$$

$$(2.9\text{-}4)$$

Since this equation has to be true for any arbitrary value of r, the coefficients of all terms have to be individually equal to zero. Or,

$$r^{-1}: \qquad\qquad a_1 = 0$$

$$r^0: 4a_2 + b^2 a_0 = 0 \qquad a_2 = -\frac{b^2}{4} a_0$$

$$r^1: 9a_3 + b^2 a_1 = 0 \qquad a_3 = -\frac{b^2}{9} a_1 = 0$$

$$r^2: 16a_4 + b^2 a_2 = 0 \qquad a_4 = -\frac{b^2}{16} a_2 = \frac{b^2}{16} \cdot \frac{b^2}{4} a_0$$

$$\vdots \qquad\qquad\qquad \vdots$$

We can see that for all odd values of i, $a_i = 0$, only the even-power terms remain. From the first few of these we can make the following generalization relating all of the subsequent terms to the first, a_0:

$$a_{2k} = \frac{(-1)^k b^{2k}}{[2^k (k!)]^2} a_0 \qquad\qquad (2.9\text{-}5)$$

so that Eq. 2.9-2 becomes

$$y = a_0 \sum_{k=0}^{\infty} \frac{(-1)^k (br/2)^{2k}}{(k!)^2}$$

$$= a_0 J_0(br), \qquad\qquad (2.9\text{-}6)$$

where $J_0(br)$ is called the *Bessel function of the first kind and of zero order*.

We note that we have only one integration constant—namely, a_0—in the solution, Eq. 2.9-6, while the differential equation, Eq. 2.9-1, is of the second order. Therefore, we know that another solution is missing. Via the method of *variation of parameters* (see Section 2.8 and Eqs. 2.8-10 through 2.8-12), the second solution can be obtained and is called *Weber's Bessel function of the second kind*:

$$Y_0(br) = \frac{2}{\pi} \left[\tilde{Y}_0(br) - (\ln 2 - \gamma) J_0(br) \right] \qquad (2.9\text{-}7)$$

where

$$\gamma = \lim_{n \to \infty} \left(1 + \tfrac{1}{2} + \tfrac{1}{3} + \cdots + \frac{1}{n} - \ln n \right) = 0.5772157\ldots,$$

is Euler's constant and

$$\tilde{Y}_0(br) = J_0(br) \int^{} \frac{dr}{r[J_0(br)]^2}$$

is Neumann's Bessel function of the second kind. From the linearity of the differential equation,* the complete solution is a linear combination of the two Bessel functions, or

$$y = C_1 J_0(br) + C_2 Y_0(br) \tag{2.9-8}$$

in which the constant a_0 has been "absorbed" into the constant C_1.

If we take the negative sign in Eq. 2.9-1, the series solution will lead to

$$y = C_3 I_0(br) + C_4 K_0(br), \tag{2.9-9}$$

where

$$I_0(br) = \sum_{k=0}^{\infty} \frac{(br/2)^{2k}}{(k!)^2} \tag{2.9-10}$$

is the modified Bessel function of the first kind and zero order and

$$K_0(br) = \frac{\pi}{2} \left[i J_0(ibr) - Y_0(ibr) \right] \tag{2.9-11}$$

is the modified Bessel function of the second kind and zero order (where $i = \sqrt{-1}$).

Equation 2.9-1 and its solutions are very important because most problems involving cylindrical geometry, such as blood flow and diffusion in a circular tube, yield this type of differential equation.[†] However, sometimes the resultant equation is slightly more complicated, in the following general form:

$$\frac{1}{r} \frac{d}{dr} \left(r \frac{dy}{dr} \right) \pm \left(b^2 \mp \frac{p^2}{r^2} \right) y = 0, \tag{2.9-12}$$

*See footnote to Eq. 2.9-1, and then follow Eq. 2.8-7.
[†]Thus, the German name for Bessel function is *Zylinderfunktion*.

where if we take the upper signs, the solution will be

$$y = C_1 J_p(br) + C_2 Y_p(br) \qquad (2.9\text{-}13)$$

and if we take the lower signs, we have

$$y = C_3 I_p(br) + C_4 K_p(br), \qquad (2.9\text{-}14)$$

where these Bessel functions are now of pth order. We shall omit their definitions except to show just one of them:

$$J_p(br) = \sum_{k=0}^{\infty} \frac{(-1)^k (br/2)^{2k+p}}{k!(k+p)!}, \qquad (2.9\text{-}15)$$

which can be obtained following the same procedure as described at the beginning of this section. Furthermore, Eq. 2.9-15 can be shown to yield Eq. 2.9-6 upon setting $p=0$. A summary of important Bessel differential equations and the solutions they yield are listed in Table 2.9-1.

In Table 2.9-1 the only difference between items (1) and (2) is the sign in front of the $k^2 x^2$ term. This is why the arguments of J_p and Y_p in the two items differ by a factor of $i = \sqrt{-1}$. The virtual equivalence of $J_p(ikx)$ to $I_p(kx)$ can be further verified by considering their series representations:

$$J_p(k'r) = \sum_{j=0}^{\infty} \frac{(-1)^j (k'r/2)^{2j+p}}{j!(j+p)!} \qquad (2.9\text{-}16)$$

$$I_p(kr) = \sum_{j=0}^{\infty} \frac{(kr/2)^{2j+p}}{j!(j+p)!}. \qquad (2.9\text{-}17)$$

If we let $k' = ik$ in Eq. 2.9-16, then

$$J_p(ikr) = \sum_{j=0}^{\infty} \frac{(-1)^j (ikr/2)^{2j+p}}{j!(j+p)!}$$

$$= \sum_{j=0}^{\infty} \frac{(-1)^j (i^2)^j i^p (kr/2)^{2j+p}}{j!(j+p)!}$$

Table 2.9-1 Important Bessel Differential Equations and Their Solutions— Bessel Functions [11, 12]

Item	Differential equation[a]	Solution
(1)	$r^2 \dfrac{d^2y}{dr^2} + r \dfrac{dy}{dr} + (k^2r^2 - p^2)y = 0$	$y = Z_p(kr) \quad (k \neq 0)$ $= C_1 J_p(kr) + C_2 Y_p(kr)$
(2)	$r^2 \dfrac{d^2y}{dr^2} + r \dfrac{dy}{dr} - (k^2r^2 + p^2)y = 0$	$y = Z_p(ikr) \quad (k \neq 0)$ $= C_1 J_p(ikr) + C_2 Y_p(ikr)$ $= C_1 I_p(kr) + C_2 K_p(kr)$
(3)	$r^2 \dfrac{d^2y}{dr^2} + r \dfrac{dy}{dr} - (\beta^2 - \alpha r^{2s})y = 0$	$y = Z_{\beta/s}\left(\dfrac{\sqrt{\alpha}}{s} r^s\right) \quad (\alpha s \neq 0)$
(4)	$\dfrac{d^2y}{dr^2} + kr^m y = 0$	$y = \sqrt{r}\, Z_{\frac{1}{m+2}}\left(\dfrac{2\sqrt{k}}{m+2} r^{\frac{m+2}{2}}\right), \quad \left(\begin{matrix} mk \neq \\ -2k \end{matrix}\right)$
(5)	$\dfrac{d}{dr}\left(r^n \dfrac{dy}{dr}\right) + kr^m y = 0$	$y = r^{ps} Z_p\left(\dfrac{\sqrt{k}}{s} r^s\right), \text{ where } s = \dfrac{m-n}{2}$ $p = \dfrac{1-n}{2s}, \quad (ks \neq 0)$
(6)	$r^2 \dfrac{d^2y}{dr^2} + r(a + 2br^t)\dfrac{dy}{dr} + [c + hr^{2s}$ $- b(1-a-t)r^t + b^2 r^{2t}]y = 0$	$y = r^{\frac{1-a}{2}} e^{-\frac{br^t}{t}} Z_p\left(\dfrac{\sqrt{h}}{s} r^s\right),$ $\text{where } p = \dfrac{1}{s}\sqrt{\left(\dfrac{1-a}{2}\right)^2 - c}$
(7)	$\dfrac{d^2y}{dr^2} + \left(\dfrac{1-2\alpha}{r}\right)\dfrac{dy}{dr} + \left[(\beta\gamma r^{\gamma-1})^2\right.$ $\left. + \dfrac{\alpha^2 - p^2\gamma^2}{r^2}\right]y = 0$	$y = r^\alpha Z_p(\beta r^\gamma)$
(8)	$\dfrac{d^2y}{dr^2} + \left(\dfrac{1-2\alpha}{r} \mp 2\beta\gamma i r^{\gamma-1}\right)\dfrac{dy}{dr}$ $+ \left[\dfrac{\alpha^2 - p^2\gamma^2}{r^2} \mp \beta\gamma(\gamma - 2\alpha)i r^{\gamma-2}\right]y = 0$	$y = r^\alpha e^{\pm i\beta r^\gamma} Z_p(\beta r^\gamma)$

[a]If p is *not* zero or a positive integer, $Z_p(kr) = J_p(kr) + J_{-p}(kr)$ and $Z_p(ikr) = I_p(kr) + I_{-p}(kr)$. That is, replace $Y_p(kr)$ with $J_{-p}(kr)$ and $K_p(kr)$ with $I_{-p}(kr)$.

62

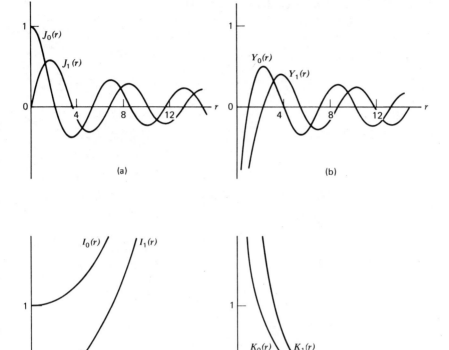

Figure 2.9-1.   Bessel functions.

$$= i^p \sum_{j=0}^{\infty} \frac{(-1)^{2j}(kr/2)^{2j+p}}{j!(j+p)!}$$

$$= i^p \sum_{j=0}^{\infty} \frac{(kr/2)^{2j+p}}{j!(j+p)!} \qquad \left\{ \text{since } (-1)^{2j} = \left[(-1)^2\right]^j = 1^j = 1 \right\}.$$

Therefore,

$$J_p(ikr) = i^p I_p(kr). \qquad (2.9\text{-}18)$$

Table 2.9-2 Relations Between Bessel Functions[a]

A. Algebraic

(1) $J_{p-1} + J_{p+1} = \dfrac{2p}{r} J_p$
(3) $I_{p-1} - I_{p+1} = \dfrac{2p}{r} I_p$

(2) $Y_{p-1} + Y_{p+1} = \dfrac{2p}{r} Y_p$
(4) $K_{p-1} - K_{p+1} = -\dfrac{2p}{r} K_p$

B. Differential

(1) $rJ_p' = pJ_p - rJ_{p+1}$
(11) $rI_p' = pI_p + rI_{p+1}$

(2) $rJ_p' = -pJ_p + rJ_{p-1}$
(12) $rI_p' = -pI_p + rI_{p-1}$

(3) $2J_p' = J_{p-1} - J_{p+1}$
(13) $2I_p' = I_{p-1} + I_{p+1}$

(4) $(r^p J_p)' = r^p J_{p-1}$
(14) $(r^p I_p)' = r^p I_{p-1}$

(5) $(r^{-p} J_p)' = -r^{-p} J_{p+1}$
(15) $(r^{-p} I_p)' = r^{-p} I_{p+1}$

(6) $rY_p' = pY_p - rY_{p+1}$
(16) $rK_p' = pK_p - rK_{p+1}$

(7) $rY_p' = -pY_p + rY_{p-1}$
(17) $rK_p' = -pK_p - rK_{p-1}$

(8) $2Y_p' = Y_{p-1} - Y_{p+1}$
(18) $2K_p' = -K_{p-1} - K_{p+1}$

(9) $(r^p Y_p)' = r^p Y_{p-1}$
(19) $(r^p K_p)' = -r^p K_{p-1}$

(10) $(r^{-p} Y_p)' = -r^{-p} Y_{p+1}$
(20) $(r^{-p} K_p)' = -r^{-p} K_{p+1}$

C. Integral

(1) $\displaystyle\int r^p J_{p-1}\, dr = r^p J_p$
(5) $\displaystyle\int r^p I_{p-1}\, dr = r^p I_p$

(2) $\displaystyle\int r^{-p} J_{p+1}\, dr = -r^{-p} J_p$
(6) $\displaystyle\int r^{-p} I_{p+1}\, dr = r^{-p} I_p$

(3) $\displaystyle\int r^p Y_{p-1}\, dr = r^p Y_p$
(7) $\displaystyle\int r^p K_{p-1}\, dr = -r^p K_p$

(4) $\displaystyle\int r^{-p} Y_{p+1}\, dr = -r^{-p} Y_p$
(8) $\displaystyle\int r^{-p} K_{p+1}\, dr = -r^{p+1} K_p$

(9) $\displaystyle\int J_p\, dr = 2 \sum_{k=0}^{\infty} J_{p+2k+1}$

(10) $\displaystyle\int r J_p(\alpha r) J_p(\beta r)\, dr = \dfrac{\beta r J_p(\alpha r) J_{p-1}(\beta r) - \alpha r J_{p-1}(\alpha r) J_p(\beta r)}{\alpha^2 - \beta^2}$

That is, the two differ by only a constant factor i^p that can be absorbed into the integration constants C_1 and C_2. Either of the two forms can be used in order to avoid the imaginary i that in some cases arises, such as in Example 6.1-3. For the special case that $p = 0$,

$$J_0(ikr) = I_0(kr). \qquad (2.9\text{-}19)$$

Table 2.9-2 (*Continued*)

D. Orthogonality

When α_i and α_j are two different positive roots of

(1) $J_p(r)=0$

(2) $J'_p(r)=0$

(3) $rJ'_p(r)+hJ_p(r)=0$, where h is a constant

then correspondingly we have the following orthogonality relations:

For (1)

$$\int_0^1 rJ_p(\alpha_i r)J_p(\alpha_j r)\,dr=\tfrac{1}{2}\big[J'_p(\alpha_i)\big]^2\delta_{ij}$$

For (2)

$$\int_0^1 rJ_p(\alpha_i r)J_p(\alpha_j r)\,dr=\tfrac{1}{2}\left(1-\frac{p^2}{\alpha_i^2}\right)\big[J_p(\alpha_i)\big]^2\delta_{ij}$$

For (3)

$$\int_0^1 rJ_p(\alpha_i r)J_p(\alpha_j r)\,dr=\tfrac{1}{2}\left(1+\frac{h^2-p^2}{\alpha_i^2}\right)[J_p(\alpha_i)]^2\delta_{ij}$$

where $\delta_{ij}=0$ if $i\neq j$ and $\delta_{ii}=1$.

---

[a]Note that the Bessel functions here are for $J_p(r),\ldots$, *not* $J_p(br),\ldots$ except as otherwise noted. Extracted from Ch.E. 224 lecture notes of Professor R. B. Bird, U. of Wisconsin (1960). See also [8] and [11].

The values of these Bessel functions are tabulated in well-known references and handbooks [11]. But it is important for us to know, at least qualitatively, how they behave. This we show in Figure 2.9-1. As will be seen in later chapters, even such sketchy information is valuable. For example, the fact that Y_0 equals $-\infty$ at $r=0$ often means that the integration constant C_2 in Eq. 2.9-8 must be equal to 0, because otherwise the solution would become infinite, contrary to physical requirement.

Another important piece of information is where these curves cross the r-axis. For example, for $J_0(x)=0$ we have $x=2.40483\ldots$, $5.52009\ldots$, $8.65373\ldots$, and so on, which we call the roots, or zeros, of the Bessel function. This usually comes from the other boundary condition, which would normally give us the other integration constant, such as C_1. To be able to use this, many relationships between Bessel functions have to be employed. These we list in Table 2.9-2; the "orthogonality" relationships will be explained in Section 2.11.

2.10 GEOMETRIC SYSTEMATICITY

Mathematics is the most logically beautiful language the human race has ever known. One example of this is the systematic variation of differential equations and their solutions as we go from one fundamental geometry to another.

Among the differential equations listed in Table 2.8-1 is the type

$$\frac{d^2y}{dx^2} \pm m^2 y = 0, \tag{2.10-1}$$

for which the solution is

$$y = C_1 e^{imx} + C_2 e^{-imx}$$

$$= C_1' \sin mx + C_2' \cos mx \tag{2.10-2}$$

if we take the plus sign and

$$y = C_3 e^{mx} + C_4 e^{-mx}$$

$$= C_3' \sinh mx + C_4' \cosh mx \tag{2.10-3}$$

if we take the minus sign. This category of equations we frequently encounter in problems involving systems of rectangular shape.

In Section 2.9 we mentioned that for cylindrical geometry, particularly infinite cylinders, the differential equation involved, Eq. 2.9-1, is

$$\frac{1}{r}\frac{d}{dr}\left(r\frac{dy}{dr}\right) \pm b^2 y = 0$$

and the solutions have been given as the Bessel functions. Now if we proceed to the spherical geometry, we will find, as will be shown in Chapter 4, that the differential equation becomes

$$\frac{1}{r^2}\frac{d}{dr}\left(r^2\frac{dy}{dr}\right) \pm b^2 y = 0, \tag{2.10-4}$$

which can be transformed, by taking $f = ry$, to

$$\frac{d^2f}{dr^2} \pm b^2 f = 0,$$

which is similar to Eq. 2.10-1. From Eqs. 2.10-2 and 2.10-3 we can see that the solution to Eq. 2.10-4 is

$$y = \frac{f}{r} = \frac{C_1}{r} e^{ibr} + \frac{C_2}{r} e^{-ibr}$$

$$= \frac{C_1'}{r} \sin br + \frac{C_2'}{r} \cos br \qquad (2.10\text{-}5)$$

if we take the plus sign and

$$y = \frac{f}{r} = \frac{C_3}{r} e^{br} + \frac{C_4}{r} e^{-br}$$

$$= \frac{C_3'}{r} \sinh br + \frac{C_4'}{r} \cosh br \qquad (2.10\text{-}6)$$

if we take the minus sign. These we have summarized in Table 2.10-1, from which we can immediately observe that they generally follow the following pattern:

$$\frac{1}{r^j} \frac{d}{dr}\left(r^j \frac{dy}{dr}\right) \pm b^2 y = 0, \qquad (2.10\text{-}7)$$

where $j = 0, 1, 2$ for rectangular, cylindrical, and spherical geometries, respectively. If we assume the same series solution as Eq. 2.9-2 and go through the same procedure as before, we will find that the first solution to Eq. 2.10-7 with the negative sign is

$$y = a_0 \left\{ 1 + \sum_{p=1}^{\infty} \left[\prod_{q=1}^{p} \frac{(-b)^2}{2q(j+2q-1)} \right] r^{2p} \right\} \qquad (2.10\text{-}8)$$

which, if we substitute the j values for the three different geometries, will give three series that are the expansions of the various trigonometric, Bessel, and hyperbolic functions listed in Table 2.10-1.* Actually this

---

*$\prod$ denotes *product* just as Σ denotes *sum*. For example, $\prod_{i=1}^{n} f(x_i) = f(x_1) \cdot f(x_2) \cdots f(x_{n-1}) \cdot f(x_n)$.

68

Table 2.10-1 Solutions for Differential Equations in the Three Basic Geometries

Geometry	Rectangular	Cylindrical	Spherical
Differential equation	$\dfrac{d^2y}{dx^2} * b^2y = 0$	$\dfrac{1}{r}\dfrac{d}{dr}\left(r\dfrac{dy}{dr}\right) * b^2y = 0$	$\dfrac{1}{r^2}\dfrac{d}{dr}\left(r^2\dfrac{dy}{dr}\right) * b^2y = 0$
Equation	$y = C_1\cos bx + C_2\sin bx$	$y = C_1 J_0(br) + C_2 Y_0(br)$	$y = \dfrac{C_1}{r}\cos br + \dfrac{C_2}{r}\sin br$
Graph			
Equation	$y = C_1 e^{bx} + C_2 e^{-bx}$ $= C_1'\cosh bx + C_2'\sinh bx$	$y = C_1 I_0(br) + C_2 K_0(br)$	$y = \dfrac{C_1}{r}e^{br} + \dfrac{C_2}{r}e^{-br}$ $= \dfrac{C_1'}{r}\cosh br + \dfrac{C_2'}{r}\sinh br$
Graph			

Solution for * = +

Solution for * = −

should be no surprise. We shall omit the second solution with the hope that the above has demonstrated the systematic mathematical variation with geometry.

2.11 ORTHOGONALITY

Two functions $f_m(x)$ and $f_n(x)$ are said to be *orthogonal* over an interval (a,b) if

$$\int_a^b f_m(x)f_n(x)\,dx = 0. \tag{2.11-1}$$

More generally, if

$$\int_a^b w(x)f_m(x)f_n(x)\,dx = 0, \tag{2.11-2}$$

Table 2.11-1 Selected Orthogonality Relationships[a] $(m,n = \text{integer})$

(A) $\displaystyle\int_0^1 \sin m\pi\xi \sin n\pi\xi\,d\xi = \begin{cases} 0 & \text{when } m \neq n \text{ or } m = n = 0 \\ \frac{1}{2} & \text{when } m = n \neq 0 \end{cases}$

(B) $\displaystyle\int_0^1 \cos m\pi\xi \cos n\pi\xi\,d\xi = \begin{cases} 0 & \text{when } m \neq n \\ \frac{1}{2} & \text{when } m = n \neq 0 \\ 1 & \text{when } m = n = 0 \end{cases}$

(C) $\displaystyle\int_0^1 \cos\gamma_m\xi \cos\gamma_n\xi\,d\xi = \begin{cases} 0 & \text{when } m \neq n \\ \frac{1}{2}\left[\dfrac{\gamma_n^2 + N + N^2}{\gamma_n^2 + N^2}\right] & \text{when } m = n \end{cases}$

where the γ_n's are the roots of $\gamma\tan\gamma = N$

(D) $\displaystyle\int_0^1 \sin\gamma_m\xi \sin\gamma_n\xi\,d\xi = \begin{cases} 0 & \text{when } m \neq n \\ \frac{1}{2}\left[\dfrac{\gamma_n^2 + N^2 - N}{\gamma_n^2 + N^2}\right] & \text{when } m = n \end{cases}$

where the γ_n's are the roots of $\gamma\cot\gamma = N$

[a]For orthogonality relationships of Bessel functions, see Part (D) of Table 2.9-2. More on sine and cosine functions can be found as footnotes on p. 468.

where $w(x)$ is a *weighting function*, they are said to be *orthogonal with respect to the weighting function w(x)* over that interval.

We shall omit all of the theories and derivations here but simply list in Table 2.11-1 a few of the functions, in addition to the Bessel functions we have mentioned in Table 2.9-2.

Even without the theories, we can see that in this table, for example, formula (A) is true. From formula (13) of Table 2.4-1, the integration table, we can see that

$$\int_0^1 \sin m\pi\xi \sin n\pi\xi \, d\xi = \left[\frac{\sin(m-n)\pi\xi}{2(m-n)\pi} - \frac{\sin(m+n)\pi\xi}{2(m+n)\pi} \right]_0^1$$

$$= \frac{\sin(m-n)\pi}{2(m-n)\pi} - \frac{\sin(m+n)\pi}{2(m+n)\pi}. \qquad (2.11\text{-}3)$$

Since m and n are integers, so are $(m+n)$ and $(m-n)$, which means that the above expression is equal to zero $(\ldots, \sin[-180°], \sin 0°, \sin 180°, \sin 360°, \ldots$ are all zero) unless $m=n$, in which case the second term still vanishes but the first term becomes $0/0$ and via L'Hopital's rule (Eq. 2.3-1) it can be found to be*

$$\lim_{m \to n} \frac{(d/dm)[\sin(m-n)\pi]}{(d/dm)[2(m-n)\pi]} = \lim_{m \to n} \frac{\pi\cos(m-n)\pi}{2\pi}$$

$$= \frac{\pi}{2\pi} = \tfrac{1}{2}.$$

The application of orthogonality is mainly in evaluating the integration constant of a series solution, as will be seen in Examples 6.1-2, 6.2-5, 6.3-2 (Case b), 6.3-3, 10.4-1, and 10.5-1.

2.12 LAPLACE TRANSFORM

One of the most powerful methods of solving *linear* differential equations, whether ordinary or partial, whether a single one or a system of simultaneous ones, is the Laplace transform method, because it reduces an ordinary differential equation, for example, to an algebraic equation that even high-school students can solve. The advantage is even more apparent

---

*This is true only for $m = n \neq 0$. If $m = n = 0$, the second term in Eq. 2.11-3 is also $0/0$, which, upon using L'Hopital's rule, can be shown to be equal to $1/2$. Thus, the entire definite integral is equal to 0, as given in formula (A) of Table 2.11-1.

when we have, for example, a system of simultaneous differential equations relating more than one dependent variable, in which case the reduction to a system of simultaneous algebraic equations can really save the day. A partial differential equation can be transformed into an ordinary differential equation, thereby making it easier to solve, by taking the Laplace transform with respect to one of its two independent variables. These schemes are illustrated in Figure 2.12-1.

The Laplace transform is only one of the many possible integral transforms. In general, a transform is one in which a function of t (or, a function defined in the t-space, as the mathematical jargon goes), after multiplied by a *kernel* $K(p,t)$ and integrated with respect to t and between limits a and b, is altered into a function of p. Symbolically we write

$$\bar{f}(p) = \int_a^b f(t)K(p,t)\,dt \qquad (2.12\text{-}1)$$

or simply $\bar{f} = Tf$.

According to this, anybody could invent any kind of transform and attach his own name to it. However, for this to be a useful tool, we must add the requirement that this be one that can be inverted and that there is a unique one-to-one correspondence between the function, $f(t)$, in the t-space and the function, $\bar{f}(p)$, in the p-space, or *image space*. That is,

$$f(t) = \int_\alpha^\beta \bar{f}(p)H(t,p)\,dp \qquad (2.12\text{-}2)$$

or $f = T^{-1}\bar{f}$. *

The Laplace transform is one of this type, the kernel being $K(p,t)=e^{-pt}$ and the limits of integration 0 and ∞. In this case we write

$$\mathcal{L}\{f(t)\} = \bar{f}(p) = \int_0^\infty f(t)e^{-pt}\,dt. \qquad (2.12\text{-}3)$$

Let us take a few simple examples.

Example 2.12-1.

$$f(t) = 1$$

$$\bar{f}(p) = \int_0^\infty e^{-pt}\,dt$$

$$= \left[-\frac{1}{p}e^{-pt}\right]_0^\infty = 0 - \left(-\frac{1}{p}\right) = \frac{1}{p}$$

*The "-1" here denotes "inverse" rather than "reciprocal"; see also footnote to Table 2.2-1 on p. 27.

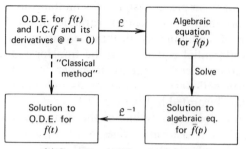

(a) To ordinary differential equation

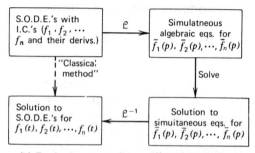

(b) To simultaneous ordinary differential equations

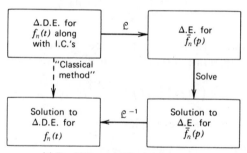

(c) To difference differential equation

Figure 2.12-1. Applications of Laplace transform.

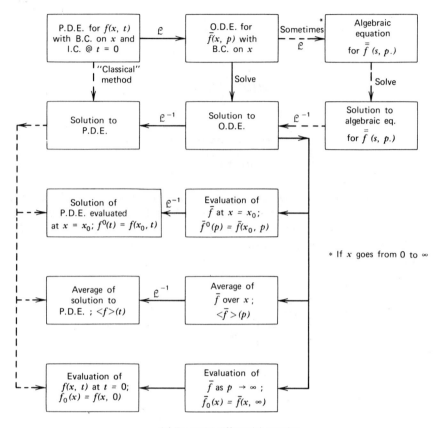

(d) To partial differential equation

Figure 2.12-1. (*Continued*)

Example 2.12-2.

$$f(t) = e^{at}$$

$$\bar{f}(p) = \int_0^\infty e^{-(p-a)t}\, dt$$

$$= -\frac{1}{p-a}\left[e^{-(p-a)t}\right]_0^\infty = \frac{1}{p-a}$$

Example 2.12-3.

$$f(t) = \cos at$$

$$\bar{f}(p) = \int_0^\infty (\cos at) e^{-pt} dt*$$

$$= \left[\frac{e^{-pt}(a \sin at - p \cos at)}{(-p)^2 + a^2} \right]_0^\infty = \frac{p}{p^2 + a^2},$$

and so on. This way we can develop an entire table of such specific functions, as shown in Part B of Table 2.12-1.† Part A of the same table deals with some general formulas such as those for df/dt and d^2f/dt^2. We shall derive these two because they are the main ingredients of most differential equations.

Example 2.12-4.

$$\mathcal{L}\left\{ \frac{df}{dt} \right\} = \int_0^\infty \frac{df}{dt} e^{-pt} dt = \int_{t=0}^{t=\infty} e^{-pt} df$$

$$= \left[fe^{-pt} \right]_{t=0}^{t=\infty} - \int_{t=0}^{t=\infty} f d(e^{-pt})^\ddagger$$

$$= -f(t=0) + p \int_0^\infty fe^{-pt} dt.$$

Noting that the definite integral in the second term is the Laplace transform of $f(t)$, we have

$$\mathcal{L}\left\{ \frac{df}{dt} \right\} = -f(0) + p\bar{f}(p).$$

Exercise 2.12-1.   Following a similar procedure, show that

$$\mathcal{L}\left\{ \frac{d^2f}{dt^2} \right\} = p^2 \bar{f}(p) - pf(0) - \frac{df(0)}{dt}.$$

*Formula 18, Table 2.4-1.

†A more complete table can be found in Erdelyi, A.; Magnus, W.; Oberhettinger, F.; and Tricomi, F. G. *Tables of Integral Transforms*, Vols. I and II. McGraw-Hill, New York, 1954.

‡Integration by parts, formula (c), Table 2.4-1.

Table 2.12-1 Selected Laplace Transforms*

Item No.	Transform	Function
(A1)	$\bar{f}(p) = \mathcal{L}\{f(t)\} = \int_0^\infty e^{-pt}f(t)\,dt$	$f(t)$
(A2)	$a\bar{f}(p) + b\bar{g}(p)$	$af(t) + bg(t)$
(A3)	$p\bar{f}(p) - f(0)$	$\dfrac{df(t)}{dt}$
(A4)	$p^2\bar{f}(p) - pf(0) - \dfrac{df(0)}{dt}$	$\dfrac{d^2f(t)}{dt^2}$
(A5)	$p^n\bar{f}(p) - \displaystyle\sum_{k=1}^{n} p^{n-k}\dfrac{d^{k-1}f(0)}{dt^{k-1}}$	$\dfrac{d^nf(t)}{dt^n}$
(A6)	$\dfrac{1}{p^n}\bar{f}(p)$	$\overbrace{\int_0^t \cdots \int_0^t}^{n\ times} f(t)\,\underbrace{dt\ldots dt}_{n\ times}$
(A7)	$(-)^n \dfrac{d^n\bar{f}(p)}{dp^n}$	$t^nf(t)$
(A8)	$\bar{f}(p-a)$	$e^{at}f(t)$
(A9)	$e^{-ap}\bar{f}(p)$	$\begin{cases} f(t-a) & \text{when} \quad t > a \\ 0 & \text{when} \quad t < a \end{cases}$
(A10)	$\bar{f}(p)\bar{g}(p)$	$\displaystyle\int_0^t f(t-s)g(s)\,ds$ or $\displaystyle\int_0^t f(s)g(t-s)\,ds$
(A11)	$\dfrac{\displaystyle\int_0^a e^{-pt}f(t)\,dt}{1 - e^{-ap}}$	$f(t+a) = f(t)$ Function that repeats itself with interval a

Table 2.12-1 *(Continued)*

Item No.	Transform $\bar{f}(p) = \int_0^\infty e^{-pt} f(t)\, dt$	Function $f(t)$
(B1)	$\dfrac{1}{p}$	1
(B2)	$\dfrac{1}{p^2}$	t
(B3)	$\dfrac{1}{p^n}, \quad n = 1, 2, 3, \ldots$	$\dfrac{t^{n-1}}{(n-1)!}$
(B4)	$\dfrac{1}{\sqrt{p}}$	$\dfrac{1}{\sqrt{\pi t}}$
(B5)	$\dfrac{1}{p^{3/2}}$	$2\sqrt{\dfrac{t}{\pi}}$
(B6)	$\dfrac{1}{p^{n+1/2}}, \quad n = 1, 2, 3, \ldots$	$\dfrac{2^n t^{n-1/2}}{[1 \cdot 3 \cdot 5 \cdots (2n-1)]\sqrt{\pi}}$
(B7)	$\dfrac{1}{p-a}$	e^{at}
(B8)	$\dfrac{1}{(p-a)^2}$	te^{at}
(B9)	$\dfrac{1}{(p-a)^n}, \quad n = 1, 2, 3, \ldots$	$\dfrac{1}{(n-1)!} t^{n-1} e^{at}$
(B10)	$\dfrac{1}{(p-a)(p-b)}, \quad a \neq b$	$\dfrac{1}{a-b}(e^{at} - e^{bt})$
(B11)	$\dfrac{p}{(p-a)(p-b)}, \quad a \neq b$	$\dfrac{1}{a-b}(ae^{at} - be^{bt})$
(B12)	$\dfrac{1}{(p-a)(p-b)(p-c)}, \ a \neq b \neq c$	$-\dfrac{(b-c)e^{at} + (c-a)e^{bt} + (a-b)e^{ct}}{(a-b)(b-c)(c-a)}$
(B13)	$\dfrac{1}{p^2 + a^2}$	$\dfrac{1}{a}\sin at$
(B14)	$\dfrac{p}{p^2 + a^2}$	$\cos at$
(B15)	$\dfrac{1}{p^2 - a^2}$	$\dfrac{1}{a}\sinh at$
(B16)	$\dfrac{p}{p^2 - a^2}$	$\cosh at$
(B17)	$\dfrac{p}{(p^2 + a^2)(p^2 + b^2)}, \quad a^2 \neq b^2$	$\dfrac{\cos at - \cos bt}{b^2 - a^2}$

Table 2.12-1 (*Continued*)

Item No.	Transform $\bar{f}(p) = \int_0^\infty e^{-pt} f(t)\, dt$	Function $f(t)$
(B18)	$\dfrac{1}{(p-a)^2 + b^2}$	$\dfrac{1}{b} e^{at} \sin bt$
(B19)	$\dfrac{p-a}{(p-a)^2 + b^2}$	$e^{at} \cos bt$
(B20)	$\dfrac{1}{\sqrt{p^2 + a^2}}$	$J_0(at)$
(B21)	$\dfrac{\left(\sqrt{p^2 + a^2} - p\right)^j}{\sqrt{p^2 + a^2}} \qquad j > 1$	$a^j J_j(at)$
(B22)	$\dfrac{1}{p} e^{-k/p}$	$J_0(2\sqrt{kt}\,)$
(B23)	$\dfrac{1}{\sqrt{p}} e^{-k/p}$	$\dfrac{1}{\sqrt{\pi t}} \cos 2\sqrt{kt}$
(B24)	$\dfrac{1}{\sqrt{p}} e^{k/p}$	$\dfrac{1}{\sqrt{\pi t}} \cosh 2\sqrt{kt}$
(B25)	$e^{-k\sqrt{p}}, \qquad k > 0$	$\dfrac{k}{2\sqrt{\pi t^3}} \exp\!\left(-\dfrac{k^2}{4t}\right)$
(B26)	$\dfrac{1}{p} e^{-k\sqrt{p}}, \qquad k \geqslant 0$	$\mathrm{erfc}\!\left(\dfrac{k}{2\sqrt{t}}\right)$
(B27)	$\dfrac{1}{\sqrt{p}} e^{-k\sqrt{p}}, \qquad k \geqslant 0$	$\dfrac{1}{\sqrt{\pi t}} \exp\!\left(-\dfrac{k^2}{4t}\right)$
(B28)	$p^{-3/2} e^{-k\sqrt{p}}, \qquad k \geqslant 0$	$2\sqrt{\dfrac{t}{\pi}}\left[\exp\!\left(-\dfrac{k^2}{4t}\right)\right] - k\,\mathrm{erfc}\!\left(\dfrac{k}{2\sqrt{t}}\right)$
(B29)	$\ln \dfrac{p-a}{p-b}$	$\dfrac{1}{t}(e^{bt} - e^{at})$

*$f(0), df(0)/dt, \ldots$ mean, respectively, $f(t), df/dt, \ldots$ evaluated at $t = 0$; a, b, c and k are constants.

As we have indicated before, being able to take the Laplace transform is only half of the work. To be able to solve a differential equation completely, we must know how to invert the solution of the algebraic equation. For simple ones this involves only consulting a table. But in most practical cases we will face a complex situation involving various types of combinations of the elementary functions listed in Part B of Table 2.12-1. This is where Part A of that table comes in. For example, when we have a sum of elementary functions, formula (A2) obviously governs. If they form a product, formula (A10), generally known as the *convolution integral*, is the choice. If we have a quotient of two functions, we use the Heaviside partial fraction expansion theorem (HPFET), which states that in

$$\mathcal{L}^{-1}\left\{ \frac{\bar{f}(p)}{\bar{g}(p)} \right\}$$

if (1) $\bar{f}(p)$ and $\bar{g}(p)$ are polynomials in p of orders m and n, respectively, both of which are integers, (2) $n > m$*, and (3) there are no multiple roots of $\bar{g}(p) = 0$,† then following the same method by which we usually expand functions in quotient form into partial fractions, we can write

$$\frac{\bar{f}(p)}{\bar{g}(p)} = \sum_{k=1}^{n} \frac{\bar{f}(a_k)}{\bar{g}'(a_k)} \cdot \frac{1}{p - a_k}$$

so that

$$\mathcal{L}^{-1}\left\{ \frac{\bar{f}(p)}{\bar{g}(p)} \right\} = \sum_{k=1}^{n} \frac{\bar{f}(a_k)}{\bar{g}'(a_k)} e^{a_k t} \tag{2.12-4}$$

where $\bar{g}'(p) = (d/dp)\bar{g}(p)$ and a_k are the roots of $\bar{g}(p) = 0$.

Specific applications of the Laplace transform are given in Examples 2.13-2, 6.1-3, 6.2-4, 6.3-2, and 10.5-1. Generally, they follow the patterns represented by the block diagrams shown in Figure 2.12-1.

2.13 PARTIAL DIFFERENTIAL EQUATIONS

When a differential equation involves a dependent variable that is a function of more than one variable, partial derivatives appear and we call this a *partial differential equation.*

---

*If not, we can divide $\bar{g}(p)$ into $\bar{f}(p)$ until this condition is satisfied.

†If so, a modified formula is used (see Wylie, C. R., Jr. *Advanced Engineering Mathematics*, 2nd ed. McGraw-Hill: New York, 1960, pp. 319-320).

The equations of change are examples of partial differential equations, except that, as in Tables 4.12-1 through 4.12-3, they are difficult to solve, either analytically or numerically. In a practical problem, however, all except two or three terms can usually be eliminated for various reasons. The remaining equation can be solved by one or more of several ways, depending on the nature of the equation.

The method of *combination of variables* is most suitable for differential equations with initial and boundary conditions* of the type given in the following example. It is sometimes called the *similarity solution* [13] or *group transformation* [14]. As an illustration of the analogy between the three types of transport, the general mathematics employed in this example can be applied to start-up flow in a semi-infinite body of fluid (Example 4.1-1 in Ref. 15), the heating of a semi-infinite solid (Example 11.1-1 in Ref. 15), and the diffusion of substance in a semi-infinite medium [16].

Example 2.13-1.

$$\frac{\partial \phi}{\partial t} = \nu \frac{\partial^2 \phi}{\partial x^2} \tag{2.13-1}$$

I.C.:   at $t \leqslant 0$, $\phi = 0$ for all $x > 0$

B.C.1:   at $x = 0$, $\phi = 1$ for all $t > 0$

B.C.2:   at $x = \infty$, $\phi = 0$ for all $t > 0$

where $\phi = \phi(x, t)$ and that ν is a constant. We can see from the above equation that in order for it to be dimensionally uniform, the group $x^2 / \nu t$ must be dimensionless. Moreover, if we combine the two independent variables x and t, to form a new independent variable, the I.C. and B.C.2 can be satisfied simultaneously. In other words, even though this is only a "hunch" that they can be combined and there is no guarantee that it will definitely work, it has passed the first test.

To completely illustrate this process, let us call this new variable

$$\eta = \frac{x}{b\sqrt{\nu t}}, \tag{2.13-2}$$

where b is a constant that may not be necessary at all. However, we shall see later on that we can choose its value so as to make mathematical manipulation easier without affecting the result.

*Abbreviated "I.C." and "B.C.", respectively.

From Eq. 2.13-2 we can write

$$\frac{\partial \phi}{\partial t} = \frac{d\phi}{d\eta} \cdot \frac{\partial \eta}{\partial t} = \frac{d\phi}{d\eta} \cdot \left(-\frac{1}{2} \right) \frac{x}{bt\sqrt{\nu t}}$$

$$= -\frac{1}{2} \cdot \frac{d\phi}{d\eta} \cdot \frac{\eta}{t} \tag{2.13-3}$$

$$\frac{\partial \phi}{\partial x} = \frac{d\phi}{d\eta} \cdot \frac{\partial \eta}{\partial x} = \frac{d\phi}{d\eta} \cdot \frac{1}{b\sqrt{\nu t}} \tag{2.13-4}$$

$$\frac{\partial^2 \phi}{\partial x^2} = \frac{\partial}{\partial x} \left(\frac{\partial \phi}{\partial x} \right) = \frac{1}{b\sqrt{\nu t}} \frac{\partial}{\partial x} \left(\frac{d\phi}{d\eta} \right)$$

$$= \frac{1}{b\sqrt{\nu t}} \cdot \frac{d}{d\eta} \left(\frac{d\phi}{d\eta} \right) \cdot \frac{\partial \eta}{\partial x}$$

$$= \left(\frac{1}{b\sqrt{\nu t}} \right)^2 \frac{d^2\phi}{d\eta^2} = \frac{\eta^2}{x^2} \cdot \frac{d^2\phi}{d\eta^2}. \tag{2.13-5}$$

Substituting Eqs. 2.13-3 and 2.13-5 into Eq. 2.13-1 gives, after proper cancellations and manipulations,

$$\frac{d^2\phi}{d\eta^2} + \tfrac{1}{2}b^2\eta \frac{d\phi}{d\eta} = 0. \tag{2.13-6}$$

If we write $\phi' = d\phi/d\eta$, the solution of this equation can be easily observed (also see formula D3, Table 2.8-1):

$$\frac{d\phi'}{d\eta} + \tfrac{1}{2}b^2\eta \, \phi' = 0. \tag{2.13-7}$$

If we further write

$$\frac{d\phi'}{\phi'} + \tfrac{1}{2}b^2\eta \, d\eta = 0,$$

it can now be easily seen that if we take $b = 2$,* the solution will be the simplest, since it carries no constant factor:

$$\ln \phi' + \eta^2 = C_1'$$

*If we work enough of this type of problem, we will find that if $\eta = x/\sqrt[\alpha]{\beta t}$, $\beta = \alpha^2$ is a good rule of thumb.

or

$$\phi' = C_1 \exp(-\eta^2) \qquad (2.13\text{-}8)$$

where $C_1 = \exp(C_1')$. Remember that $\phi' = d\phi/d\eta$ and that we still have to solve this once more. Thus,

$$\frac{d\phi}{d\eta} = C_1 \exp(-\eta^2) \qquad (2.13\text{-}9)$$

Interestingly enough this seemingly simple expression *cannot* be integrated analytically.† We thus write

$$\phi = C_1 \int e^{-\eta^2} d\eta + C_2 \qquad (2.13\text{-}10)$$

and apply the boundary conditions, which have now become

B.C.1′: at $\eta = 0$, $\phi = 1$

B.C.2′: at $\eta = \infty$ $\phi = 0$

so that we can generate two equations from Eq. 2.13-10:

$$1 = C_1 \int \exp(-\eta^2) \, d\eta \Big|_{\eta=0} + C_2 \qquad (2.13\text{-}11)$$

$$0 = C_1 \int \exp(-\eta^2) \, d\eta \Big|_{\eta=\infty} + C_2. \qquad (2.13\text{-}12)$$

Eliminating C_2 from these two last equations yields:

$$-1 = C_1 \int_0^\infty \exp(-\eta^2) \, d\eta$$

The definite integral $\int_0^\infty \exp(-\eta^2)d\eta$ can be evaluated by expanding the

†Except in series form, since

$$e^{-\eta^2} = 1 - \eta^2 + \frac{\eta^4}{2!} - \frac{\eta^6}{3!} + \cdots$$

[see formula (1), Table 2.5-1] so that

$$\int e^{-\eta^2} d\eta = \eta - \frac{\eta^3}{3 \cdot 1!} + \frac{\eta^5}{5 \cdot 2!} - \frac{\eta^7}{7 \cdot 3!} + \cdots + C.$$

integrand in infinite series and shown to be exactly equal to $\sqrt{\pi}/2$. Thus, we have

$$C_1 = -\frac{2}{\sqrt{\pi}} \qquad (2.13\text{-}13)$$

Eliminating C_2 by subtracting Eq. 2.13-11 from Eq. 2.13-10 yields

$$\phi = 1 + C_1 \int_0^\eta \exp(-\eta^2)\, d\eta$$

Upon substitution of Eq. 2.13-13, we have, finally

$$\phi = 1 - \frac{2}{\sqrt{\pi}} \int_0^\eta \exp(-\eta^2)\, d\eta, \qquad (2.13\text{-}14)$$

where $(2/\sqrt{\pi})\int_0^\eta \exp(-\eta^2)\, d\eta$ is a well-tabulated function called the *error function*, or erf(η). Its values are given in Table 2.13-1. The entire expression $[1 - \text{erf}(\eta)]$ is called the *complementary error function*, denoted by erfc(η).

We also note that in Eq. 2.13-14 the η in the integrand and differential can be replaced by any symbol without changing the result of the problem, because in the final analysis, it will be replaced by the integration limits anyway:

$$\phi = 1 - \frac{2}{\sqrt{\pi}} \int_0^\eta \exp(-\eta^2)\, d\eta = 1 - \frac{2}{\sqrt{\pi}} \int_0^\eta \exp(-\xi^2)\, d\xi = \cdots . \qquad (2.13\text{-}15)$$

Such a variable is called a *dummy variable* because it does not have any meaning itself in this context.

It appears that some intuition or experience is required in choosing the right combination of variables and that there are few, if any, "sure-fire" formulas. However, its ultimate justification is that it works, although the boundary and initial conditions must first pass the compatibility test, as shown earlier in the above example.

The second often-used method in solving partial differential equations is the *separation of variables* method. Instead of combining the two independent variables, the dependent variable is separated into two or more functions each of which is a function of only one independent variable. Applications of this are given in Examples 6.1-2, 6.1-4, 6.2-4, 6.2-5, 6.2-6, 6.3-2, 6.3-3, 10.3-2, 10.4-1, and 10.5-1. In others, a function of as many as four independent variables can be separated (see Problem 11.K_3 in Ref. 15).

Table 2.13-1 Values of the error function, erf (x)

x	0.00	0.01	0.02	0.03	0.04	0.05	0.06	0.07	0.08	0.09
0.0	0.00000	0.01128	0.02256	0.03384	0.04511	0.05637	0.06762	0.07886	0.09008	0.10128
0.1	0.11246	0.12362	0.13476	0.14587	0.15695	0.16800	0.17901	0.18999	0.20094	0.21184
0.2	0.22270	0.23352	0.24430	0.25502	0.26570	0.27633	0.28690	0.29742	0.30788	0.31828
0.3	0.32863	0.33891	0.34913	0.35928	0.36936	0.37938	0.38933	0.39921	0.40901	0.41874
0.4	0.42839	0.43797	0.44747	0.45689	0.46623	0.47548	0.48466	0.49375	0.50275	0.51167
0.5	0.52050	0.52924	0.53790	0.54646	0.55494	0.56332	0.57162	0.57982	0.58792	0.59594
0.6	0.60386	0.61168	0.61941	0.62705	0.63459	0.64203	0.64938	0.65663	0.66378	0.67084
0.7	0.67780	0.68467	0.69143	0.69810	0.70468	0.71116	0.71754	0.72382	0.73001	0.73610
0.8	0.74210	0.74800	0.75381	0.75952	0.76514	0.77067	0.77610	0.78144	0.78669	0.79184
0.9	0.79691	0.80188	0.80677	0.81156	0.81627	0.82089	0.82542	0.82987	0.83423	0.83851
1.0	0.84270	0.84681	0.85084	0.85478	0.85865	0.86244	0.86614	0.86977	0.87333	0.87680
1.1	0.88021	0.88353	0.88679	0.88997	0.89308	0.89612	0.89910	0.90200	0.90484	0.90761
1.2	0.91031	0.91296	0.91553	0.91805	0.92051	0.92290	0.92524	0.92751	0.92973	0.93190
1.3	0.93401	0.93606	0.93807	0.94002	0.94191	0.94376	0.94556	0.94731	0.94902	0.95067
1.4	0.95229	0.95385	0.95538	0.95686	0.95830	0.95970	0.96105	0.96237	0.96365	0.96490
1.5	0.96611	0.96728	0.96841	0.96952	0.97059	0.97162	0.97263	0.97360	0.97455	0.97546
1.6	0.97635	0.97721	0.97804	0.97884	0.97962	0.98038	0.98110	0.98181	0.98249	0.98315
1.7	0.98379	0.98441	0.98500	0.98558	0.98613	0.98667	0.98719	0.98769	0.98817	0.98864
1.8	0.98909	0.98952	0.98994	0.99035	0.99074	0.99111	0.99147	0.99182	0.99216	0.99248
1.9	0.99279	0.99309	0.99338	0.99366	0.99392	0.99418	0.99443	0.99466	0.99489	0.99511
2.0	0.99532									

In some of these cases the separation is into product form and appears to be arbitrary. Again the ultimate justification of this method is its success. Theoretically, the dependent variable in any partial differential equation is separable; however, the lack of suitable initial and boundary conditions often drives us to a dead end and makes it necessary for us to turn to other methods. In other cases, such as Examples 6.2-6 and 10.4-1, the dependent variable is the sum of the separated functions, based on theoretical foundation.

As mentioned in Section 2.12, the Laplace transform method can be quite useful in solving partial differential equations. It reduces a partial differential equation into an ordinary one. In fact, it often makes the procedure much simpler, as shown in the following example.

Example 2.13-2.   The differential equation in Example 2.13-1,

$$\frac{\partial \phi}{\partial t} = \nu \frac{\partial^2 \phi}{\partial x^2},$$
(2.13-1)

can be solved by taking the Laplace transform of this equation with respect to t, noting the initial condition that at $t=0$, $\phi=0$, and using formula (A3) in Table 2.12-1:

$$p\bar{\phi}(x,p) = \nu \frac{d^2 \bar{\phi}}{dx^2}.$$
(2.13-16)

Note that from $\phi(x,t)$ we have changed it such that the transform is a function of x and p. Since p is not a variable with respect to which any derivative is taken, we have now an ordinary differential equation for $\bar{\phi}$ instead. Following formula (C3) of Table 2.8-1, the general solution to Eq. 2.13-16 is

$$\bar{\phi}(x,p) = A e^{\sqrt{p/\nu}\, x} + B e^{-\sqrt{p/\nu}\, x}$$
(2.13-17)

with the *transformed* boundary conditions

$$\text{B.C.1: at } x=0, \qquad \bar{\phi}=1/p$$

$$\text{B.C.2: at } x=\infty, \qquad \bar{\phi}=0.$$

Note here that the boundary conditions are for $\bar{\phi}$ instead of ϕ. Hence, we cannot write $\phi=1$ but the transform of both sides. According to formula (B1) in Table 2.12-1, the transform of 1 is $1/p$.

Using B.C.2 we can see that A must be zero; otherwise, the solution would not be finite. Using B.C.1 then yields

$$B = \frac{1}{p},$$

so that the specific solution to Eq. 2.13-16 is

$$\bar{\phi}(x,p) = \frac{1}{p} e^{-\sqrt{p/\nu}\, x}. \tag{2.13-18}$$

By using formula (B26) of Table 2.12-1, Eq. 2.13-18 can be inverted to give the solution to Eq. 2.13-1:

$$\phi(x,t) = \text{erfc}\left(\frac{x}{2\sqrt{\nu t}}\right), \tag{2.13-19}$$

which is identical to that obtained by combination of variables in Example 2.13-1.

One note of interest here is that since Eq. 2.13-16 is an ordinary differential equation, it should be solvable by the Laplace transform method itself. In the present problem, this is indeed the case because the variable, x, to be transformed, also ranges from 0 to ∞, as indicated by the two boundary conditions; thus, the Laplace transform is well defined.*

As a matter of interest, let us now again take the Laplace transform, this time of Eq. 2.13-16 and with respect to x, thus forming an algebraic equation with a *double Laplace transform*;

$$\frac{p}{\nu}\bar{\bar{\phi}}(s,p) = s^2\bar{\bar{\phi}}(s,p) - s\bar{\phi}(x=0) + \left.\frac{d\bar{\phi}}{dx}\right|_{x=0}. \tag{2.13-20}$$

It should be noted here that to avoid confusion between the two transforms we use the symbol s to represent the second one. Moreover, we do not know the value of $d\bar{\phi}/dx$ at $x=0$. But, we know that it should be a constant, which we will call $\bar{\phi}_0'$ now and evaluate later. This plus B.C.1 yields

$$\left(s^2 - \frac{p}{\nu}\right)\bar{\bar{\phi}} = \frac{s}{p} - \bar{\phi}_0'.$$

*In some cases, problems with a variable not ranging from 0 to ∞ can be solved by Laplace transform by imagining its extension and then "cutting out" the part desired. However, these should be treated as exceptions rather than the rule.

Solving for $\bar{\bar{\phi}}$, we have

$$\bar{\bar{\phi}}(s,p) = \frac{s}{p(s^2 - p/\nu)} - \frac{\bar{\phi}_0'}{s^2 - p/\nu}. \qquad (2.13\text{-}21)$$

Taking the inverse Laplace transform of this equation with respect to s, we obtain, by formulas (B15) and (B16) of Table 2.12-1,

$$\bar{\phi}(x,p) = \frac{1}{p}\cosh\sqrt{\frac{p}{\nu}}\,x - \bar{\phi}_0'\sqrt{\frac{\nu}{p}}\,\sinh\sqrt{\frac{p}{\nu}}\,x. \qquad (2.13\text{-}22)$$

Applying B.C.2 and noting that $\cosh y$ approaches $\sinh y$ as y approaches infinity, we can evaluate $\bar{\phi}_0'$:

$$\bar{\phi}_0' = \frac{1}{\sqrt{\nu p}},$$

so that Eq. 2.13-22 becomes

$$\bar{\phi}(x,p) = \frac{1}{p}\left(\cosh\sqrt{\frac{p}{\nu}}\,x - \sinh\sqrt{\frac{p}{\nu}}\,x\right)^{*}$$

$$= \frac{1}{p}e^{-\sqrt{p/\nu}\,x},$$

which is exactly the same as Eq. 2.13-18, whereupon we can take the inverse Laplace transform again to obtain $\phi(x,t)$ as before.

It might be of interest to mention that theoretically, we could have taken the inverse Laplace transform with respect to p before taking that with respect to s. That is to say, the order of taking the Laplace transform and the inverse Laplace transform is immaterial. However, in actural practice, one usually finds one way more convenient than the other.

As an extension of Figure 2.12-1, the Laplace transform method is perhaps most useful in solving simultaneous partial differential equations by converting them into simultaneous ordinary differential equations (if there are only two independent variables), whereupon we can solve it the "classical" way or by further converting them into simultaneous algebraic equations via the Laplace transform.

---

*Remember that $\sinh a = (e^a - e^{-a})/2$ and $\cosh a = (e^a + e^{-a})/2$.

2.14 NONLINEAR PROBLEMS AND NUMERICAL METHODS

Almost all of the methods discussed so far are for solving linear differential equations, which appear most frequently in research journals and scientific texts. However, many realistic problems are nonlinear by nature and can be described by linear differential equations only after approximations. Frequently approximations introduce only small errors or yield at least qualitatively useful information.

When approximations are impermissible, we normally try to make appropriate variable transformation in order to linearize the equation. But, this does not always work, and frequently we have to make simplifying assumptions to achieve this or employ numerical methods. Since nonlinear differential equations belong to a separate subject that would require extensive space to discuss, we shall omit them and, whenever possible, select only problems that yield linear differential equations. Using nonlinear problems might illustrate mathematical elegance but would contribute very little to the understanding of transport phenomena. We shall only cite two references here for those who want to pursue this subject [17, 18].

Numerical methods, in addition to being able to solve nonlinear differential equations, can solve almost any problem for which we cannot find an analytic solution. With the rapid development in computer software and hardware, they have become and are becoming increasingly popular. But a few words of caution are in order here. Many people, because of the easiness of obtaining numerical solutions, prematurely abandon their effort to obtain analytic solutions. This is rather unfortunate because the analytic solution is usually easier to use in the long run and offers more insight than the numerical solution. The latter does not produce a functional relationship in a form whereby one can readily observe the important features of the system solved. Because of this and the fact that discussion of numerical methods would again occupy valuable space here, we shall attempt to use examples with analytic solutions as much as possible. A reference is cited here [19] for those who have the need for some knowledge in this subject. In cases where examples with numerical solutions must be used to illustrate a certain theory, references to the particular method(s) used will be given.

2.15 VECTOR AND ITS OPERATIONS

In elementary physics we learned that a physical quantity is a *scalar* if it has only magnitude, whereas a *vector* is one that comprises both magni-

tude and direction. Time, energy, temperature, and pressure are examples
of the scalar, while force, acceleration, velocity, and momentum those of
the vector. We will wait till Section 2.16 to explain what a *tensor* is, but the
following notation is used throughout the book:

Quantity	Typeface	Suggested Handwritten Form	Product of an Operation
Scalar	s (lightface italic)	s	( ) (parentheses)
Vector	$\mathbf{v}$ (boldface)	$\underline{v}$	[ ] (brackets)
Tensor	$\boldsymbol{\tau}$ (boldface Greek)	$\underline{\underline{\tau}}$	{ } (braces)

We are already familiar with the geometrical representation of the
addition and subtraction of two vectors as depicted in Figure 2.15-1.

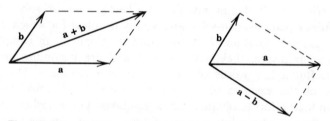

Figure 2.15-1.   Addition and subtraction of two vectors.

Any vector in a three-dimensional space can be projected into components in the three directions, such as x, y, z in the Cartesian coordinate
system. However, to be more general and to enable us to use the more
convenient Σ (summation) notations, we shall in this chapter use $1, 2, 3$
instead. Then, any vector can be completely defined by giving the magnitudes of these components, which are scalar quantities measured on the
respective coordinate axes. Analytically, we write

$$\mathbf{v} = v_1 \mathbf{d}_1 + v_2 \mathbf{d}_2 + v_3 \mathbf{d}_3 \tag{2.15-1}$$

where v_1, v_2, and v_3 are components of the vector $\mathbf{v}$ and $\mathbf{d}_1$, $\mathbf{d}_2$, and $\mathbf{d}_3$ are

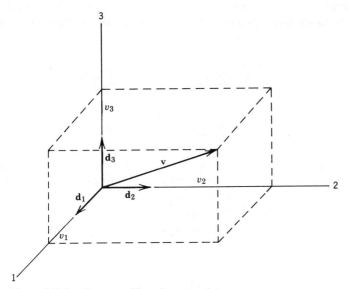

Figure 2.15-2. Decomposition of a vector into components.

the *unit vectors* in the directions indicated by the subscripts.* Since the unit vector has a dimensionless length of 1, it does not change the magnitude of a component but merely "vectorizes" it. Graphically, these quantities are shown in Figure 2.15-2.

With the concept of unit vectors, addition and subtraction of vectors can be performed in a more systematic manner:

$$\mathbf{a} + \mathbf{b} = (a_1 + b_1)\mathbf{d}_1 + (a_2 + b_2)\mathbf{d}_2 + (a_3 + b_3)\mathbf{d}_3$$

$$= \sum_{i=1}^{3} (a_i + b_i)\mathbf{d}_i \qquad (2.15\text{-}2)$$

$$\mathbf{a} - \mathbf{b} = (a_1 - b_1)\mathbf{d}_1 + (a_2 - b_2)\mathbf{d}_2 + (a_3 - b_3)\mathbf{d}_3$$

$$= \sum_{i=1}^{3} (a_i - b_i)\mathbf{d}_i. \qquad (2.15\text{-}3)$$

*In most mathematics books the unit vector in the x, y, and z-directions are more frequently represented by $\mathbf{i}, \mathbf{j}$, and $\mathbf{k}$, respectively, but here we use $\mathbf{d}_i$ for the same reason we use the $1, 2, 3$ directions. Furthermore, the "dyadic product" (Section 2.16) $\mathbf{d}_i\mathbf{d}_j$ is closely associated with the unit tensor $\boldsymbol{\delta}$ and Kronecker delta δ_{ij}, although the reader may simply view the $\mathbf{d}$ as denoting "direction," and, therefore, a vector. The unit vectors $\mathbf{i}, \mathbf{j}, \mathbf{k}$ in the x, y, z-directions are sometimes specifically known as the *base vectors*.

The advantage of using unit vectors is especially evident when addition or subtraction, or any operation for that matter, involving more than two vectors is desired:

$$\sum_{j=1}^{n} \mathbf{a}_j = \sum_{j=1}^{n} a_{j_1}\mathbf{d}_1 + \sum_{j=1}^{n} a_{j_2}\mathbf{d}_2 + \sum_{j=1}^{n} a_{j_3}\mathbf{d}_3$$

$$= \sum_{i=1}^{3}\sum_{j=1}^{n} a_{j_i}\mathbf{d}_i, \tag{2.15-4}$$

where all the $\mathbf{a}_j$ with $j = 1,2,\dots,n$ are the n vectors to be summed; that is, the vectors are first decomposed and their components on the same axis summed or subtracted, resulting in three components, which are then combined to form the resultant vector.

The magnitude of a vector in terms of its components in Cartesian space is

$$|\mathbf{v}| = \sqrt{v_1^2 + v_2^2 + v_3^2} = \sqrt{\sum_{i=1}^{3} v_i^2} \tag{2.15-5}$$

and its direction in relation to the Cartesian $(1,2,3)$ coordinate axes is given by the direction cosines

$$\cos\alpha_1 = \frac{v_1}{|\mathbf{v}|}, \qquad \cos\alpha_2 = \frac{v_2}{|\mathbf{v}|}, \qquad \cos\alpha_3 = \frac{v_3}{|\mathbf{v}|}, \tag{2.15-6}$$

where α_1, α_2, and α_3 are angles between the vector and the $1,2,3$ axes, respectively.

The multiplication of a vector $\mathbf{v}$ by a scalar s yields a vector in exactly the same direction except that now its magnitude is s times that of the original vector $\mathbf{v}$.

The *dot product* of two vectors $\mathbf{a}$ and $\mathbf{b}$ is defined by

$$(\mathbf{a}\cdot\mathbf{b}) = ab\cos\phi_{ab}, \tag{2.15-7}$$

where ϕ_{ab} is the angle ($< 180°$) between vectors $\mathbf{a}$ and $\mathbf{b}$. Since the product is a scalar, denoted by the parentheses, it is also called the *scalar product*.

Geometrically, the dot product gives the area of a rectangle formed by either of the vectors and the projection of the other on it (see Figure 2.15-3). The dot product of a vector with itself is the square of its magnitude:

$$(\mathbf{a}\cdot\mathbf{a}) = aa\cos 0° = a^2 = |\mathbf{a}|^2. \tag{2.15-8}$$

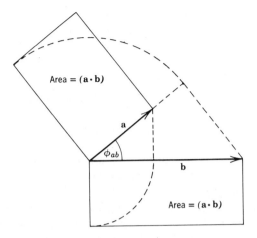

Figure 2.15-3. Geometrical representation of the dot product of two vectors.

On the other hand, the dot product of *any* two vectors 90° apart is *always* 0, regardless of their magnitudes, since $\cos 90° = 0$.

With these in mind, the dot multiplication can be performed in an analytical manner using the component concept:

$$(\mathbf{a} \cdot \mathbf{b}) = [\, a_1\mathbf{d}_1 + a_2\mathbf{d}_2 + a_3\mathbf{d}_3\,] \cdot [\, b_1\mathbf{d}_1 + b_2\mathbf{d}_2 + b_3\mathbf{d}_3\,]$$

$$= a_1 b_1 (\mathbf{d}_1 \cdot \mathbf{d}_1) + a_1 b_2 (\mathbf{d}_1 \cdot \mathbf{d}_2) + \cdots + a_3 b_3 (\mathbf{d}_3 \cdot \mathbf{d}_3) \qquad (2.15\text{-}9)$$

and these can be more simply expressed as

$$(\mathbf{a} \cdot \mathbf{b}) = \left[\, \sum_{i=1}^{3} a_i\mathbf{d}_i \,\right] \cdot \left[\, \sum_{j=1}^{3} b_j\mathbf{d}_j \,\right]$$

$$= \sum_{i=1}^{3} \sum_{j=1}^{3} a_i b_j (\mathbf{d}_i \cdot \mathbf{d}_j). \qquad (2.15\text{-}10)$$

Since the unit vectors are mutually perpendicular, all dot products between $\mathbf{d}_1, \mathbf{d}_2,$ and $\mathbf{d}_3$ in Eq. 2.15-9 disappear except $(\mathbf{d}_1 \cdot \mathbf{d}_1), (\mathbf{d}_2 \cdot \mathbf{d}_2),$ and $(\mathbf{d}_3 \cdot \mathbf{d}_3),$ which are equal to $|\mathbf{d}|^2$ or unity. This actually means that in Eq. 2.15-10 all terms where $i \neq j$ are nonexistent. Therefore,

$$(\mathbf{a} \cdot \mathbf{b}) = a_1 b_1 + a_2 b_2 + a_3 b_3$$

$$= \sum_{i=1}^{3} a_i b_i. \tag{2.15-11}$$

The *cross product* of two vectors $\mathbf{a}$ and $\mathbf{b}$ is defined as

$$[\mathbf{a} \times \mathbf{b}] = (ab \sin \phi_{ab}) \mathbf{n}_{ab}, \tag{2.15-12}$$

where $\mathbf{n}_{ab}$ is a unit vector normal to the plane formed by $\mathbf{a}$ and $\mathbf{b}$ pointing in the direction that a right-hand screw will advance if turned from $\mathbf{a}$ toward $\mathbf{b}$ via the shortest route (i.e., less than 180°). Since the product is a vector, as indicated by the brackets, it is also called the *vector product*.

Geometrically, the magnitude of the cross product is equal to the area of the parallelogram formed by the two vectors, as shown in Figure 2.15-4. It also follows that the cross product of a vector with itself is 0, since $\sin 0° = 0$, while two vectors 90° apart and their cross product are mutually perpendicular and, therefore, form a rectangular parallelepiped whose volume is equal to $|\mathbf{a} \times \mathbf{b}|^2$.

Analytically, the cross product can be obtained the same way we developed the dot product via Eqs. 2.15-9 through 2.15-11. That is,

$$[\mathbf{a} \times \mathbf{b}] = \left[\sum_{i=1}^{3} a_i \mathbf{d}_i \right] \times \left[\sum_{j=1}^{3} b_j \mathbf{d}_j \right]$$

$$= \sum_{i=1}^{3} \sum_{j=1}^{3} a_i b_j [\mathbf{d}_i \times \mathbf{d}_j]. \tag{2.15-13}$$

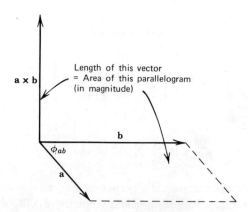

Figure 2.15-4. Geometrical representation of the cross product of two vectors.

From the above discussion we know that all the $[\mathbf{d}_i \times \mathbf{d}_i]$ ($i = 1, 2, 3$) disappear; only those $[\mathbf{d}_i \times \mathbf{d}_j]$ with $i \neq j$ remain. That is,

$$[\mathbf{d}_1 \times \mathbf{d}_1] = [\mathbf{d}_2 \times \mathbf{d}_2] = [\mathbf{d}_3 \times \mathbf{d}_3] = 0 \qquad (2.15\text{-}14)$$

$$[\mathbf{d}_1 \times \mathbf{d}_2] = -[\mathbf{d}_2 \times \mathbf{d}_1] = \mathbf{d}_3 \qquad (2.15\text{-}15)$$

$$[\mathbf{d}_2 \times \mathbf{d}_3] = -[\mathbf{d}_3 \times \mathbf{d}_2] = \mathbf{d}_1 \qquad (2.15\text{-}16)$$

$$[\mathbf{d}_3 \times \mathbf{d}_1] = -[\mathbf{d}_1 \times \mathbf{d}_3] = \mathbf{d}_2, \qquad (2.15\text{-}17)$$

since $\mathbf{d}_1, \mathbf{d}_2$, and $\mathbf{d}_3$ are unit vectors on a set of three mutually perpendicular right-hand coordinate axes. Using the determinant expression, the non-vanishing terms in Eq. 2.15-13 can be neatly written

$$[\mathbf{a} \times \mathbf{b}] = \begin{vmatrix} \mathbf{d}_1 & \mathbf{d}_2 & \mathbf{d}_3 \\ a_1 & a_2 & a_3 \\ b_1 & b_2 & b_3 \end{vmatrix}. \qquad (2.15\text{-}18)$$

It is obvious from Eqs. 2.15-15 through 2.15-18 that the order of cross multiplication of vectors is important, while for dot multiplication the order is immaterial. Since $\sin \phi_{ab} = \sin(-\phi_{ba}) = -\sin \phi_{ba}$ but $\cos \phi_{ab} = \cos(-\phi_{ba}) = \cos \phi_{ba}$.]

If a scalar variable s is a function of the position coordinates (x_1, x_2, x_3)* in the Cartesian space, then

$$\nabla s = \frac{\partial s}{\partial x_1} \mathbf{d}_1 + \frac{\partial s}{\partial x_2} \mathbf{d}_2 + \frac{\partial s}{\partial x_3} \mathbf{d}_3$$

$$= \sum_{i=1}^{3} \frac{\partial s}{\partial x_i} \mathbf{d}_i \qquad (2.15\text{-}19)$$

is called the *gradient of the scalar field s*. In a way, we can envision the "del" (∇) as

$$\nabla = \frac{\partial}{\partial x_1} \mathbf{d}_1 + \frac{\partial}{\partial x_2} \mathbf{d}_2 + \frac{\partial}{\partial x_3} \mathbf{d}_3$$

$$= \sum_{i=1}^{3} \frac{\partial}{\partial x_i} \mathbf{d}_i, \qquad (2.15\text{-}20)$$

*More commonly this is (x, y, z), but remember that we have been using 1, 2, 3 for the three directions; thus, the use of (x_1, x_2, x_3) as a position follows.

which "operates," or "does a differential operation," on the scalar. Since it has the characteristics of a vector, the del is called a *vector differential operator*. The reader should keep in mind that the del has no physical meaning whatsoever by itself since it is *not* a physical quantity. It merely *behaves like* a vector and *must* have something to operate on.

If **v** is a vector variable with v_1, v_2, and v_3 as its components in the 1,2, and 3 directions and is a function of the position coordinates (x_1, x_2, x_3), then

$$\frac{\partial v_1}{\partial x_1} + \frac{\partial v_2}{\partial x_2} + \frac{\partial v_3}{\partial x_3}$$

is called the *divergence of the vector field* **v**. Following the discussions earlier in this section, the reader can easily write

$$(\nabla \cdot \mathbf{v}) = \left[\frac{\partial}{\partial x_1} \mathbf{d}_1 + \frac{\partial}{\partial x_2} \mathbf{d}_2 + \frac{\partial}{\partial x_3} \mathbf{d}_3 \right] \cdot [v_1 \mathbf{d}_1 + v_2 \mathbf{d}_2 + v_3 \mathbf{d}_3]$$

$$= \frac{\partial v_1}{\partial x_1} + \frac{\partial v_2}{\partial x_2} + \frac{\partial v_3}{\partial x_3}$$

$$= \sum_{i=1}^{3} \frac{\partial v_i}{\partial x_i} \qquad (2.15\text{-}21)$$

while keeping in mind that this does not mean the dot product of the del and vector **v**. It is another type of differential operation. The divergence of vector **v** is sometimes written "div **v**."

The *Laplacian operator* is

$$\nabla^2 = (\nabla \cdot \nabla) = \frac{\partial^2}{\partial x_1^2} + \frac{\partial^2}{\partial x_2^2} + \frac{\partial^2}{\partial x_3^2}$$

$$= \sum_{i=1}^{3} \frac{\partial^2}{\partial x_i^2} . \qquad (2.15\text{-}22)$$

Again, this is not a quantity but a second-order differential operator that acts like a scalar.

The Laplacian can operate on a vector as well as a scalar:

$$(\nabla^2 s) = (\nabla \cdot \nabla) s = \frac{\partial^2 s}{\partial x_1^2} + \frac{\partial^2 s}{\partial x_2^2} + \frac{\partial^2 s}{\partial x_3^2}$$

$$= \sum_{i=1}^{3} \frac{\partial^2 s}{\partial x_i^2} \qquad (2.15\text{-}23)$$

Table 2.15-1 Characteristics of Operations Involving Vectors

	Commutative Law	Associative Law	Distributive Law
Addition	$\mathbf{a}+\mathbf{b}=\mathbf{b}+\mathbf{a}$	$(\mathbf{a}+\mathbf{b})+\mathbf{c}=\mathbf{a}+(\mathbf{b}+\mathbf{c})$	
Subtraction	$\mathbf{a}-\mathbf{b}=(-\mathbf{b})+\mathbf{a}$	No: $\therefore (\mathbf{a}-\mathbf{b})-\mathbf{c}=\mathbf{a}-(\mathbf{b}+\mathbf{c})$	
Multiplication of vector by scalar	$s\mathbf{v}=\mathbf{v}s$	$r(s\mathbf{v})=(rs)\mathbf{v}$	$(r+s)\mathbf{v}=r\mathbf{v}+s\mathbf{v}$
Dot multiplication	$(\mathbf{a}\cdot\mathbf{b})=(\mathbf{b}\cdot\mathbf{a})$	No: $(\mathbf{a}\cdot\mathbf{b})\mathbf{c}\neq\mathbf{a}(\mathbf{b}\cdot\mathbf{c})$	$(\mathbf{a}\cdot[\mathbf{b}+\mathbf{c}])=$ $(\mathbf{a}\cdot\mathbf{b})+(\mathbf{a}\cdot\mathbf{c})$
Cross multiplication	No: $\therefore [\mathbf{a}\times\mathbf{b}]=-[\mathbf{b}\times\mathbf{a}]$	No: $[[\mathbf{a}\times\mathbf{b}]\times\mathbf{c}]\neq[\mathbf{a}\times[\mathbf{b}\times\mathbf{c}]]$	$[\mathbf{a}\times[\mathbf{b}+\mathbf{c}]]=$ $[\mathbf{a}\times\mathbf{b}]+[\mathbf{a}\times\mathbf{c}]$
Gradient	No: $\nabla s\neq s\nabla$	No: $\therefore \nabla(rs)=r(\nabla s)+s(\nabla r)$	$\nabla(r+s)=\nabla r+\nabla s$
Divergence	No: $(\nabla\cdot\mathbf{v})\neq(\mathbf{v}\cdot\nabla)$	No: $\therefore (\nabla\cdot s\mathbf{v})=s(\nabla\cdot\mathbf{v})+(\mathbf{v}\cdot\nabla s)$	$(\nabla\cdot[\mathbf{v}+\mathbf{w}])=$ $(\nabla\cdot\mathbf{v})+(\nabla\cdot\mathbf{w})$
Dyadic multiplication	No: $\{\mathbf{ab}\}\neq\{\mathbf{ba}\}$	No: $[\mathbf{a}\cdot\{\mathbf{bc}\}]\neq[\{\mathbf{ab}\}\cdot\mathbf{c}]$	$\{\mathbf{a}(\mathbf{b}+\mathbf{c})\}=$ $\{\mathbf{ab}\}+\{\mathbf{ac}\}$

$$(\nabla^2 \mathbf{v}) = (\nabla \cdot \nabla)\mathbf{v} = \frac{\partial^2 \mathbf{v}}{\partial x_1^2} + \frac{\partial^2 \mathbf{v}}{\partial x_2^2} + \frac{\partial^2 \mathbf{v}}{\partial x_3^2}$$

$$= \sum_{i=1}^{3} \frac{\partial^2 \mathbf{v}}{\partial x_i^2} = \sum_{i=1}^{3} \sum_{j=1}^{3} \frac{\partial^2 v_j}{\partial x_i^2} \mathbf{d}_j$$

$$= \nabla^2 \sum_{j=1}^{3} v_j \mathbf{d}_j. \qquad (2.15\text{-}24)$$

Note that Eq. 2.15-24 is not correct for non-Cartesian coordinates. It should also be noted with care that not all of the commutative, associative, and distributive laws apply to all of the vector operations. The summary in Table 2.15-1 should give a quick and easy reference.

In the last line of Table 2.15-1 we have included the dyadic product, whose definition and operation will require the understanding of tensors and thus are discussed in the following section. We put it here merely to make the table more complete.

2.16 TENSOR AND ITS OPERATIONS

Just as a vector $\mathbf{v}$ is specified by giving a set of three components v_1, v_2, and v_3 in a three-dimensional space, a tensor τ is specified by giving the nine (three times three) elements $\tau_{11}, \tau_{12}, \tau_{13}, \tau_{21}, \ldots, \tau_{33}$. It is more specifically known as the *second-order tensor*, because of the extra "generation" of subscripts but is loosely called *tensor* because the first-order tensor is generally known as *vector*.

A tensor can be expressed in the form of a matrix, which is an array of numbers or functions obeying certain laws of operation [20]:

$$\tau = \left\{ \begin{array}{ccc} \tau_{11} & \tau_{12} & \tau_{13} \\ \tau_{21} & \tau_{22} & \tau_{23} \\ \tau_{31} & \tau_{32} & \tau_{33} \end{array} \right\}. \qquad (2.16\text{-}1)$$

In addition, a tensor must also obey a set of laws of coordinate transformation, just as a vector must obey similar ones, examples of which are given in Section 2.17. The stress with its nine components (see Figure 4.4-1) acting on a body or on an element of fluid is an example of the tensor. In fact, it is the origin of the term *tensor*.

The tensor or matrix should not be confused with the determinant, which is a scalar obtainable from an array of numbers or functions (i.e., matrix) through an operation obeying certain rules.

When two vectors are placed adjacent to each other without any operational sign, they form the *dyadic product*, or *dyad*, a special second-order tensor the elements of which consist of all possible products between components of the two vectors as follows:

$$\{\mathbf{ab}\} = \begin{Bmatrix} a_1b_1 & a_1b_2 & a_1b_3 \\ a_2b_1 & a_2b_2 & a_2b_3 \\ a_3b_1 & a_3b_2 & a_3b_3 \end{Bmatrix}. \tag{2.16-2}$$

That is, the ij-element of the dyad $\{\mathbf{ab}\}$ is the product of the i-component of vector $\mathbf{a}$ and the j-component of vector $\mathbf{b}$.

The *transpose* τ^T of tensor τ is one formed by interchanging the rows and columns (i.e., the indices on each of the elements) of τ. That is,

$$\tau^T = \begin{Bmatrix} \tau_{11} & \tau_{21} & \tau_{31} \\ \tau_{12} & \tau_{22} & \tau_{32} \\ \tau_{13} & \tau_{23} & \tau_{33} \end{Bmatrix}. \tag{2.16-3}$$

A *symmetric tensor* is one that is identical to its own transpose, (i.e., all $\tau_{ij} = \tau_{ji}$). The *unit tensor* δ is one in which all diagonal elements are unity and all other elements (nondiagonals) zero, or

$$\delta = \begin{Bmatrix} 1 & 0 & 0 \\ 0 & 1 & 0 \\ 0 & 0 & 1 \end{Bmatrix} \tag{2.16-4}$$

The elements of the unit tensor are the *Kronecker delta*, which is defined as

$$\delta_{ij} = \begin{cases} 0 \text{ when } i \neq j \\ 1 \text{ when } i = j. \end{cases} \tag{2.16-5}$$

The dyadic product of any two unit vectors is called a *unit dyad*. There are nine of them:

$$\mathbf{d}_1\mathbf{d}_1 = \begin{Bmatrix} 1 & 0 & 0 \\ 0 & 0 & 0 \\ 0 & 0 & 0 \end{Bmatrix} \quad \mathbf{d}_1\mathbf{d}_2 = \begin{Bmatrix} 0 & 1 & 0 \\ 0 & 0 & 0 \\ 0 & 0 & 0 \end{Bmatrix} \quad \mathbf{d}_1\mathbf{d}_3 = \begin{Bmatrix} 0 & 0 & 1 \\ 0 & 0 & 0 \\ 0 & 0 & 0 \end{Bmatrix}$$

$$\mathbf{d}_2\mathbf{d}_1 = \begin{Bmatrix} 0 & 0 & 0 \\ 1 & 0 & 0 \\ 0 & 0 & 0 \end{Bmatrix} \quad \mathbf{d}_2\mathbf{d}_2 = \begin{Bmatrix} 0 & 0 & 0 \\ 0 & 1 & 0 \\ 0 & 0 & 0 \end{Bmatrix} \quad \mathbf{d}_2\mathbf{d}_3 = \begin{Bmatrix} 0 & 0 & 0 \\ 0 & 0 & 1 \\ 0 & 0 & 0 \end{Bmatrix}$$

$$\mathbf{d}_3\mathbf{d}_1 = \begin{Bmatrix} 0 & 0 & 0 \\ 0 & 0 & 0 \\ 1 & 0 & 0 \end{Bmatrix} \quad \mathbf{d}_3\mathbf{d}_2 = \begin{Bmatrix} 0 & 0 & 0 \\ 0 & 0 & 0 \\ 0 & 1 & 0 \end{Bmatrix} \quad \mathbf{d}_3\mathbf{d}_3 = \begin{Bmatrix} 0 & 0 & 0 \\ 0 & 0 & 0 \\ 0 & 0 & 1 \end{Bmatrix}.$$

$$\tag{2.16-6}$$

It can be immediately observed that the ij-element of $\{\mathbf{d}_i\mathbf{d}_m\}$ is equal to the product of two Kronecker deltas δ_{li} and δ_{mj}.

The unit dyads and unit tensor may seem to be merely trivial cases of tensors; however, they do occupy important positions in the world of tensors. For example, a comparison of Eq. 2.16-6 with Eq. 2.15-1 clearly shows that the unit dyads play a similar role in tensor operations as the unit vectors in vector operations, for they "tensorize" the elements. After the nine elements are all tensorized by the unit dyads, they can be combined to form the tensor. To illustrate, we write this in the reverse direction to see how the tensor is "decomposed" and expressed analytically:

$$
\tau = \begin{Bmatrix} \tau_{11} & \tau_{12} & \tau_{13} \\ \tau_{21} & \tau_{22} & \tau_{23} \\ \tau_{31} & \tau_{32} & \tau_{33} \end{Bmatrix}
$$

$$
= \tau_{11} \begin{Bmatrix} 1 & 0 & 0 \\ 0 & 0 & 0 \\ 0 & 0 & 0 \end{Bmatrix} + \tau_{12} \begin{Bmatrix} 0 & 1 & 0 \\ 0 & 0 & 0 \\ 0 & 0 & 0 \end{Bmatrix} + \tau_{13} \begin{Bmatrix} 0 & 0 & 1 \\ 0 & 0 & 0 \\ 0 & 0 & 0 \end{Bmatrix}
$$

$$
+ \tau_{21} \begin{Bmatrix} 0 & 0 & 0 \\ 1 & 0 & 0 \\ 0 & 0 & 0 \end{Bmatrix} + \cdots + \tau_{33} \begin{Bmatrix} 0 & 0 & 0 \\ 0 & 0 & 0 \\ 0 & 0 & 1 \end{Bmatrix}
$$

$$
= \sum_{i=1}^{3}\sum_{j=1}^{3} \tau_{ij}\mathbf{d}_i\mathbf{d}_j. \tag{2.16-7}
$$

The reader will appreciate this advantage as he proceeds further into the following discussions.

The sum of two tensors is formed by adding corresponding elements of them, resulting in another:

$$
\{\sigma + \tau\} = \begin{Bmatrix} \sigma_{11}+\tau_{11} & \sigma_{12}+\tau_{12} & \sigma_{13}+\tau_{13} \\ \sigma_{21}+\tau_{21} & \sigma_{22}+\tau_{22} & \sigma_{23}+\tau_{23} \\ \sigma_{31}+\tau_{31} & \sigma_{32}+\tau_{32} & \sigma_{33}+\tau_{33} \end{Bmatrix}. \tag{2.16-8}
$$

In the shorthand Σ form, it can be written as

$$
\{\sigma + \tau\} = \sum_{i=1}^{3}\sum_{j=1}^{3}(\sigma_{ij}+\tau_{ij})\mathbf{d}_i\mathbf{d}_j. \tag{2.16-9}
$$

The advantage of using the unit dyads will be even more evident if the addition of more than two tensors is to be performed. The difference between two tensors can be obtained via the same procedure.

Similar to the multiplication of a vector by a scalar, the multiplication of a tensor τ by a scalar s yields a tensor whose elements are s times those of τ; that is,

$$\{s\tau\} = \begin{Bmatrix} s\tau_{11} & s\tau_{12} & s\tau_{13} \\ s\tau_{21} & s\tau_{22} & s\tau_{23} \\ s\tau_{31} & s\tau_{32} & s\tau_{33} \end{Bmatrix}$$

$$= \sum_{i=1}^{3} \sum_{j=1}^{3} (s\tau_{ij}) \mathbf{d}_i \mathbf{d}_j. \tag{2.16-10}$$

The *single-dot product* of tensors $\boldsymbol{\sigma}$ and $\boldsymbol{\tau}$, in that order, is a tensor whose ijth element is equal to the sum of products between corresponding elements of the ith row of $\boldsymbol{\sigma}$ and jth column of $\boldsymbol{\tau}$. Diagrammatically,

$$\{\boldsymbol{\sigma}\cdot\boldsymbol{\tau}\} = \begin{Bmatrix} \sigma_{11} & \sigma_{12} & \sigma_{13} \\ \sigma_{21} & \sigma_{22} & \sigma_{23} \\ \sigma_{31} & \sigma_{32} & \sigma_{33} \end{Bmatrix} \begin{Bmatrix} \tau_{11} & \tau_{12} & \tau_{13} \\ \tau_{21} & \tau_{22} & \tau_{23} \\ \tau_{31} & \tau_{32} & \tau_{33} \end{Bmatrix}$$

$$= \begin{Bmatrix} \sigma_{11}\tau_{11}+\sigma_{12}\tau_{21}+\sigma_{13}\tau_{31} & \sigma_{11}\tau_{12}+\sigma_{12}\tau_{22}+\sigma_{13}\tau_{32} & \sigma_{11}\tau_{13}+\sigma_{12}\tau_{23}+\sigma_{13}\tau_{33} \\ \sigma_{21}\tau_{11}+\sigma_{22}\tau_{21}+\sigma_{23}\tau_{31} & \sigma_{21}\tau_{12}+\sigma_{22}\tau_{22}+\sigma_{23}\tau_{32} & \sigma_{21}\tau_{13}+\sigma_{22}\tau_{23}+\sigma_{23}\tau_{33} \\ \sigma_{31}\tau_{11}+\sigma_{32}\tau_{21}+\sigma_{33}\tau_{31} & \sigma_{31}\tau_{12}+\sigma_{32}\tau_{22}+\sigma_{33}\tau_{32} & \sigma_{31}\tau_{13}+\sigma_{32}\tau_{23}+\sigma_{33}\tau_{33} \end{Bmatrix}$$

In effect, we have

$$\{\boldsymbol{\sigma}\cdot\boldsymbol{\tau}\}_{ij} = \sum_{k=1}^{3} \sigma_{ik}\tau_{kj}$$

or

$$\{\boldsymbol{\sigma}\cdot\boldsymbol{\tau}\} = \sum_{i=1}^{3} \sum_{j=1}^{3} \left(\sum_{k=1}^{3} \sigma_{ik}\tau_{kj} \right) \mathbf{d}_i \mathbf{d}_j. \tag{2.16-11}$$

Since the product is a tensor, it is also known as the *tensor product*.

If we view the vector as a tensor that has only a single row or column (thus the term *first-order tensor*), the dot multiplication of a tensor with a vector is merely a special case of that between two tensors. Thus,

$$[\boldsymbol{\tau} \cdot \mathbf{v}] = \left\{ \begin{bmatrix} \tau_{11} & \tau_{12} & \tau_{13} \\ \tau_{21} & \tau_{22} & \tau_{23} \\ \tau_{31} & \tau_{32} & \tau_{33} \end{bmatrix} \right\} \cdot \begin{bmatrix} v_1 \\ v_2 \\ v_3 \end{bmatrix}$$

$$= \begin{bmatrix} \tau_{11}v_1 + \tau_{12}v_2 + \tau_{13}v_3 \\ \tau_{21}v_1 + \tau_{22}v_2 + \tau_{23}v_3 \\ \tau_{31}v_1 + \tau_{32}v_2 + \tau_{33}v_3 \end{bmatrix}$$

$$= \sum_{i=1}^{3} \left(\sum_{k=1}^{3} \tau_{ik}v_k \right) \mathbf{d}_i. \tag{2.16-12}$$

The last expression of Eq. 2.16-12 can also be reached by considering all possible combinations of the terms as follows:

$$[\boldsymbol{\tau} \cdot \mathbf{v}] = \left\{ \sum_{i=1}^{3} \sum_{j=1}^{3} \tau_{ij}\mathbf{d}_i\mathbf{d}_j \right\} \cdot \left[\sum_{k=1}^{3} v_k \mathbf{d}_k \right]$$

$$= \sum_{i=1}^{3} \sum_{j=1}^{3} \sum_{k=1}^{3} \tau_{ij}v_k [\mathbf{d}_i\mathbf{d}_j \cdot \mathbf{d}_k]$$

$$= \sum_{i=1}^{3} \sum_{j=1}^{3} \sum_{k=1}^{3} \tau_{ij}v_k \mathbf{d}_i \delta_{jk}$$

$$= \sum_{i=1}^{3} \left(\sum_{j=1}^{3} \tau_{ij}v_j \right) \mathbf{d}_i. \tag{2.16-13}$$

Here, between the third and fourth expressions, we have used the following identity:

$$[\mathbf{d}_i\mathbf{d}_j \cdot \mathbf{d}_k] = \mathbf{d}_i \delta_{jk} \tag{2.16-14}$$

for which we simply first perform the operation using the dot (or one of the dots in the case of a double-dot product as in Eq. 2.16-17) and the two quantities immediately adjacent to it and then perform whatever that is left (in the case of Eq. 2.16-14, nothing). This rule extends to other operations:

$$[\mathbf{d}_i \cdot \mathbf{d}_j \mathbf{d}_k] = \delta_{ij} \mathbf{d}_k \qquad (2.16\text{-}15)$$

$$[\mathbf{d}_i \mathbf{d}_j \cdot \mathbf{d}_k \mathbf{d}_l] = \delta_{jk} \mathbf{d}_i \mathbf{d}_l \qquad (2.16\text{-}16)$$

$$(\mathbf{d}_i \mathbf{d}_j : \mathbf{d}_k \mathbf{d}_l) = \delta_{jk} (\mathbf{d}_i \cdot \mathbf{d}_l)$$

$$= \delta_{jk} \delta_{il}. \qquad (2.16\text{-}17)$$

The application of this basic rule in Eq. 2.16-17 is especially important, for without it the double-dot product of two tensors would be difficult to perform using the expansion (see the following paragraph). The reader can also go back and verify Eqs. 2.16-11 and 2.16-12 using this rule.

The double-dot product of two tensors is written

$$(\boldsymbol{\sigma} : \boldsymbol{\tau}) = \left(\left\{ \sum_{i=1}^{3} \sum_{j=1}^{3} \sigma_{ij} \mathbf{d}_i \mathbf{d}_j \right\} : \left\{ \sum_{k=1}^{3} \sum_{l=1}^{3} \tau_{kl} \mathbf{d}_k \mathbf{d}_l \right\} \right)$$

$$= \sum_{i=1}^{3} \sum_{j=1}^{3} \sum_{k=1}^{3} \sum_{l=1}^{3} \sigma_{ij} \tau_{kl} (\mathbf{d}_i \mathbf{d}_j : \mathbf{d}_k \mathbf{d}_l).$$

According to Eq. 2.16-17, we have

$$(\boldsymbol{\sigma} : \boldsymbol{\tau}) = \sum_{i=1}^{3} \sum_{j=1}^{3} \sum_{k=1}^{3} \sum_{l=1}^{3} \sigma_{ij} \tau_{kl} \delta_{il} \delta_{jk}$$

$$= \sum_{i=1}^{3} \sum_{j=1}^{3} \sigma_{ij} \tau_{ji}, \qquad (2.16\text{-}18)$$

since k and l have to be equal to j and i, respectively, for the terms to exist. Since the double-dot product is a scalar it is also called the *scalar product*.

By considering the differential operator ∇ as a vector, the differential operation can be most easily performed. For example, one would expect $\nabla \mathbf{v}$ to be a sort of "dyadic product" as

$$\{\nabla \mathbf{v}\} = \begin{Bmatrix} \dfrac{\partial v_1}{\partial x_1} & \dfrac{\partial v_2}{\partial x_1} & \dfrac{\partial v_3}{\partial x_1} \\[2mm] \dfrac{\partial v_1}{\partial x_2} & \dfrac{\partial v_2}{\partial x_2} & \dfrac{\partial v_3}{\partial x_2} \\[2mm] \dfrac{\partial v_1}{\partial x_3} & \dfrac{\partial v_2}{\partial x_3} & \dfrac{\partial v_3}{\partial x_3} \end{Bmatrix}$$

$$= \sum_{i=1}^{3} \sum_{j=1}^{3} \frac{\partial}{\partial x_i} v_j \mathbf{d}_i \mathbf{d}_j$$

or

$$\{\nabla \mathbf{v}\} = \left[\sum_{i=1}^{3} \frac{\partial}{\partial x_i} \mathbf{d}_i \right] \left[\sum_{j=1}^{3} v_j \mathbf{d}_j \right]$$

$$= \sum_{i=1}^{3} \sum_{j=1}^{3} \frac{\partial}{\partial x_i} v_j \mathbf{d}_i \mathbf{d}_j. \qquad (2.16\text{-}19)$$

Other operations can be similarly done by following this principle and the basic rule mentioned earlier on p. 100.

2.17 COORDINATE TRANSFORMATION

Because of the fact that in most cases, the system or apparatus involved in the transport of material, energy, and momentum is either cylindrical or spherical in configuration, the expression of the problem in cylindrical and spherical coordinates is most important. To do this, we can start with the generalized orthogonal coordinates and then substitute specific characteristic parameters for the particular system involved [21]. However, this requires much advanced mathematical background. The benefit derived from doing this does not justify the effort.

A simpler, although less general, way is sufficient for our purpose. It consists only of basic trigonometric conversions and differential operations, as summarized in Table 2.17-1.

Differential operations in cylindrical and spherical coordinate systems can best be related to those in the rectangular coordinate system by using the chain rule of partial differentiation—for example, if we regard

$$v_x = v_x(r, \theta, z)$$

$$= v_x[r(x,y,z), \theta(x,y,z), z(x,y,z)]$$

$$dv_x = \left(\frac{\partial v_x}{\partial r} \right)_{\theta,z} dr + \left(\frac{\partial v_x}{\partial \theta} \right)_{r,z} d\theta + \left(\frac{\partial v_x}{\partial z} \right)_{r,\theta} dz. \qquad (2.17\text{-}28)$$

"Dividing through" by dx while holding y and z constant, we have

$$\left(\frac{\partial v_x}{\partial x} \right)_{y,z} = \left(\frac{\partial v_x}{\partial r} \right)_{\theta,z} \left(\frac{\partial r}{\partial x} \right)_{y,z} + \left(\frac{\partial v_x}{\partial \theta} \right)_{r,z} \left(\frac{\partial \theta}{\partial x} \right)_{y,z}. \qquad (2.17\text{-}29)$$

In this case the term $(\partial v_x / \partial z)_{r,\theta} (\partial z / \partial x)_{y,z}$ disappears because in the

Table 2.17-1 Coordinate Transformation

A. *Position Variables*

Pictorial Description

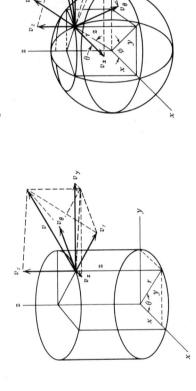

Rectangular and Cylindrical
Coordinates

Rectangular and Spherical
Coordinates

Coordinates Transformation

Formulas

Rectangular to Cylindrical

(1) $x = r\cos\theta$

(2) $y = r\sin\theta$

(3) $z = z$

Rectangular to Spherical

(4) $x = r\sin\theta\cos\phi$

(5) $y = r\sin\theta\sin\phi$

(6) $z = r\cos\theta$

Table 2.17-1 *(Continued)*

Cylindrical to Rectangular

$$(7) \quad r = \sqrt{x^2 + y^2}$$

$$(8) \quad \theta = \tan^{-1}\frac{y}{x}$$

$$(9) \quad z = z$$

Spherical to Rectangular

$$(10) \quad r = \sqrt{x^2 + y^2 + z^2}$$

$$(11) \quad \theta = \tan^{-1}\frac{\sqrt{x^2 + y^2}}{z}$$

$$(12) \quad \phi = \tan^{-1}\frac{y}{x}$$

Line Segment

$$(13) \quad (ds)^2 = (dx)^2 + (dy)^2 + (dz)^2$$

$$(14) \quad (ds)^2 = (dr)^2 + r^2(d\theta)^2 + (dz)^2 \qquad (15) \quad (ds)^2 = (dr)^2 + r^2(d\theta)^2 + r^2\sin^2\theta\,(d\phi)^2$$

B. *Vector Components*

Pictorial Description

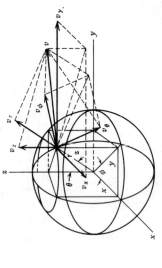

Rectangular and Spherical
Coordinates

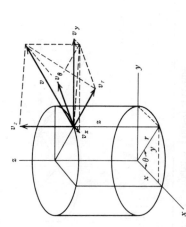

Rectangular and Cylindrical
Coordinates

104

Transformation of Vector Components

Formulas

Rectangular to Cylindrical

$$(16) \quad v_x = v_r \cos\theta - v_\theta \sin\theta$$

$$(17) \quad v_y = v_r \sin\theta + v_\theta \cos\theta$$

$$(18) \quad v_z = v_z$$

Cylindrical to Rectangular

$$(22) \quad v_r = v_x \cos\theta + v_y \sin\theta$$

$$(23) \quad v_\theta = -v_x \sin\theta + v_y \cos\theta$$

$$(24) \quad v_z = v_z$$

Rectangular to Spherical

$$(19) \quad v_x = v_r \sin\theta \cos\phi + v_\theta \cos\theta \cos\phi - v_\phi \sin\phi$$

$$(20) \quad v_y = v_r \sin\theta \sin\phi + v_\theta \cos\theta \sin\phi + v_\phi \cos\phi$$

$$(21) \quad v_z = v_r \cos\theta - v_\theta \sin\theta$$

Spherical to Rectangular

$$(25) \quad v_r = v_x \sin\theta \cos\phi + v_y \sin\theta \sin\phi + v_z \cos\theta$$

$$(26) \quad v_\theta = v_x \cos\theta \cos\phi + v_y \cos\theta \sin\phi - v_z \sin\theta$$

$$(27) \quad v_\phi = -v_x \sin\phi + v_y \cos\phi$$

second quantity z in cylindrical coordinates is independent of x. Similarly,

$$\left(\frac{\partial v_y}{\partial y}\right)_{x,z} = \left(\frac{\partial v_y}{\partial r}\right)_{\theta,z}\left(\frac{\partial r}{\partial y}\right)_{x,z} + \left(\frac{\partial v_y}{\partial \theta}\right)_{r,z}\left(\frac{\partial \theta}{\partial y}\right)_{x,z} \qquad (2.17\text{-}30)$$

$$\left(\frac{\partial v_z}{\partial z}\right)_{x,y} = \left(\frac{\partial v_z}{\partial r}\right)_{\theta,z}\left(\frac{\partial r}{\partial z}\right)_{x,y} + \left(\frac{\partial v_z}{\partial \theta}\right)_{r,z}\left(\frac{\partial \theta}{\partial z}\right)_{x,y} + \left(\frac{\partial v_z}{\partial z}\right)_{r,\theta}\left(\frac{\partial z}{\partial z}\right)_{x,y}$$

$$= \left(\frac{\partial v_z}{\partial z}\right)_{r,\theta} \qquad (2.17\text{-}31)$$

since r and θ depend only on x and y. So if we held *both* x and y constant, $\partial r/\partial z$ and $\partial \theta/\partial z$ naturally went to zero. Note that in Eq. 2.17-31 although $(\partial v_z/\partial z)_{x,y} = (\partial v_z/\partial z)_{r,\theta}$ this is only an exception and that these two terms actually have different meanings. On the left-hand side, v_z is regarded as a function of x, y, z, but on the right, a function of r, θ, z.

The importance of specifying the subscript for partial differentiation has been demonstrated in Section 2.6, in which we have found that the apparent shortcut $\partial r/\partial x = 1/(\partial x/\partial r)$ is actually a booby trap. Speaking of booby traps, another one is $(\nabla \cdot \mathbf{v}) = (\partial v_r/\partial r) + (\partial v_\theta/\partial \theta) + (\partial v_z/\partial z)$. But if we carefully substitute appropriate quantities for the terms in Eqs. 2.17-29 through 2.17-31 we will find it untrue. From Eqs. (16) through (18) of Table 2.17-1 we have,

$$\left(\frac{\partial v_x}{\partial r}\right)_{\theta,z} = \frac{\partial v_r}{\partial r}\cos\theta - \frac{\partial v_\theta}{\partial r}\sin\theta \qquad (2.17\text{-}32)$$

$$\left(\frac{\partial v_x}{\partial \theta}\right)_{r,z} = -v_r\sin\theta + \frac{\partial v_r}{\partial \theta}\cos\theta - v_\theta\cos\theta - \frac{\partial v_\theta}{\partial \theta}\sin\theta \qquad (2.17\text{-}33)$$

$$\left(\frac{\partial v_y}{\partial r}\right)_{\theta,z} = \frac{\partial v_r}{\partial r}\sin\theta + \frac{\partial v_\theta}{\partial r}\cos\theta \qquad (2.17\text{-}34)$$

$$\left(\frac{\partial v_y}{\partial \theta}\right)_{r,z} = \frac{\partial v_r}{\partial \theta}\sin\theta + v_r\cos\theta + \frac{\partial v_\theta}{\partial \theta}\cos\theta - v_\theta\sin\theta. \qquad (2.17\text{-}35)$$

From Eqs. (7) and (8) of Table 2.17-1, we have,

$$\left(\frac{\partial r}{\partial x}\right)_{y,z}=\left(\frac{\partial}{\partial x}\sqrt{x^2+y^2}\right)_{y,z}=\frac{1}{2}\cdot\frac{2x}{\sqrt{x^2+y^2}}$$

$$=\frac{x}{r}=\cos\theta \tag{2.17-36}$$

$$\left(\frac{\partial\theta}{\partial x}\right)_{y,z}=\left(\frac{\partial}{\partial x}\tan^{-1}\frac{y}{x}\right)_{y,z}=\frac{1}{1+(y/x)^2}\cdot y\left(-\frac{1}{x^2}\right)$$

$$=-\frac{y}{r^2}=-\frac{\sin\theta}{r} \tag{2.17-37}$$

$$\left(\frac{\partial r}{\partial y}\right)_{x,z}=\left(\frac{\partial}{\partial y}\sqrt{x^2+y^2}\right)_{x,z}=\frac{1}{2}\cdot\frac{2y}{\sqrt{x^2+y^2}}$$

$$=\frac{y}{r}=\sin\theta \tag{2.17-38}$$

$$\left(\frac{\partial\theta}{\partial y}\right)_{x,z}=\left(\frac{\partial}{\partial y}\tan^{-1}\frac{y}{x}\right)_{x,z}=\frac{1}{1+(y/x)^2}\cdot\frac{1}{x}$$

$$=\frac{x}{r^2}=\frac{\cos\theta}{r}. \tag{2.17-39}$$

Substituting Eqs. 2.17-32 through 2.17-39 into Eqs. 2.17-29 through 2.17-31 and subsequently into the following

$$(\nabla\cdot\mathbf{v})=\frac{\partial v_x}{\partial x}+\frac{\partial v_y}{\partial y}+\frac{\partial v_z}{\partial z},$$

we obtain

$$(\nabla\cdot\mathbf{v})=\left(\frac{\partial v_r}{\partial r}\cos\theta-\frac{\partial v_\theta}{\partial r}\sin\theta\right)\cos\theta$$

$$+\left(-v_r\sin\theta+\frac{\partial v_r}{\partial\theta}\cos\theta-v_\theta\cos\theta-\frac{\partial v_\theta}{\partial\theta}\sin\theta\right)$$

$$\cdot\left(-\frac{\sin\theta}{r}\right)+\left(\frac{\partial v_r}{\partial r}\sin\theta+\frac{\partial v_\theta}{\partial r}\cos\theta\right)\sin\theta$$

$$+\left(\frac{\partial v_r}{\partial \theta}\sin\theta + v_r\cos\theta + \frac{\partial v_\theta}{\partial \theta}\cos\theta - v_\theta\sin\theta\right)\frac{\cos\theta}{r} + \left(\frac{\partial v_z}{\partial z}\right)_{r,z}$$

$$= \frac{\partial v_r}{\partial r}(\cos^2\theta + \sin^2\theta) + \frac{v_r}{r}(\sin^2\theta + \cos^2\theta)$$

$$+ \frac{\partial v_\theta}{\partial \theta}\cdot\frac{\sin^2\theta + \cos^2\theta}{r} + \left(\frac{\partial v_z}{\partial z}\right)_{r,z}$$

$$= \frac{\partial v_r}{\partial r} + \frac{v_r}{r} + \frac{1}{r}\frac{\partial v_\theta}{\partial \theta} + \frac{\partial v_z}{\partial z}$$

$$= \frac{1}{r}\frac{\partial}{\partial r}(rv_r) + \frac{1}{r}\frac{\partial v_\theta}{\partial \theta} + \frac{\partial v_z}{\partial z}. \tag{2.17-40}$$

Therefore,

$$(\nabla\cdot\mathbf{v}) = \frac{\partial v_x}{\partial x} + \frac{\partial v_y}{\partial y} + \frac{\partial v_z}{\partial z}$$

$$= \frac{1}{r}\frac{\partial}{\partial r}(rv_r) + \frac{1}{r}\frac{\partial v_\theta}{\partial \theta} + \frac{\partial v_z}{\partial z}$$

$$\neq \frac{\partial v_r}{\partial r} + \frac{\partial v_\theta}{\partial \theta} + \frac{\partial v_z}{\partial z}!$$

A similar procedure can be applied to obtain the equivalent in spherical coordinates.

Table 2.17-2 lists some of the important differential quantities to be encountered in this book, in the three types of coordinate systems.

PROBLEMS

2.A Exercise in Differentiation

(a) Prove Eq. e in Table 2.2-1 from Eq. 2.2-5 and the first footnote to Table 2.2-1.

(b) Find the derivative of $(d/dx)(1/\sqrt{f(x)})$ from the following methods:

(i) Eq. d Table 2.2-1

(ii) Eq. 2.2-5.

Table 2.17-2 Differential Operations in Various Coordinates Systems

Quantity	Rectangular		
	x-Component	y-Component	z-Component
∇s	$\dfrac{\partial s}{\partial x}$	$\dfrac{\partial s}{\partial y}$	$\dfrac{\partial s}{\partial z}$
$(\nabla \cdot \mathbf{v})$	$\dfrac{\partial v_x}{\partial x} + \dfrac{\partial v_y}{\partial y} + \dfrac{\partial v_z}{\partial z}$		
$(\nabla^2 s)$	$\dfrac{\partial^2 s}{\partial x^2} + \dfrac{\partial^2 s}{\partial y^2} + \dfrac{\partial^2 s}{\partial z^2}$		
$[\nabla^2 \mathbf{v}]$	$\dfrac{\partial^2 v_x}{\partial x^2} + \dfrac{\partial^2 v_x}{\partial y^2}$ $+ \dfrac{\partial^2 v_x}{\partial z^2}$	$\dfrac{\partial^2 v_y}{\partial x^2} + \dfrac{\partial^2 v_y}{\partial y^2}$ $+ \dfrac{\partial^2 v_y}{\partial z^2}$	$\dfrac{\partial^2 v_z}{\partial x^2} + \dfrac{\partial^2 v_z}{\partial y^2}$ $+ \dfrac{\partial^2 v_z}{\partial z^2}$
$[\nabla \cdot \tau]$	$\dfrac{\partial \tau_{xx}}{\partial x} + \dfrac{\partial \tau_{xy}}{\partial y}$ $+ \dfrac{\partial \tau_{xz}}{\partial z}$	$\dfrac{\partial \tau_{yx}}{\partial x} + \dfrac{\partial \tau_{yy}}{\partial y}$ $+ \dfrac{\partial \tau_{yz}}{\partial z}$	$\dfrac{\partial \tau_{zx}}{\partial x} + \dfrac{\partial \tau_{zy}}{\partial y}$ $+ \dfrac{\partial \tau_{zz}}{\partial z}$
$[\mathbf{v} \cdot \nabla \mathbf{v}]$	$v_x \dfrac{\partial v_x}{\partial x} + v_y \dfrac{\partial v_x}{\partial y}$ $+ v_z \dfrac{\partial v_x}{\partial z}$	$v_x \dfrac{\partial v_y}{\partial x} + v_y \dfrac{\partial v_y}{\partial y}$ $+ v_z \dfrac{\partial v_y}{\partial z}$	$v_x \dfrac{\partial v_z}{\partial x} + v_y \dfrac{\partial v_z}{\partial y}$ $v_z \dfrac{\partial v_z}{\partial z}$

Table 2.17-2 (*Continued*)

Quantity	Cylindrical		
	r-Component	θ-Component	z-Component
∇s	$\dfrac{\partial s}{\partial r}$	$\dfrac{1}{r}\dfrac{\partial s}{\partial \theta}$	$\dfrac{\partial s}{\partial z}$
$(\nabla \cdot \mathbf{v})$	$\dfrac{1}{r}\dfrac{\partial}{\partial r}(rv_r) + \dfrac{1}{r}\dfrac{\partial v_\theta}{\partial \theta} + \dfrac{\partial v_z}{\partial z}$		
$(\nabla^2 s)$	$\dfrac{1}{r}\dfrac{\partial}{\partial r}\left(r\dfrac{\partial s}{\partial r}\right) + \dfrac{1}{r^2}\dfrac{\partial^2 s}{\partial \theta^2} + \dfrac{\partial^2 s}{\partial z^2}$		
$[\nabla^2 \mathbf{v}]$	$\dfrac{\partial}{\partial r}\left(\dfrac{1}{r}\dfrac{\partial}{\partial r}(rv_r)\right)$ $+\dfrac{1}{r^2}\dfrac{\partial^2 v_r}{\partial \theta^2}$ $-\dfrac{2}{r^2}\dfrac{\partial v_\theta}{\partial \theta} + \dfrac{\partial^2 v_r}{\partial z^2}$	$\dfrac{\partial}{\partial r}\left(\dfrac{1}{r}\dfrac{\partial}{\partial r}(rv_\theta)\right)$ $+\dfrac{1}{r^2}\dfrac{\partial^2 v_\theta}{\partial \theta^2}$ $+\dfrac{2}{r^2}\dfrac{\partial v_r}{\partial \theta} + \dfrac{\partial^2 v_\theta}{\partial z^2}$	$\dfrac{1}{r}\dfrac{\partial}{\partial r}\left(r\dfrac{\partial v_z}{\partial r}\right)$ $+\dfrac{1}{r^2}\dfrac{\partial^2 v_z}{\partial \theta^2} + \dfrac{\partial^2 v_z}{\partial z^2}$
$[\nabla \cdot \boldsymbol{\tau}]$	$\dfrac{1}{r}\dfrac{\partial}{\partial r}(r\tau_{rr})$ $+\dfrac{1}{r}\dfrac{\partial \tau_{r\theta}}{\partial \theta}$ $-\dfrac{1}{r}\tau_{\theta\theta} + \dfrac{\partial \tau_{rz}}{\partial z}$	$\dfrac{1}{r}\dfrac{\partial \tau_{\theta\theta}}{\partial \theta} + \dfrac{\partial \tau_{r\theta}}{\partial r}$ $+\dfrac{2}{r}\tau_{r\theta} + \dfrac{\partial \tau_{\theta z}}{\partial z}$	$\dfrac{1}{r}\dfrac{\partial}{\partial r}(r\tau_{rz})$ $+\dfrac{1}{r}\dfrac{\partial \tau_{\theta z}}{\partial \theta}$ $+\dfrac{\partial \tau_{zz}}{\partial z}$
$[\mathbf{v}\cdot\nabla \mathbf{v}]$	$v_r\dfrac{\partial v_r}{\partial r} + \dfrac{v_\theta}{r}\dfrac{\partial v_r}{\partial \theta}$ $-\dfrac{v_\theta^2}{r} + v_z\dfrac{\partial v_r}{\partial z}$	$v_r\dfrac{\partial v_\theta}{\partial r} + \dfrac{v_\theta}{r}\dfrac{\partial v_\theta}{\partial \theta}$ $+\dfrac{v_r v_\theta}{r} + v_z\dfrac{\partial v_\theta}{\partial z}$	$v_r\dfrac{\partial v_z}{\partial r} + \dfrac{v_\theta}{r}\dfrac{\partial v_z}{\partial \theta}$ $+v_z\dfrac{\partial v_z}{\partial z}$

Table 2.17-2 (*Continued*)

Quantity	Spherical		
	r-Component	θ-Component	ϕ-Component
∇s	$\dfrac{\partial s}{\partial r}$	$\dfrac{1}{r}\dfrac{\partial s}{\partial \theta}$	$\dfrac{1}{r\sin\theta}\dfrac{\partial s}{\partial \phi}$
$(\nabla \cdot \mathbf{v})$	$\dfrac{1}{r^2}\dfrac{\partial}{\partial r}(r^2 v_r) + \dfrac{1}{r\sin\theta}\dfrac{\partial}{\partial \theta}(v_\theta \sin\theta) + \dfrac{1}{r\sin\theta}\dfrac{\partial v_\phi}{\partial \phi}$		
$(\nabla^2 s)$	$\dfrac{1}{r^2}\dfrac{\partial}{\partial r}\left(r^2 \dfrac{\partial s}{\partial r}\right) + \dfrac{1}{r^2 \sin\theta}\dfrac{\partial}{\partial \theta}\left(\sin\theta \dfrac{\partial s}{\partial \theta}\right) + \dfrac{1}{r^2 \sin^2\theta}\dfrac{\partial^2 s}{\partial \phi^2}$		
$[\nabla^2 \mathbf{v}]$	$\nabla^2 v_r - \dfrac{2v_r}{r^2} - \dfrac{2}{r^2}\dfrac{\partial v_\theta}{\partial \theta}$ $- \dfrac{2v_\theta \cot\theta}{r^2}$ $- \dfrac{2}{r^2 \sin\theta}\dfrac{\partial v_\phi}{\partial \phi}$	$\nabla^2 v_\theta + \dfrac{2}{r^2}\dfrac{\partial v_r}{\partial \theta}$ $- \dfrac{v_\theta}{r^2 \sin^2\theta}$ $- \dfrac{2\cos\theta}{r^2 \sin^2\theta}\dfrac{\partial v_\phi}{\partial \phi}$	$\nabla^2 v_\phi - \dfrac{v_\phi}{r^2 \sin^2\theta}$ $+ \dfrac{2}{r^2 \sin\theta}\dfrac{\partial v_r}{\partial \phi}$ $+ \dfrac{2\cos\theta}{r^2 \sin^2\theta}\dfrac{\partial v_\theta}{\partial \phi}$
$[\nabla \cdot \boldsymbol{\tau}]$	$\dfrac{1}{r^2}\dfrac{\partial}{\partial r}(r^2 \tau_{rr})$ $+ \dfrac{1}{r\sin\theta}\dfrac{\partial}{\partial \theta}(\tau_{r\theta}\sin\theta)$ $+ \dfrac{1}{r\sin\theta}\dfrac{\partial \tau_{r\phi}}{\partial \phi}$ $- \dfrac{\tau_{\theta\theta} + \tau_{\phi\phi}}{r}$	$\dfrac{1}{r^2}\dfrac{\partial}{\partial r}(r^2 \tau_{r\theta})$ $+ \dfrac{1}{r\sin\theta}\dfrac{\partial}{\partial \theta}(\tau_{\theta\theta}\sin\theta)$ $+ \dfrac{1}{r\sin\theta}\dfrac{\partial \tau_{\theta\phi}}{\partial \phi}$ $+ \dfrac{\tau_{r\theta}}{r} - \dfrac{\cot\theta}{r}\tau_{\phi\phi}$	$\dfrac{1}{r^2}\dfrac{\partial}{\partial r}(r^2 \tau_{r\phi})$ $+ \dfrac{1}{r}\dfrac{\partial \tau_{\theta\phi}}{\partial \theta} + \dfrac{\tau_{r\phi}}{r}$ $+ \dfrac{1}{r\sin\theta}\dfrac{\partial \tau_{\phi\phi}}{\partial \phi}$ $+ \dfrac{2\cot\theta}{r}\tau_{\theta\phi}$
$[\mathbf{v} \cdot \nabla \mathbf{v}]$	$v_r \dfrac{\partial v_r}{\partial r} + \dfrac{v_\theta}{r}\dfrac{\partial v_r}{\partial \theta}$ $+ \dfrac{v_\phi}{r\sin\theta}\dfrac{\partial v_r}{\partial \phi}$ $- \dfrac{v_\phi^2 + v_\theta^2}{r}$	$v_r \dfrac{\partial v_\theta}{\partial r} + \dfrac{v_\theta}{r}\dfrac{\partial v_\theta}{\partial \theta}$ $+ \dfrac{v_\phi}{r\sin\theta}\dfrac{\partial v_\theta}{\partial \phi}$ $+ \dfrac{v_r v_\theta}{r} - \dfrac{v_\phi^2 \cot\theta}{r}$	$v_r \dfrac{\partial v_\phi}{\partial r} + \dfrac{v_\theta}{r}\dfrac{\partial v_\phi}{\partial \theta}$ $+ \dfrac{v_\phi}{r\sin\theta}\dfrac{\partial v_\phi}{\partial \phi}$ $+ \dfrac{v_\phi v_r}{r} + \dfrac{v_\theta v_\phi \cot\theta}{r}$

2.B Partial Differentiation: Stream Function in Renal Flow [22]

In Problem 6.B (a) we can obtain (as in Example 6.1-4) from the equations of continuity and motion

$$\frac{1}{r}\frac{\partial}{\partial r}(rv_r) + \frac{\partial v_z}{\partial z} = 0 \qquad (2.B-1)$$

$$\frac{1}{\mu}\frac{\partial p}{\partial r} = \frac{\partial^2 v_r}{\partial r^2} + \frac{1}{r}\frac{\partial v_r}{\partial r} - \frac{v_r}{r^2} + \frac{\partial^2 v_r}{\partial z^2} \qquad (2.B-2)$$

$$\frac{1}{\mu}\frac{\partial p}{\partial z} = \frac{\partial^2 v_z}{\partial r^2} + \frac{1}{r}\frac{\partial v_z}{\partial r} + \frac{\partial^2 v_z}{\partial z^2}. \qquad (2.B-3)$$

By differentiating Eqs. 2.B-2 and 2.B-3 with respect to z and r, respectively, and the definitions for *stream function** ψ such that

$$\frac{1}{r}\frac{\partial \psi}{\partial z} \equiv v_r \qquad (2.B-4)$$

$$-\frac{1}{r}\frac{\partial \psi}{\partial r} \equiv v_z, \qquad (2.B-5)$$

show that

(a)   Eq. 2.B-1 is automatically satisfied by Eqs. 2.B-4 and 2.B-5.

(b) $E^2(E^2\psi) = 0$ $\qquad\qquad\qquad\qquad\qquad\qquad$ (2.B-6)

where the *operator* E^2 is, for cylindrical coordinates,

$$E^2 = \frac{\partial^2}{\partial r^2} - \frac{1}{r}\frac{\partial}{\partial r} + \frac{\partial^2}{\partial z^2}. \qquad (2.B-7)$$

2.C Partial Differentiation: Velocity Distributions in the Hemispheric Portion of the Pseudopod of an Amoeba [23]

In order to obtain the velocity distributions v_r and v_θ in the hemisphere of the pseudopod of an amoeba (see Problem 6.D), it is necessary to define the stream function for spherical geometry

$$v_r = -\frac{1}{r^2 \sin\theta}\frac{\partial \psi}{\partial \theta} \qquad (2.C-1)$$

*See Problem 6.B about the meaning of stream function.

$$v_\theta = \frac{1}{r \sin\theta} \frac{\partial\psi}{\partial r}, \qquad (2.\text{C-}2)$$

while assuming ψ to be of the form

$$\psi = \left(\frac{A}{r} + Br + Cr^2 + Dr^4\right) \sin^2\theta \qquad (2.\text{C-}3)$$

with the boundary conditions

> B.C.1: at $r=0$, v_r and v_θ must be finite
>
> B.C.2: at $\theta = \frac{\pi}{2}$, $v_r = 0$
>
> B.C.3: at $r = R$, $v_r = 0$
>
> B.C.4: at $r = R$ and $\theta = \frac{\pi}{2}$, $v_\theta = V$,

where V is the uniform velocity of the outer portion of the pseudopod. Staats and Wasan [23] gave the following solutions:

$$v_r = V\left(1 - \frac{r^2}{R^2}\right)\cos\theta, \qquad (2.\text{C-}4)$$

$$v_\theta = V\left(2\frac{r^2}{R^2} - 1\right)\sin\theta. \qquad (2.\text{C-}5)$$

Are they correct? Show whether the solutions satisfy the differential equation *and* the boundary conditions.

2.D Exercise in Partial Differentiation and Differentiation of an Integral: Correlation of Blood-Flow Data

The general expression for the volumetric flow rate of the laminar flow of a Casson fluid as shown in Eq. 9.1-20 can be rewritten as

$$Q = \frac{8\pi}{(\Delta p/L)^3} \int_{\tau_y}^{\tau_R} \tau_{rz}^2 \dot\gamma \, d\tau_{rz} \qquad (2.\text{D-}1)$$

where

$$\dot\gamma = -\frac{dv_z}{dr} \qquad \text{(function of } \tau_{rz}\text{)}$$

$$\tau_R = \frac{\Delta PR}{2L}$$

$$\tau_y = \text{yield stress} \qquad \text{(a constant)}.$$

Show that

$$\dot{\gamma}_R = \frac{Q}{\pi R^3}\left[3+\left(\frac{\partial \ln Q}{\partial \ln \Delta P}\right)_{R,L}\right]. \qquad (2.\text{D-}2)$$

(In doing this problem, you will have the opportunity to use the Leibnitz formula for the differentiation of integrals, Eq. 2.6-15.)

2.E Exercise in Integration: Time-Averaged Volumetric Rate of Pulsatile Blood Flow

In Eq. 6.1-51 (with Eq. 6.1-46) the volumetric flow rate Q is found to be a function of t; that is, it is also periodic in nature but with a time lag behind the periodic pressure gradient. Show that over a long period of time, the time-averaged flow rate is

$$\bar{Q} = \frac{\displaystyle\int_0^{2\pi/\omega} Q(t)\,dt}{\displaystyle\int_0^{2\pi/\omega} dt}$$

$$= \frac{\pi R^4}{8\mu}\left(\frac{\Delta P}{L}+\frac{16A\lambda^2}{\pi}\sum_{k=1}^{\infty}\frac{1}{\alpha_k^4}\cdot\frac{1-\exp(-2\pi\alpha_k^2/\lambda)}{\alpha_k^4+\lambda^2}\right), \qquad (2.\text{E-}1)$$

where $\lambda = R^2\omega/\nu$.

2.F Integration of Partial Derivatives: Pressure Distributions in Renal Tubule [24]

Following the same procedure as in Example 6.1-4 (Eqs. 6.1-83 through 6.1-85) and in Appendix C, show that the pressure distribution in the renal tubule with exponentially declining bulk flow (see Problem 6.B) is

$$p(r,z)-p(0,0) = -\left(\frac{8Q_0\mu}{\pi R^4}-\frac{16v_0\mu}{\alpha R^3}\right)z$$

$$-\frac{2v_0\mu}{R}\left[\frac{J_0(\alpha R)J_0(\alpha r)}{J_1^2(\alpha R)-J_2(\alpha R)J_0(\alpha R)}\right](1-e^{-\alpha z}), \qquad (2.\text{F-}1)$$

where $p(0,0)$ is the pressure at $r=0, z=0$, while the other terms are defined and the pertinent equations to be used can be found in Problem 6.B. That is, from Eqs. 2.B-2, 2.B-3, 6.B-9, and 6.B-10, we can obtain the expressions for $\partial p/\partial r$ and $\partial p/\partial z$. Does it matter whether we integrate via $(0,0)\rightarrow(r,0)\rightarrow(r,z)$ or via $(0,0)\rightarrow(0,z)\rightarrow(r,z)$?

2.G Infinite Series

Using infinite series, or Eqs. (2.5-8) and (2.5-9) with the definitions of hyperbolic functions in Table 2.1-1, show that

$$e^{ix} = i\sin x + \cos x \qquad (2.G-1)$$

and

$$e^{2\pi ki} = 1, \qquad (2.G-2)$$

where k are all possible integers; that is, $k = 0, \pm 1, \pm 2, \dots, \pm\infty$; and $i = \sqrt{-1}$.

2.H Solution to Partial Differential Equations: Transient Absorption and Diffusion of γ-Globulin by Lung Tissues

Solve Eq. 6.3-11 to obtain Eq. 6.3-12 by the following methods:
(a) Separation-of-variables method, by assuming solution of the type

$$\Gamma(\xi,\tau) = f(\xi)\cdot g(\tau).$$

(b) Laplace transform method, by first taking the Laplace transform of c_A (or Γ) with respect to t (or τ), and then solving the resulting ordinary differential equation for $\bar{c}_A$ (or $\bar{\Gamma}$) followed by inversion of the solution.

2.I Vectors

Show that, by writing vectors a, b, and c in the form of Eq. 2.15-1,

$$(\mathbf{a}\cdot\mathbf{b})\mathbf{c} \neq \mathbf{a}(\mathbf{b}\cdot\mathbf{c}).$$

Actually there is an obvious reason why the two sides are not equal, even without performing the mathematics. What is this argument?

2.J Differential Operation Involving Vectors

By writing the operator ∇ and the vector $\mathbf{v}$ in the forms of Eqs. 2.15-20 and 2.15-1, respectively, show that

$$(\nabla\cdot\rho\mathbf{v}) = \rho(\nabla\cdot\mathbf{v}) + (\mathbf{v}\cdot\nabla\rho),$$

where ρ is a scalar variable.

2.K Dyadic Products

Prove

$$[\nabla \cdot \rho vv] = [v \cdot \nabla \rho v] + \rho v(\nabla \cdot v)$$

either by writing out the components as in Problems 2.I and 2.J or by first proving the tensor identity

$$[\nabla \cdot vw] = [v \cdot \nabla w] + w(\nabla \cdot v).$$

2.L Differential Operation Involving Tensors

Prove that

$$(\nabla \cdot [\tau \cdot v]) = (v \cdot [\nabla \cdot \tau]) + (\tau : \nabla v).$$

REFERENCES

1. Einstein, A. *Ann. Phys.*, **19**, 289 (1906).

2. Fåhraeus, R., and Lindqvist, T. *Am. J. Physiol.*, **96**, 562 (1931).

3. Vand, V. *J. Phys. Coll. Chem.*, **62**, 277 (1948).

4. Macey, R. I. *Bull. Math. Biophys.*, **25**(1), 1 (1963).

5. Aiba, S.; Humphrey, A. E.; and Millis, N. F. *Biochemical Engineering.* Academic: New York, 1965, Section 4.1.3.

6. Abramowitz, M., and Stegun, I. A. *Handbook of Mathematical Functions with Formulas, Graphs, and Mathematical Tables*, NBS-AMS 55. Government Printing Office: Washington, D. C., 1964.

7. *CRC Mathematical Tables.* Chemical Rubber Publishing: Cleveland, 1962.

8. Speigel, M. R., *Mathematical Handbook of Formulas and Tables*, Schaum's Outline Series, McGraw-Hill: New York, 1968.

9. Lih, M. M. *Bull. Math. Biophys.*, **31**(1), 143 (1969).

10. Siddique, M. R. Private communication, 1969.

11. Jahnke, E.; Emde, F.; and Lösch, F. *Table of Higher Functions*, 6th ed. McGraw-Hill: New York, 1960.

12. Hildebrand, F. B. *Advanced Calculus for Applications*, 2nd ed. Prentice-Hall: Englewood Cliffs, N.J., 1964.

13. Hansen, A. G. *Similarity Analysis of Boundary-Value Problems in Engineering.* Prentice-Hall: Englewood Cliffs, N.J., 1965.

14. Birkhoff, G. *Hydrodynamics*, rev. ed. Princeton Univ. Press, Princeton, N.J., 1961.

15. Bird, R. B., Stewart, W. E.; and Lightfoot, E. N. *Transport Phenomena.* Wiley: New York, 1960.

16. Kelman, R. B. *Bull. Math Biophys.*, **27** (1), 53 (1965).

17. Ames, W. F. *Nonlinear Ordinary Differential Equations in Transport Processes.* Academic: New York, 1968.

18. Ames, W. F. *Nonlinear Partial Differential Equations in Engineering.* Academic: New York, 1965.

19. Southworth, R. W., and DeLeeuw, S. L. *Digital Computation and Numerical Methods.* McGraw-Hill: New York, 1965.

20. Hildebrand, F. B. *Methods of Applied Mathematics*, 2nd ed. Prentice-Hall: Englewood Cliffs, N.J., 1965.

21. Aris, R. *Vectors, Tensors, and the Basic Equations in Fluid Mechanics.* Prentice-Hall: Englewood Cliffs, N.J., 1962.

22. Macey, R. I. *Bull. Math. Biophys.*, **27**, 117 (1965).

23. Staats, W. R. and Wasan, D. T. In *Chemical Engineering in Medicine, Chem. Eng. Prog. Symp. Series No. 66*, Vol. LXII, E. F. Leonard, Ed. American Institute of Chemical Engineers: New York, 1966, p. 132.

24. Kozinski, A. A.; Schmidt, F. P.; and Lightfoot, E. N. *Ind. Eng. Chem. Fund.*, **9**(3), 502 (1970).

CHAPTER THREE

Fundamental Principles

In this chapter we begin to discuss our subject proper—transport phenomena. Since the organization of the material largely depends on analogies, we shall first state the general law of transport. This simple law may seem abstract until we apply it to the three types of transports—mass, momentum, and energy, whereupon it takes on practical meanings.

It should not be forgotten, however, that mathematical analogies are merely a simple way to "model" the real situation; it would be a mistake to interpret them as physical realities. For example, the mass flux, a vector, has three components, while the momentum flux, a second-order tensor, has nine. In addition, there are a host of non-Newtonian fluids, some of which are quite relevant in physiological work. On the other hand departures from Fourier's and Fick's laws are believed to be of much smaller variety. Another example is mass transport of which diffusion is only one of many modes—others being convective, active, and facilitated transports. Even in diffusion alone we have coupling effects with energy transport. These indicate that the analogies end somewhere. These exceptions will also be discussed, whenever physiologically related, so as to provide background for some of the special-interest topics later.

3.1 GENERAL PHENOMENOLOGICAL LAW

The general law governing the transport of physical entities can be stated as follows: *The flux of a physical entity (momentum, energy, or mass) is directly proportional to, and in the opposite direction of, the gradient of the potential (velocity, temperature, or concentration, respectively) causing the movement,* where *flux* is defined as "time rate of flow per unit area" and *gradient* the "rate of increase per unit length."

In other words, "everything goes downhill"—a phenomenon that we all have been familiar with since childhood days. In short-hand form we can write

$$\text{flux} = -(\text{coefficient}) \times (\text{gradient}). \qquad (3.1\text{-}1)$$

The following sections will illustrate the specific meaning of this general statement in the three different subject areas.

3.2 MOMENTUM TRANSPORT

If we put a deck of playing cards, all stacked up neatly, on the palm of one hand and use the other hand to press the deck slightly while giving it horizontal push, we would find that the side of the deck forms a parallelogram, as shown in Figure 3.2-1. Elementary physics tells us that between the sliding cards there exists *friction*. If the deck of cards is replaced by a solid block that is subjected to the same force and attempted motion, we know that *shear stress* is produced whose magnitude is defined as the shear force divided by the area upon which the force is exerted. In the case of the playing cards, for example, the surface area of the playing card should be used.

Now consider a fluid—make it a liquid for the sake of convenience in observation and appreciation—between two large* parallel, horizontal flat plates of surface area A, separated by a distance Y. The bottom plate is held stationary while at time $t = 0$, the upper one is given a horizontal motion with a constant velocity V. Figure 3.2-2 shows the sequence of events with (b) showing the fluid at rest before the motion takes place and (c) showing that at the precise moment $t = 0$, the top plate is set in motion while the fluid is still at rest. Sketch (d) shows that a short time later the

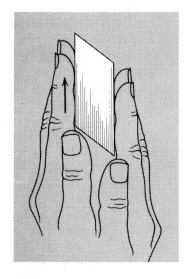

Figure 3.2-1. A deck of cards under shear.

*Large enough so that no fluid flows in the sideways direction within the area under consideration.

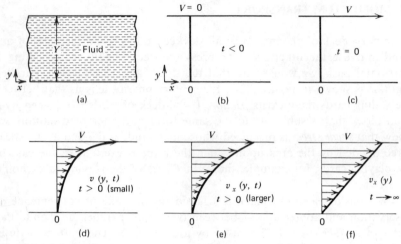

Figure 3.2-2. Sequence of events in setting in motion the upper of two parallel plates.

motion is, in layman's terms, "transmitted through" a small depth of fluid. As time elapses, we find that the motion is transmitted through an increasing depth of fluid (e) until finally, a sufficiently long time later, the velocity profile between the two plates takes a linear form (f) and no longer changes. We say that a "steady state" has been attained: that is, the velocity distribution is unchanged with respect to time.

A constant force F is required to overcome the friction between fluid layers (analogous to the friction between playing cards) and to maintain the motion of the upper plate. The following relationship represents an experimental fact, first observed by Sir Isaac Newton:

$$\frac{F}{A} \propto \frac{V}{Y}^{*}$$

or

$$\frac{F}{A} = \mu \frac{V}{Y}, \tag{3.2-1}$$

where μ is the proportionality constant and is called the *viscosity* of the fluid.

In general, linear velocity distribution such as that depicted in Figure 3.2-2(f) cannot be assumed, in which case Eq. 3.2-1 takes on a differential

*The symbol $\propto$ stands for "proportional to".

form by considering only an infinitesimally thin layer of the fluid,

$$\tau_{yx} = -\mu \frac{dv_x}{dy}, \tag{3.2-2}$$

which is known as the *Newton's law of viscosity*. The existence of the minus sign will be explained later. The derivative dv_x/dy is the *velocity gradient*, or *shear rate*. It is obvious that in Eq. 3.2-2 we represent F/A by τ_{yx}, the symbol for shear stress, with subscripts yx indicating that the x-directed shear force is acting on an area perpendicular to the y-axis.

Furthermore, τ_{yx} also has the meaning of "flux of x momentum in the y-direction." Since momentum is directly proportional to velocity, the fluid element having a larger v_x has a larger x-directed momentum, and this momentum is transferred to the fluid element in the next layer, which has a lower v_x, in the sense that the former "drags" the latter along. From the definition of flux, momentum flux has the unit of

$$\frac{\text{g-cm/sec}}{\text{sec-cm}^2} = \frac{\text{g-cm}}{\text{sec}^2\text{cm}^2}$$

$$= \frac{\text{dynes}}{\text{cm}^2}$$

or*

$$\tau_{yx}[=]\text{dynes/cm}^2$$

or

$$\frac{\text{lb}_\text{m}\text{-ft/sec}}{\text{sec-ft}^2} = \frac{\text{lb}_\text{m}\text{-ft}}{\text{sec}^2\text{ft}^2}$$

$$= \frac{1}{32.174}\text{lb}_\text{f}/\text{ft}^{2\dagger}$$

$$\tau_{yx}[=]\text{lb}_\text{f}/\text{ft}^2.^\ddagger$$

The same units can be arrived at from the stress point of view. Thus, we can use a wavy arrow to show the transfer of momentum as in Figure 3.2-3.

*The symbol [=] stands for "has the units of."

†The author does not believe it is necessary to include the conversion factor g_c for British units as long as one remembers (which he should) that 32.174 lb_m-ft/sec^2 is 1 lb-force. It does not seem logical that whether an equation has a certain term depends on whether it is in cgs or British units. Furthermore, the author has found students easily confused by g_c and g. Thus, down with the g_c factor!

‡Hereafter we shall drop the British units in favor of the cgs system for the sake of brevity.

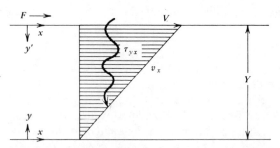

Figure 3.2-3. Velocity gradient and momentum flux.

The transition from Eq. 3.2-1 to Eq. 3.2-2 can be more easily understood if we label the upper plate and the lower plate. Then,

$$\frac{V}{Y} = \frac{V-0}{Y-0} = \frac{v_{x_1} - v_{x_0}}{y_1 - y_0} = \frac{\Delta v_x}{\Delta y},$$

and if we let Δy and Δv_x approach zero, then $\Delta v_x / \Delta y$ approach dv_x / dy (Section 2.2). The minus sign in Eq. 3.2-2 arises from the fact that momentum "flows downhill." In the $x-y$ system of Figure 3.2-2 dv_x / dy has a positive value, but τ_{yx} is in the negative y-direction. Since the viscosity is a scalar (always positive), there must be a minus sign to "balance" the signs. Note also that the existence of this minus sign does *not* depend on the particular coordinate system chosen. If we use the $x-y'$ system as depicted in Figure 3.2-2, dv_x / dy' is negative, while the momentum flux $\tau_{y'x}$ is now in the positive y'-direction. Thus, we still need the minus sign to balance the equation.

3.3 ENERGY TRANSPORT

Suppose we have a slab of finite thickness Y and a sufficiently large surface area A, as shown in Figure 3.3-1 (a). Initially the slab is maintained at a temperature of T_0 throughout (b). At time $t = 0$, we suddenly raise the temperature of the upper surface to T_1 (c). A short while later the temperature adjacent to the upper surface begins to rise (d) because of the heat flow created by the temperature difference. The temperature rise continues and spreads toward the lower surface (e) until eventually a steady-state linear temperature profile is established (f).

A constant heat flow Q is required to maintain the upper surface

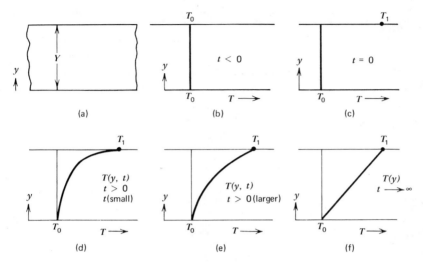

Figure 3.3-1.   Sequence of events in raising the temperature of one surface of a slab.

temperature at T_1 and the linear temperature distribution.* Analogous to Eq. 3.2-1 in the case of shearing of a fluid, the following is true:

$$\frac{Q}{A} \propto \frac{T_1 - T_0}{Y}$$

or

$$\frac{Q}{A} = k \frac{T_1 - T_0}{Y}, \tag{3.3-1}$$

where k is the thermal conductivity of the material. If we consider a curved geometry and/or other factors that might give a nonlinear temperature distribution, only an infinitesimally thin layer is taken. We then have

$$q_y = -k \frac{dT}{dy}, \tag{3.3-2}$$

which is *Fourier's law of heat conduction*. In this equation, q_y is the heat flux in the y-direction and is equal to Q/A, the rate of heat flow per unit area; dT/dy is the temperature gradient, or driving potential of heat flow, in the y-direction. The negative sign exists here for the same reason it exists in the Newton's law of viscosity: heat flows in the direction of decreasing

*This implies that the same constant heat flow Q must also be "drained off" at the low-temperature surface in order to maintain this steady-state temperature distribution.

temperature. The reader may wish to go through the mathematical formality himself.

At the beginning of this section, we assumed a "sufficiently large" surface area A, so that there is no temperature gradient (and heat flow) in the x- and z-directions. However, when physical dimensions are such that this assumption cannot be made, we do have to consider heat flow in one or both of the other directions:*

$$q_x = -k \frac{\partial T}{\partial x} \tag{3.3-3}$$

$$q_z = -k \frac{\partial T}{\partial z}, \tag{3.3-4}$$

while Eq. 3.3-2 becomes

$$q_y = -k \frac{\partial T}{\partial y} \tag{3.3-5}$$

According to Section 2.15, these equations can be vectorially combined to form a single vector equation. That is, if we multiply Eqs. 3.3-3, 3.3-4, and 3.3-5 by the unit vectors in the three corresponding directions and sum them together, we have

$$\mathbf{q} = -k \nabla T, \tag{3.3-6}$$

which amounts to a shorthand way of expressing the three equations. In this equation, $\mathbf{q}$ is the vectorial heat flux and ∇T the temperature gradient.

3.4 MASS TRANSPORT

There are several modes of mass transfer, but the one that is mathematically analogous to heat conduction and momentum transfer discussed above is called *diffusion*. When diffusion is coupled with fluid motion due to external means, we have *convective mass transfer*. Other higher order mass-transfer modes include *active transport* and *facilitated transport*, both of which are of physiological interest. We shall devote special sections to them later. Simply explained, facilitated transport is diffusion aided by a carrier, while active transport goes against concentration gradients, with energy provided by such means as biological or biochemical reactions.

On the contrary, diffusion, as we generally know it, is a passive process that can further be classified in terms of the types and relative sizes of the diffusing molecules and diffusion medium. The kinetic theory of gases tells

*Note the change from total to partial derivatives.

us that molecules travel in highly irregular zigzag patterns. If the mean free path (average distance traveled between collisions) of a species is small compared with the size of the medium, the statistical average of the actions of the diffusing molecules may be very slow. We call this *bulk diffusion*. On the other hand, if the mean free path is of the same or larger order of magnitude as the medium (such as in the fine pores of a solid medium), the individual action of the zigzagging molecules can be "felt"; we call this *Knudson diffusion*.

Furthermore, diffusion can be caused by concentration and pressure gradients as well as by electrolytic and thermal effects. It can also produce one or more of these effects. This is especially true with multicomponent diffusion, wherein such coupling effects can be fully explained via non-equilibrium thermodynamics. However, we will discuss them only when the occasion arises. Here we shall introduce only the simplest mode of diffusion that is analogous to the momentum and energy transport we have just discussed in the last two sections. It is more precisely known as *ordinary diffusion*, but the word *ordinary* is usually omitted unless such omission would create ambiguity or confusion.

If we broke a bottle of perfume at one corner of the room, its fragrance would migrate through the room, or, more technically, its molecules would diffuse through the air across the room. In such a case, people standing along the path of diffusion at a certain instant smell the fragrance at different strengths, with the person closest to the perfume source getting the greatest satisfaction. On the other hand, a person standing at the far end of the room smells a fragrance that is increasing with respect to time. This is of course because of the cencentration gradient of the perfume along the path and the increasing level of concentration as time elapses, similar to the situation described in Figures 3.2-2 and 3.3-1.

It has indeed been found that the phenomenon of concentration diffusion is similar to that of heat conduction and that Figure (3.3-1) can be changed to represent the sequence of events in initiating a one-dimensional diffusion process (i.e., infinite extent in the other two directions) by simply replacing T with ρ_A, where ρ_A denotes mass concentration of species A (see Figure 3.4-1). Furthermore, the phenomenological law governing the gradient and flux is also analogous:

$$j_{A_y} = - \mathcal{D}_{AB} \frac{d\rho_A}{dy}, \qquad (3.4\text{-}1)$$

where j_{A_y} is the mass flux of A in the y-direction and $\mathcal{D}_{AB}$ is the *binary diffusivity*. This is called *Fick's (first) law of diffusion*. It appears in many other forms (e.g., in terms of molar and mass fluxes and concentrations,

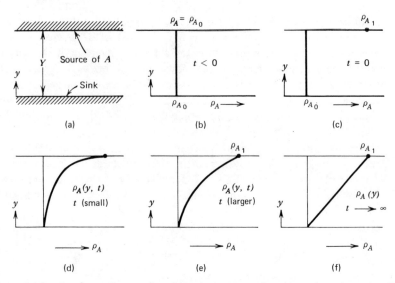

Figure 3.4-1. Sequence of events in raising the concentration at one boundary of a diffusional medium.

both total and diffusional). These involve a massive array of definitions and terminology and shall be dealt with in Chapter 5.

Similar to Eqs. 3.3-3, 3.3-4, and 3.3-5, if ρ_A also changes in the x- and z-directions, Fick's law (Eq. 3.4-1) can be written in all three directions:

$$j_{A_x} = -\mathcal{D}_{AB} \frac{\partial \rho_A}{\partial x} \tag{3.4-2}$$

$$j_{A_y} = -\mathcal{D}_{AB} \frac{\partial \rho_A}{\partial y} \tag{3.4-3}$$

$$j_{A_z} = -\mathcal{D}_{AB} - \frac{\partial \rho_A}{\partial z}. \tag{3.4-4}$$

In fact, $j_{A_x}, j_{A_y},$ and j_{A_z} are the three components of the mass flux vector $\mathbf{j}_A$. Consequently, by combining Eqs. 3.4-2, 3.4-3, and 3.4-4 vectorially, we obtain the vector form of Fick's law of diffusion:

$$\mathbf{j}_A = -\mathcal{D}_{AB} \nabla \rho_A, \tag{3.4-5}$$

where $\nabla \rho_A$ is simply the mass-concentration gradient.

We can see the analogy between this last equation and Eq. (3.3-3). However, when we attempt to carry this analogy over to momentum transport, one difficulty develops: both τ_{yx} and v_x have one more subscript than, for example, q_y and T, respectively. This is because v_x, v_y, and v_z are already components of a vector ($\mathbf{v}$) whereas T or ρ_A is just a scalar with no "components" to speak of (only one, itself). This makes a quantity such as τ_{yx} "one degree more complex" than components of $\mathbf{q}$ and $\mathbf{j}_A$ because it is only one of nine components (six shear stresses and three normal stresses acting in the three directions). The analogy involved is also more complex.

3.5 VISCOSITY, DIFFUSIVITY, AND THERMAL CONDUCTIVITY

If we plot the shear stress τ_{yx} against the negative velocity gradient, or shear rate, $-dv_x/dy$, for a common fluid such as water, a straight line passing through the origin results (Figure 3.5-1, –·–·– line). According to Newton's law of viscosity, the slope of this line, being constant, is the viscosity μ of this fluid. It is obvious that if the slope is large, we have to apply a large shear stress to achieve a certain velocity gradient, in which case we say that this fluid is highly viscous. A fluid obeying such a linear relationship is called a *Newtonian fluid*.

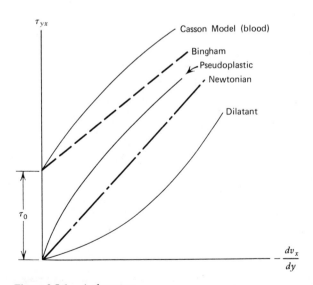

Figure 3.5-1. A rheogram.

There are also other types of fluids that deviate from this behavior. Some exhibit curvatures on this "rheogram" and are called *pseudoplastics* or *dilatant fluids*, depending on which way the lines curve. Another type of fluid, *Bingham plastic*, displays a yield-stress phenomenon; that is, below a certain value τ_0 of shear stress the fluid behaves like a solid body, but when this yield-stress value is exceeded, the fluid flows like a Newtonian fluid.

This latter type of fluid is of particular interest in biomedical work, since blood exhibits this yield-stress phenomenon—except that the behavior in the range immediately above τ_0 is somewhat more complex than the pure linearity of a Bingham fluid. This can be represented by the Casson model [1] and is briefly discussed in Chapter 9. The reader is referred to the classic treatise of Whitmore [2], in which the rheology of still another type of fluid, the viscoelastic fluid, is also discussed.* A viscoelastic fluid is one which only partially returns to its orginal form when the applied stress is removed. Synovial fluid is an example of this. Those fluids that deviate from Newtonian behavior are categorically called *non-Newtonian* fluids.

Although the physical properties of a substance are intrinsic, in actual application the physical parameters of the system may influence them. For example, carefully carried out experiments have shown the following:

1.   Serum (plasma without fibrinogen) is a Newtonian fluid.
2.   Plasma (serum plus fibrinogen) is also a Newtonian fluid.
3.   Whole blood (plasma plus erythrocytes, leukocytes, etc.) is non-Newtonian only in medium-size vessels, such as arterioles.

The currently accepted theory is that erythrocytes, with fibrinogen molecules adhering on their surfaces, tend to stick together forming "rouleaux," which in turn give rise to networks, thus causing non-Newtonian behavior. However, in large vessels such as arteries, the blood flows so rapidly and the shear rate is so high that the yield-stress phenomenon is negligible, thus exhibiting Newtonian behavior again.

In very small vessels, of a diameter approximating that of a red blood cell, it makes no sense to talk about the viscosity of "plasma plus red cells," because there will be so few red blood cells flowing abreast. Hence, in the microcirculation, we can no longer consider blood as a uniform fluid.[†] We are dealing with erythrocytes (and white blood cells) and "foreign" particles floating in plasma, thus constituting a heterogeneous fluid. We shall discuss blood rheology in Chapter 9.

Rheology can be defined as "the science of the deformation and flow of matter ranging between liquid and solids"[3].

[†]Although a suspension of red blood cells in plasma, blood is generally considered as a continuous, homogeneous fluid but with non-Newtonian behavior in larger vessels.

Similar to the viscosity in Newton's law, the $\mathcal{D}_{AB}$ in Fick's law is called the *diffusional coefficient*, or simply *diffusivity*, and the k in Fourier's law the *thermal conductivity*. Like Newtonian viscosity, both diffusivity and thermal conductivity are usually constant at constant temperature and pressure. Expressions for theoretically calculating them are given in standard texts [4,5]. Perhaps it is important at least to keep in mind some qualitative behaviors. For example, with liquids, the viscosity decreases with increasing temperature (as evidenced by heating cooking oil in a frying pan) and is relatively insensitive to pressure change, whereas with gases the viscosity increases with both temperature and pressure because of the more vigorous molecular motions and closer proximity of molecules, respectively. For simple descriptions of the pertinent formulas the reader is advised to consult Chapters 1, 8, and 16 of Bird, et al.[6].

3.6 PHENOMENOLOGICAL ANALOGIES

The reader has perhaps detected some similarities between Eqs. 3.2-2, 3.3-2, and 3.4-1 and the meanings of the terms in them, as we have glimpsed in Eq. 3.1-1. In Eq. 3.4-1, the total mass concentration, or density, ρ, of the mixture of species A and B was assumed constant,* where $\rho = \rho_A + \rho_B$ with $\rho_A = \rho\omega_A$ and ω_A is the mass fraction of A in the mixture (see Section 5.1). If in Eqs. 3.2-2 and 3.3-2, ρ, μ, and $\hat{C}_p$†, the heat capacity per unit mass, are also kept constant, they can be moved in and out of the differentiation operation at will, and these two equations become, respectively,

$$\tau_{yx} = -\frac{\mu}{\rho} \frac{d}{dy}(\rho v_x) \qquad (3.6\text{-}1)$$

$$q_y = -\frac{k}{\rho\hat{C}_p} \frac{d}{dy}(\rho\hat{C}_p T). \qquad (3.6\text{-}2)$$

Compared with Eq. 3.4-1,

$$j_{Ay} = -\mathcal{D}_{AB} \frac{d}{dy}(\rho_A),$$

we can see that we are differentiating, with respect to y (the d/dy operation), mass per unit volume (ρ_A), momentum per unit volume (ρv_x),

---

*Otherwise Eq. 3.4-1 would be $j_{Ay} = -\rho\mathcal{D}_{AB}(d\omega_A/dy)$. Also, for simplicity in explanation we are here talking only about binary diffusion. Multicomponent systems will be discussed in Chapter 5.
†The symbol ^ on top of a quantity means "per unit mass" of that quantity.

and energy per unit volume $(\rho \hat{C}_p T)$ in these equations, as one can easily verify by considering the units of these quantities. For example, mv_x (where m is mass) is momentum; that is why ρv_x (where ρ is density, or mass per unit volume) is momentum per unit volume. These derivatives are then actually the mass, momentum, and energy gradients (per unit volume) and are directly proportional to the corresponding mass, momentum, and energy fluxes on the left-hand sides of these equations.

The proportionality constant $\mathcal{D}_{AB}$ is the mass diffusivity, as we have explained earlier, having the units of square centimeters per second. The constant for Eq. 3.6-1, μ/ρ, is called the *kinematic viscosity* and is usually designated by the symbol ν. By analogy and by considering its units, which are also square centimeters per second, one would expect that we should be able to regard it as the "momentum diffusivity." Indeed we can.

It then follows that we can call $k/\rho\hat{C}_p$ the *thermal diffusivity*; it is usually denoted by α. This quantity, not the thermal conductivity k alone, governs the rate of penetration of heat in a substance, for k represents only the efficiency of conduction, the transfer of energy from one point to a neighboring one. It does not tell how much heat the material "absorbs." If the material has a high density ρ and/or high heat capacity per unit mass $\hat{C}_p$, each unit volume of it will absorb a great deal of energy for each unit of temperature increase, thus leaving relatively little energy to be passed on to its neighbor. In other words, high thermal conductivity can be at least partially offset by high heat capacity per unit volume $(\rho\hat{C}_p)$ to yield a low thermal diffusivity.

Similarly, considering the momentum diffusivity $\nu = \mu/\rho$ again, we can see that a high viscosity μ, which should normally mean a fast transmission of motion, can be compensated by a high density ρ, to yield a low-momentum penetration rate, because the large inertia produced by this high mass per unit volume makes movement difficult.

In conclusion, we can summarize the three phenomenological laws,

$$j_{A_y} = - \mathcal{D}_{AB} \frac{d}{dy}(\rho_A)$$

$$\tau_{yx} = - \nu \frac{d}{dy}(\rho v_x)$$

$$q_y = - \alpha \frac{d}{dy}(\rho \hat{C}_p T),$$

into a general expression,

$$\binom{\text{Flux of entity}}{\text{being transported}} = - (\text{Diffusivity}) \times \binom{\text{Gradient of}}{\text{said entity}},$$

Table 3.6-1 Analogies in Phenomenological Laws

	Mass	Momentum	Energy	Remarks
Flux (cgs units)	j_{Ay}, g/cm²-sec	τ_{yx}, $\text{g-cm/sec}^2\text{-cm}^2 = \dfrac{\text{g-cm}}{\text{sec}}\Big/\text{cm}^2\text{-sec}$	q_y, cal/cm²-sec	Note that it is always entity per unit area-time.
Diffusivity (cgs units)	$\mathcal{D}_{AB}$, cm²/sec	ν, $\dfrac{\text{g}}{\text{cm-sec}}\Big/\dfrac{\text{g}}{\text{cm}^3} =$ cm²/sec	α, $\dfrac{(\text{cal/sec-cm-}°\text{K})}{(\text{g/cm}^3)(\text{cal/g-}°\text{K})} =$ cm²/sec	Note that it is always area per unit time.
Entity (per unit volume, cgs units)	ρ_A, g/cm³	ρv_x, $\dfrac{\text{g}}{\text{cm}^3}\cdot\dfrac{\text{cm}}{\text{sec}} = \dfrac{\text{g-cm}}{\text{sec}}\Big/\text{cm}^3$	$\rho\hat{C}_p T$, $\dfrac{\text{g}}{\text{cm}^3}\cdot\dfrac{\text{cal}}{\text{g-}°\text{K}}\,°\text{K} =$ cal/cm³	Note that it is always entity per unit volume.

which we have already stated in Section 3.1 without elaboration. The various analogous quantities included in this generalized phenomenological law are given in Table 3.6-1.

REFERENCES

1. Casson, N. In *Rheology of Disperse Systems*, C. C. Mill, Ed. Pergamon: New York, 1959, p. 84.
2. Whitmore, R. L. *Rheology of the Circulation*. Pergamon: New York, 1968.
3. Fredrickson, A. G. *Principles and Applications of Rheology*. Prentice-Hall: Englewood Cliffs, N. J., 1964.
4. Hirschfelder, J. O.; Curtiss, C. F.; and Bird, R. B. *Molecular Theory of Gases and Liquids*, rev. ed. Wiley: New York, 1964.
5. Chapman, S., and Cowling, T. G. *Mathematical Theory of Non-Uniform Gases*, 2nd ed. Cambridge Univ. Press: Cambridge, 1951.
6. Bird, R. B.; Stewart, W. E.; and Lightfoot, E. N. *Transport Phenomena*, Wiley: New York, 1960.

CHAPTER FOUR

The Equations of Change

In solving any problem in the physical sciences, one always starts with two types of equations. One, the *constitutive equation*, or *equation of state*, describes the physical properties of the *constituent* or material involved. The other, the equation of change, describes the system. These two combined form the basic equation to be solved.

Since there are three transport phenomena involved, there are three equations in each of the two above-mentioned categories. In Chapter 3, we have already introduced the three constitutive equations in the forms of the three phenomenological laws. In this chapter we will develop the three equations of change—equations of continuity, motion, and energy—so called because they deal respectively with changes in the concentration, velocity, and temperature of the system. They are also called the conservation equations, because they are based on the principles of conservation of, respectively, mass, momentum, and energy.

We will first develop the equations of change for pure substances. These will enable us to solve problems of fluid flow and heat transfer in single-component systems. Since the fluids in biological systems are mostly multicomponent, the concepts of mixture, concentration, and the like will then be introduced in Chapter 5 so that we can develop the equations of change for multicomponent systems. These will allow us to solve various types of problems, including mass transfer in multicomponent systems.

4.1 TIME DERIVATIVES

Before we start developing the equations of change, the various types of time derivatives commonly used have to be illustrated. This can best be done with a bird-watching example.*

Since birds are in motion, the concentration c (i.e., the number of birds, say, per cubic foot in the sky) is a function of position (i.e., the x,y,z coordinates) as well as time. This we write as $c(x,y,z,t)$, where t stands for time.

*Think of the recent problems facing the residents of Graceham, Maryland where a large flock of birds streamed into the community every evening at an estimated maximum rate of 25,000 per minute over a mere 60 acres of pine grove area.

133

If the bird-watcher stays on the ground and looks up to the sky at a definite spot, the rate of change of bird concentration he observes is the partial derivative of c with respect to t, or $\partial c / \partial t$, since the x, y, and z coordinates are held fixed, or constant. As mentioned in Section 2.6, to avoid misunderstanding, one writes

$$\left(\frac{\partial c}{\partial t} \right)_{x,y,z},$$

meaning "rate of change of c with respect to t while holding x, y, and z constant."

Now if the observer rides in a small airplane that flies at its own velocity, then the bird concentration change he observes will be the *total time derivative*, dc/dt, which according to the chain rule is mathematically defined as

$$\frac{dc}{dt} = \frac{\partial c}{\partial t} + \frac{\partial c}{\partial x}\frac{dx}{dt} + \frac{\partial c}{\partial y}\frac{dy}{dt} + \frac{\partial c}{\partial z}\frac{dz}{dt}, \tag{4.1-1}$$

where $\partial c / \partial x$ means $(\partial c / \partial x)_{y,z,t}$ and so on and (x, y, z) are the coordinates of the observer at any time, and changing.* Then dx/dt, dy/dt, and dz/dt are components of the velocity of the airplane; and $\partial c / \partial x$, $\partial c / \partial y$, and $\partial c / \partial z$ are components of the concentration change with respect to the airplane's position *at a certain time*. Therefore, the total time derivatives reflect the concentration change with respect to both time and the observer's position.

Now, instead of making the observer ride on a plane, let us imagine that he is a weightless creature floating in the air with the local air velocity $\mathbf{v}$. Then for this particular case dx/dt, dy/dt, and dz/dt become v_x, v_y, and v_z, respectively, the components of local velocity of the air. This special case of total time derivative is called the *substantial time derivative*, or "time derivative following motion", denoted by Dc/Dt. Mathematically, we write

$$\frac{Dc}{Dt} = \frac{\partial c}{\partial t} + v_x \frac{\partial c}{\partial x} + v_y \frac{\partial c}{\partial y} + v_z \frac{\partial c}{\partial z}. \tag{4.1-2}$$

In vector notations, this is

$$\frac{Dc}{Dt} = \frac{\partial c}{\partial t} + (\mathbf{v} \cdot \nabla c). \tag{4.1-3}$$

*Note the difference here between this equation and Eq. 2.6-9. In Eq. 2.6-9, f is a function of t only through $x(t)$, $y(t)$, and $z(t)$, that is, indirectly. Here, in addition to the indirect dependence c is a direct function of t as well. Thus, we have the extra term $(\partial c / \partial t)_{x,y,z}$.

4.2 THE EULERIAN AND LAGRANGIAN APPROACHES

The equations of change can be derived from basic physical principles using either of two approaches. In the first, the system under consideration is a tiny fixed element in space, which we can view as an empty three-dimensional frame of constant volume. Then we apply the principles of mass, momentum, and energy conservation to obtain the equations of continuity, motion, and energy, respectively. The conservation of mass, for example, dictates that the rate of mass entering the system must equal that leaving plus that accumulated, because it simply cannot be created or destroyed.* This approach is associated with the Swiss mathematician and physicist Leonhard Euler and is therefore regarded as the Eulerian point of view.

In the second approach, the system studied is a tiny element of fluid following the motion of the main body of the fluid. The shape and volume of the element and density of the material in it may change as it flows along, but the mass within the element should remain constant. With this in mind, we can derive the equation of continuity; applying the Newton's second law of motion and the first law of thermodynamics gives us the equations of motion and energy. This approach was extensively used by the French mathematician and astronomer J. L. Lagrange and hence is called the Lagrangian approach.

Although detailed derivations of these equations will be given in the following sections, the summary in Table 4.2-1 of the principles used and the forms of the resulting equations will give the reader a preview of the general picture.

In order to show both of these approaches, we shall derive the equation of continuity using the Eulerian approach, the equation of motion via the Lagrangian approach, and then the equation of energy using both methods.

Another interesting approach is to first derive a single general equation of change based on the conservation of a general entity. Then by making this entity mass, momentum, and energy we obtain the equations of continuity, motion, and energy, respectively [1]. This in principle has the obvious advantage of condensing the procedure; however, the formal vector notations have yet to be worked out.

---

*In multicomponent systems this statement has to be modified because material of a species can be converted to another by virtue of chemical reaction.

Table 4.2-1   Principles Used in, and Forms of, the Equation of Change[a]

Equation of:	Eulerian Approach: Fixed element in space		Lagrangian Approach: Fluid element following motion
Continuity	Conservation of mass $$\frac{\partial \rho}{\partial t} + (\nabla \cdot \rho v) = 0$$	$\longrightarrow$	"Isolation of mass" $$\frac{D\rho}{Dt} + \rho(\nabla \cdot \mathbf{v}) = 0$$
Motion	Conservation of momentum $$\frac{\partial}{\partial t} v = -[\nabla \cdot \rho vv] - \nabla p - [\nabla \cdot \tau] + \rho g$$	$\longleftrightarrow$	Newton's second law of motion $$\rho \frac{Dv}{Dt} = -\nabla p - [\nabla \cdot \tau] + \rho g$$
Energy	Conservation of energy $$\frac{\partial}{\partial t} \rho \hat{E} = -(\nabla \cdot \rho \hat{E} v) - (\nabla \cdot \mathbf{g}) - (\nabla \cdot p v) - (\nabla \cdot [\tau \cdot v])$$	$\longrightarrow$	First law of thermodynamics $$\rho \frac{D\hat{E}}{Dt} = -(\nabla \cdot \mathbf{g}) - (\nabla p v) - (\nabla \cdot [\tau \cdot v])$$

[a] $\leftrightarrow$ indicates mathematical equivalence, or one being obtainable from the other by purely mathematical manipulations.

4.3 THE EQUATION OF CONTINUITY (EULERIAN DERIVATION)

Basically, the equation of continuity is the principle of conservation of mass applied to an infinitesimal fixed element in space. Let us consider such an element in a rectangular coordinate system, as shown in Figure 4.3-1.

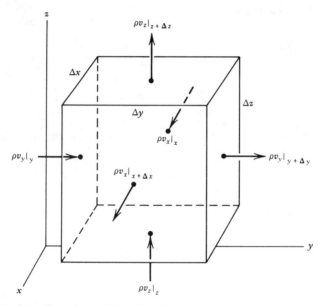

Figure 4.3-1.   Mass fluxes into and out of an element fixed in space.

The mass conservation principle says that in a period of time the net accumulation of mass in a system must be equal to the gain minus the loss during that period of time, since it can be neither created nor destroyed; or

$$\{\text{Accumulation of mass}\} = \{\text{Gain in mass}\} - \{\text{loss of mass}\}. \quad (4.3\text{-}1)$$

Let us define ρ as the density of the fluid; v_x, v_y, and v_z as the x, y, and z components of the fluid velocity; and $\Delta x, \Delta y$, and Δz as the dimensions of the cubic element. For simplicity, let us also assume that the fluid enters the cube at x, y, z and leaves at $x + \Delta x, y + \Delta y, z + \Delta z$.* For accounting

---

*Actually the reader can prove that the fluid could, for example, enter at $x, y + \Delta y, z$ and leave at $x + \Delta x, y, z + \Delta z$, and the result would still hold.

purpose, we have, for a certain time period Δt,

$$\text{Accumulation in mass} = \Delta(\rho\,\Delta x\,\Delta y\,\Delta z) = \Delta x\,\Delta y\,\Delta z\,\Delta\rho. \quad (4.3\text{-}2)$$

The latter part is true, since the fixed element (or "frame") also has a fixed volume.

To account for the mass gained and lost during the same time period Δt, we should realize that quantities such as ρv_x and ρv_y are mass fluxes, as evidenced by their units:

$$\rho v_x [=]\frac{g}{cm^3} \times \frac{cm}{sec}$$

$$[=]\frac{g}{cm^2\text{-sec}}.$$

Multiplying these fluxes with the cross-sectional areas through which they flow, we will have the mass flow rates, which, upon multiplication by the time period, give the total mass in the various input and output streams during the said time period.

We thus have the following:

	x-direction	y-direction	z-direction

Mass Gain:

$$\rho v_x|_x\Delta y\,\Delta z\,\Delta t \qquad \rho v_y|_y\Delta x\,\Delta z\,\Delta t \qquad \rho v_z|_z\Delta x\,\Delta y\,\Delta t \qquad (4.3\text{-}3)$$

Mass Loss:

$$\rho v_x|_{x+\Delta x}\Delta y\,\Delta z\,\Delta t \qquad \rho v_y|_{y+\Delta y}\Delta x\,\Delta z\,\Delta t \qquad \rho v_z|_{z+\Delta z}\Delta x\,\Delta y\,\Delta t \quad (4.3\text{-}4)$$

where $v_x|_x$ means v_x evaluated at x, and so on. Originally each term should be $\rho v_x \Delta y\,\Delta z\,\Delta t|_x$, $\rho v_y \Delta x\,\Delta z\,\Delta t|_{y+\Delta y}$, and so on, but since $\Delta x, \Delta y, \Delta z$, and Δt are independent of the position coordinates (x, y, z), they can be taken to the right of the "|". Substituting Eqs. 4.3-2 through 4.3-4 into Eq. 4.3-1, we have

$$\Delta x\,\Delta y\,\Delta z\,\Delta\rho = (\rho v_x|_x - \rho v_x|_{x+\Delta x})\Delta y\,\Delta z\,\Delta t$$

$$+ (\rho v_y|_y - \rho v_y|_{y+\Delta y})\Delta x\,\Delta z\,\Delta t$$

$$+ (\rho v_z|_z - \rho v_z|_{z+\Delta z})\Delta x\,\Delta y\,\Delta t.$$

Dividing the entire equation by $\Delta x \Delta y \Delta z \Delta t$ and collecting all the terms into the left-hand side, we have

$$\frac{\Delta \rho}{\Delta t} + \frac{\rho v_x|_{x+\Delta x} - \rho v_x|_x}{\Delta x} + \frac{\rho v_y|_{y+\Delta y} - \rho v_y|_y}{\Delta y}$$

$$+ \frac{\rho v_z|_{z+\Delta z} - \rho v_z|_z}{\Delta z} = 0.$$

Letting $\Delta x, \Delta y, \Delta z$, and Δt approach zero, the following differential equation results, following the definitions of derivatives given in Sections 2.2 and 2.6:

$$\frac{\partial \rho}{\partial t} + \frac{\partial}{\partial x}(\rho v_x) + \frac{\partial}{\partial y}(\rho v_y) + \frac{\partial}{\partial z}(\rho v_z) = 0. \qquad (4.3\text{-}5)$$

The partial time derivative is used here because the element is fixed in space. Using vector notation, Eq. 4.3-5 becomes

$$\frac{\partial \rho}{\partial t} + (\nabla \cdot \rho \mathbf{v}) = 0. \qquad (4.3\text{-}6)$$

This is the equation of continuity for a pure substance. In addition to the meaning of $\rho \mathbf{v}$ explained earlier, $(\nabla \cdot \rho \mathbf{v})$ has the meaning of *net rate of mass outflow per unit volume* because

$$(\nabla \cdot \rho \mathbf{v})[=]\frac{1}{\text{cm}} \cdot \frac{\text{g}}{\text{cm}^3} \cdot \frac{\text{cm}}{\text{sec}}$$

$$[=]\frac{\text{g}}{\text{sec}}/\text{cm}^3.$$

Therefore, the continuity equation simply states that *the time rate of increase of density within a small element in space is equal to the net rate of mass inflow divided by its volume*, since

$$\frac{\partial \rho}{\partial t} = -(\nabla \cdot \rho \mathbf{v}). \qquad (4.3\text{-}6a)$$

To show the mathematical interchangeability between the Eulerian and Lagrangian forms of the equation of continuity, we can expand $(\nabla \cdot \rho \mathbf{v})$ as follows:*

*See Problem 2.J, and Table 2.15-1.

$$(\nabla \cdot \rho \mathbf{v}) = \rho(\nabla \cdot \mathbf{v}) + (\mathbf{v} \cdot \nabla \rho). \qquad (4.3\text{-}7)$$

Substituting this into Eq. 4.3-6, we have

$$\frac{\partial \rho}{\partial t} + (\mathbf{v} \cdot \nabla \rho) + \rho(\nabla \cdot \mathbf{v}) = 0.$$

By definition, the first two terms of this equation constitute the substantial derivative of ρ (see Eq. 4.1-3). Thus,

$$\frac{D\rho}{Dt} + \rho(\nabla \cdot \mathbf{v}) = 0. \qquad (4.3\text{-}8)$$

For incompressible fluids, the density ρ is constant. Both Eqs. 4.3-6 and 4.3-8 then become

$$(\nabla \cdot \mathbf{v}) = 0, \qquad (4.3\text{-}9)$$

which is the equation of continuity usually employed for liquids.

4.4 THE EQUATION OF MOTION (LAGRANGIAN DERIVATION)

To derive the equation of motion via the Lagrangian approach, Newton's second law of motion is applied to an element of fluid following the bulk fluid motion. Let us consider such an element, as pictured in Figure 4.4-1. Newton's second law states that the acceleration is equal to the resultant sum of forces divided by the mass and is in the same direction as the resultant force.

Since both force and acceleration are vectors, this law can be separately applied to the x, y, and z directions. For the x direction, for example, we can write

$$(\text{Mass}) \times (x\text{-acceleration}) = \Sigma x\text{-forces}. \qquad (4.4\text{-}1)$$

For the element under consideration

$$\text{Mass} = \rho \, \Delta x \, \Delta y \, \Delta z \qquad (4.4\text{-}2)$$

and

$$x\text{-acceleration} = \frac{Dv_x}{Dt}. \qquad (4.4\text{-}3)$$

Here the substantial derivative is used because the element is moving with local bulk velocity.

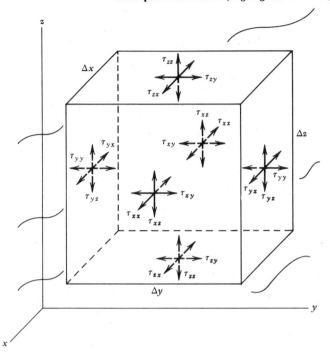

Figure 4.4-1. Viscous forces (shear and normal stresses) acting on the surfaces of an element following motion.

There are several contributions to the total x-directed force. The pressure force is simply equal to the pressure drop in the direction of flow multiplied by the cross-sectional area of the element in the same direction. The body force, so named because it acts on each and every "chunk" of material in the element, includes, most frequently, gravitational force. In special cases, magnetic, electrostatic, electrolytic, and buoyancy forces may also play their roles. The viscous surface force, arising from the relative motion between this flowing element and the surrounding fluid, is equal to the shear or normal stress, whichever the case may be, multiplied by the surface area on which the stress occurs. These forces are graphically represented in Figure 4.4-1. To summarize, all these various forces can be tabulated as follows:

Body force:

$$\rho \, \Delta x \, \Delta y \, \Delta z \, g_x \qquad\qquad (4.4\text{-}4)$$

Pressure force:

$$In\ (+) \qquad\qquad Out\ (-)$$

$$p|_x \Delta y\,\Delta z \qquad\qquad p|_{x+\Delta x}\Delta y\,\Delta z \qquad\qquad (4.4\text{-}5)$$

Viscous surface force:

$$\tau_{xx}|_x \Delta y\,\Delta z \qquad \tau_{xx}|_{x+\Delta x}\Delta y\,\Delta z \qquad\qquad (4.4\text{-}6)$$

$$\tau_{yx}|_y \Delta x\,\Delta z \qquad \tau_{yx}|_{y+\Delta y}\Delta x\,\Delta z \qquad\qquad (4.4\text{-}7)$$

$$\tau_{zx}|_z \Delta x\,\Delta y \qquad \tau_{zx}|_{z+\Delta z}\Delta x\,\Delta y. \qquad\qquad (4.4\text{-}8)$$

where p denotes pressure and g_x the x-directed body force per unit mass.

If we substitute all these quantities (Eqs. 4.4-2 through 4.4-8) into Eq. 4.4-1 and divide the entire equation by $\Delta x\,\Delta y\,\Delta z$ while letting Δx, Δy, and Δz approach zero, a differential equation results:

$$\rho\frac{Dv_x}{Dt} = -\frac{\partial p}{\partial x} - \left(\frac{\partial \tau_{xx}}{\partial x} + \frac{\partial \tau_{yx}}{\partial y} + \frac{\partial \tau_{zx}}{\partial z}\right) + \rho g_x. \qquad (4.4\text{-}9)$$

Similar expressions can be obtained for the y- and z-directions by the same procedure or simply permuting x, y, and z:

$$\rho\frac{Dv_y}{Dt} = -\frac{\partial p}{\partial y} - \left(\frac{\partial \tau_{xy}}{\partial x} + \frac{\partial \tau_{yy}}{\partial y} + \frac{\partial \tau_{zy}}{\partial z}\right) + \rho g_y \qquad (4.4\text{-}10)$$

$$\rho\frac{Dv_z}{Dt} = -\frac{\partial p}{\partial z} - \left(\frac{\partial \tau_{xz}}{\partial x} + \frac{\partial \tau_{yz}}{\partial y} + \frac{\partial \tau_{zz}}{\partial z}\right) + \rho g_z. \qquad (4.4\text{-}11)$$

A vector equation can be obtained with Eqs. 4.4-9, 4.4-10, and 4.4-11 as its components. That is, if we multiply them by their respective unit vectors and subsequently sum them together, we have

$$\rho\frac{D\mathbf{v}}{Dt} = -\nabla p - [\nabla\cdot\boldsymbol{\tau}] + \rho\mathbf{g}. \qquad (4.4\text{-}12)$$

The physical meaning of this equation is

$$
\left\{
\begin{array}{c}
\text{Mass (per unit volume)} \\
\text{times acceleration} \\
\text{(following motion)}
\end{array}
\right\}
$$

$$
\rho \frac{D\mathbf{v}}{Dt}
$$

$$
= \left\{
\begin{array}{c}
\text{Pressure force} \\
\text{(per unit volume)}
\end{array}
\right\}
+
\left\{
\begin{array}{c}
\text{Viscous force} \\
\text{(per unit volume)}
\end{array}
\right\}
+
\left\{
\begin{array}{c}
\text{Body force} \\
\text{(per unit volume)}
\end{array}
\right\}
$$

$$
-\nabla p \qquad\qquad -[\nabla\cdot\boldsymbol{\tau}] \qquad\qquad \rho\mathbf{g}
$$

In order to show how Eq. 4.4-12 can be converted into the Eulerian form, we use the expansion

$$
\frac{D}{Dt}(\rho\mathbf{v}) = \rho\frac{D\mathbf{v}}{Dt} + \mathbf{v}\frac{D\rho}{Dt} \tag{4.4-13}
$$

and the definition of the substantial derivative (Eq. 4.1-3)

$$
\frac{D}{Dt}(\rho\mathbf{v}) = \frac{\partial}{\partial t}\rho\mathbf{v} + (\mathbf{v}\cdot\nabla)\rho\mathbf{v} \tag{4.4-14}
$$

or*

$$
\frac{D}{Dt}(\rho\mathbf{v}) = \frac{\partial}{\partial t}\rho\mathbf{v} + [\mathbf{v}\cdot\nabla\rho\mathbf{v}]. \tag{4.4-15}
$$

From these we have

$$
\rho\frac{D\mathbf{v}}{Dt} = \frac{\partial}{\partial t}\rho\mathbf{v} + [\mathbf{v}\cdot\nabla\rho\mathbf{v}] - \mathbf{v}\frac{D\rho}{Dt}. \tag{4.4-16}
$$

Using the equation of continuity (Eq. 4.3-8), Eq. 4.4-16 becomes

$$
\rho\frac{D\mathbf{v}}{Dt} = \frac{\partial}{\partial t}\rho\mathbf{v} + [\mathbf{v}\cdot\nabla\rho\mathbf{v}] + \rho\mathbf{v}(\nabla\cdot\mathbf{v}). \tag{4.4-17}
$$

Using the identity proven in Problem 2.K, we have

$$
\rho\frac{D\mathbf{v}}{Dt} = \frac{\partial}{\partial t}\rho\mathbf{v} + [\nabla\cdot\rho\mathbf{v}\mathbf{v}]. \tag{4.4-18}
$$

*The reader may wish to prove that $(\mathbf{v}\cdot\nabla)\rho\mathbf{v}=[\mathbf{v}\cdot\nabla\rho\mathbf{v}]$.

Substituting Eq. 4.4-18 into Eq. 4.4-12 results in the Eulerian form of the equation of motion:

$$\frac{\partial}{\partial t}\rho\mathbf{v} = -[\nabla\cdot\rho\mathbf{vv}] - \nabla p - [\nabla\cdot\boldsymbol{\tau}] + \rho\mathbf{g}. \qquad (4.4\text{-}19)$$

Those who are unfamiliar with vector and tensor notations are advised to review Section 2.15 and 2.16.

The physical meaning of Eq. 4.4-19 in terms of the principle of conservation of momentum applied to an elemental frame fixed in space, is

$$\left\{\begin{array}{c}\text{Rate of increase}\\ \text{of momentum (per}\\ \text{unit volume)}\end{array}\right\} = \left\{\begin{array}{c}\text{Net rate of momentum gained (in–out)}\\ \text{by convection (per unit volume)}\end{array}\right\}$$

$$\frac{\partial}{\partial t}\rho\mathbf{v} \qquad\qquad -[\nabla\cdot\rho\mathbf{vv}]$$

$$+\left\{\begin{array}{c}\text{Pressure force on}\\ \text{element (per unit}\\ \text{volume)}\end{array}\right\} + \left\{\begin{array}{c}\text{Rate of momentum gained}\\ \text{by viscous transfer}\\ \text{(per unit volume)}\end{array}\right\} + \left\{\begin{array}{c}\text{Body force on}\\ \text{element (per}\\ \text{unit volume)}\end{array}\right\}$$

$$-\nabla p \qquad\qquad -[\nabla\cdot\boldsymbol{\tau}] \qquad\qquad \rho\mathbf{g}$$

The $\mathbf{vv}$ in the term $-[\nabla\cdot\rho\mathbf{vv}]$ is a "dyadic product," which makes it a tensor (see Section 2.16). For the moment we need not concern ourselves with this except to observe that since $\rho\mathbf{v}$ is momentum (per unit volume), multiplying it by $\mathbf{v}$ (velocity) and then differentiating the product with respect to distance (the ∇ operation) give the meaning of "per unit time." Thus, the entire term is time rate of change of momentum (per unit volume).

For an *incompressible Newtonian* fluid, ρ is constant and*

$$\tau_{xx} = -2\mu\frac{\partial v_x}{\partial x} \qquad (4.4\text{-}20\mathrm{a})$$

---

*In Chapter 3, we had, for example $\tau_{yx} = -\mu\partial v_x/\partial y$, which is for one-dimensional flow. The expressions here are more general.

$$\tau_{yy}^{\cdot} = -2\mu \frac{\partial v_y}{\partial y} \tag{4.4-20b}$$

$$\tau_{zz} = -2\mu \frac{\partial v_z}{\partial z} \tag{4.4-20c}$$

$$\tau_{xy} = \tau_{yx} = -\mu \left(\frac{\partial v_x}{\partial y} + \frac{\partial v_y}{\partial x} \right) \tag{4.4-20d}$$

$$\tau_{yz} = \tau_{zy} = -\mu \left(\frac{\partial v_y}{\partial z} + \frac{\partial v_z}{\partial y} \right) \tag{4.4-20e}$$

$$\tau_{zx} = \tau_{xz} = -\mu \left(\frac{\partial v_z}{\partial x} + \frac{\partial v_x}{\partial z} \right), \tag{4.4-20f}$$

where μ is also constant. Thus, Eqs. 4.4-9, 4.4-10, 4.4-11, and 4.4-12 become[†]

$$\rho \frac{Dv_x}{Dt} = -\frac{\partial p}{\partial x} + \mu \left(\frac{\partial^2 v_x}{\partial x^2} + \frac{\partial^2 v_x}{\partial y^2} + \frac{\partial^2 v_x}{\partial z^2} \right) + \rho g_x \tag{4.4-21}$$

$$\rho \frac{Dv_y}{Dt} = -\frac{\partial p}{\partial y} + \mu \left(\frac{\partial^2 v_y}{\partial x^2} + \frac{\partial^2 v_y}{\partial y^2} + \frac{\partial^2 v_y}{\partial z^2} \right) + \rho g_y \tag{4.4-22}$$

$$\rho \frac{Dv_z}{Dt} = -\frac{\partial p}{\partial z} + \mu \left(\frac{\partial^2 v_z}{\partial x^2} + \frac{\partial^2 v_z}{\partial y^2} + \frac{\partial^2 v_z}{\partial z^2} \right) + \rho g_z \tag{4.4-23}$$

or

$$\rho \frac{D\mathbf{v}}{Dt} = -\nabla p + \mu \nabla^2 \mathbf{v} + \rho \mathbf{g}. \tag{4.4-24}$$

The corresponding Eulerian forms are

$$\rho \left(\frac{\partial v_x}{\partial t} + v_x \frac{\partial v_x}{\partial x} + v_y \frac{\partial v_x}{\partial y} + v_z \frac{\partial v_x}{\partial z} \right) = -\frac{\partial p}{\partial x} + \mu \left(\frac{\partial^2 v_x}{\partial x^2} + \frac{\partial^2 v_x}{\partial y^2} + \frac{\partial^2 v_x}{\partial z^2} \right) + \rho g_x$$

$$\tag{4.4-25}$$

[†]In addition to substitution and differentiation, the derivational procedure includes using the equation of continuity for incompressible fluids (see Eq. 4.3-9):

$$\frac{\partial v_x}{\partial x} + \frac{\partial v_y}{\partial y} + \frac{\partial v_z}{\partial z} = 0$$

$$\rho\left(\frac{\partial v_y}{\partial t}+v_x\frac{\partial v_y}{\partial x}+v_y\frac{\partial v_y}{\partial y}+v_z\frac{\partial v_y}{\partial z}\right)=-\frac{\partial p}{\partial y}+\mu\left(\frac{\partial^2 v_y}{\partial x^2}+\frac{\partial^2 v_y}{\partial y^2}+\frac{\partial^2 v_y}{\partial z^2}\right)+\rho g_y$$

(4.4-26)

$$\rho\left(\frac{\partial v_z}{\partial t}+v_x\frac{\partial v_z}{\partial x}+v_y\frac{\partial v_z}{\partial y}+v_z\frac{\partial v_z}{\partial z}\right)=-\frac{\partial p}{\partial z}+\mu\left(\frac{\partial^2 v_z}{\partial x^2}+\frac{\partial^2 v_z}{\partial y^2}+\frac{\partial^2 v_z}{\partial z^2}\right)+\rho g_z,$$

(4.4-27)

which can be vectorially combined to form*

$$\rho\frac{\partial \mathbf{v}}{\partial t}+[\rho\mathbf{v}\cdot\boldsymbol{\nabla}\mathbf{v}]=-\boldsymbol{\nabla}p+\mu\nabla^2\mathbf{v}+\rho\mathbf{g}.$$

(4.4-28)

Equations 4.4-25, 4.4-26, and 4.4-27 are most convenient to use in solving fluid-flow problems, because they have already combined the equations of continuity, motion, and state. They are generally known as the Navier–Stokes equations. But it should be emphasized that they are good only for fluids that are both incompressible and Newtonian. The derivation of their counterparts in the other coordinate systems are discussed in Section 4.11 and the full set of equations is given in Table 4.12-2.

4.5 THE EQUATION OF ENERGY (EULERIAN DERIVATION)

For an element fixed in space, the principle of energy conservation can be written as follows:

$$\left\{\begin{array}{c}\text{Rate of}\\\text{accumulation}\\\text{of IE+KE}\end{array}\right\}=\left\{\begin{array}{c}\text{Net rate of IE}\\\text{and KE input}\\\text{by flow}\end{array}\right\}$$
$$+\left\{\begin{array}{c}\text{Net rate of energy}\\\text{input by heat}\\\text{conduction}\end{array}\right\}-\left\{\begin{array}{c}\text{Net rate of work}\\\text{done by system}\\\text{on surroundings}\end{array}\right\},\qquad(4.5\text{-}1)$$

where IE is the abbreviation for *internal energy* and KE for *kinetic energy*. Note that we have no PE (*potential energy*) term because the potential

---

*Equation 4.4-28 can also be obtained directly from Eq. 4.4-24 via the same mathematical manipulation by which we obtained Eq. 4.4-19 from Eq. 4.4-12.

energy at a fixed position remains constant. The rate of accumulation of internal energy is

$$\frac{\partial}{\partial t}\rho\,\Delta x\,\Delta y\,\Delta z\,\hat{U},$$

where U is the internal energy and the caret above it signifies "per unit mass" as footnoted in Section 3.6. The rate of accumulation of kinetic energy is

$$\frac{\partial}{\partial t}\rho\,\Delta x\,\Delta y\,\Delta z\cdot\tfrac{1}{2}v^2.$$

Therefore, since $\Delta x\,\Delta y\,\Delta z$ is constant,

$$\left\{\begin{array}{c}\text{Rate of accumulation}\\ \text{of IE}+\text{KE}\end{array}\right\}=\Delta x\,\Delta y\,\Delta z\frac{\partial}{\partial t}\rho\left(\hat{U}+\tfrac{1}{2}v^2\right).\qquad(4.5\text{-}2)$$

The rate of internal and kinetic energies by flow can again be broken down into three directional components:

Rate of IE + KE by Flow

	$In(+)$	$Out\,(-)$
x-direction	$\rho v_x\left(\hat{U}+\tfrac{1}{2}v^2\right)\big|_x\Delta y\,\Delta z$	$\rho v_x\left(\hat{U}+\tfrac{1}{2}v^2\right)\big|_{x+\Delta x}\Delta y\,\Delta z$
y-direction	$\rho v_y\left(\hat{U}+\tfrac{1}{2}v^2\right)\big|_y\Delta x\,\Delta z$	$\rho v_y\left(\hat{U}+\tfrac{1}{2}v^2\right)\big|_{y+\Delta y}\Delta x\,\Delta z$
z-direction	$\rho v_z\left(\hat{U}+\tfrac{1}{2}v^2\right)\big|_z\Delta x\,\Delta y$	$\rho v_z\left(\hat{U}+\tfrac{1}{2}v^2\right)\big|_{z+\Delta z}\Delta x\,\Delta y\,.$

$$(4.5\text{-}3)$$

The rate of energy entering and leaving by heat conduction can be itemized similarly:

Rate of Energy Flow by Heat Conduction

	$In\,(+)$	$Out\,(-)$	
x-direction	$q_x\big|_x\Delta y\,\Delta z$	$q_x\big|_{x+\Delta x}\Delta y\,\Delta z$	
y-direction	$q_y\big|_y\Delta x\,\Delta z$	$q_y\big|_{y+\Delta y}\Delta x\,\Delta z$	$(4.5\text{-}4)$
z-direction	$q_z\big|_z\Delta x\,\Delta y$	$q_z\big|_{z+\Delta z}\Delta x\,\Delta y\,,$	

where q_x, q_y, and q_z are the x, y, and z components, respectively, of the heat flux (vector) $\mathbf{q}$.

Since work is equal to force times distance, the rate of work is then equal to force times distance per unit time, or force times velocity. Corresponding to the forces listed in the derivation of the equation of motion, the rate-of-work terms are itemized below. Note that the "In" terms represent the work done by surroundings on the system and are hence negative:

Rate of work done
by system against: *In* $(-)$ *Out* $(+)$

Pressure Force

x-direction $pv_x|_x \Delta y\, \Delta z$ $pv_x|_{x+\Delta x} \Delta y\, \Delta z$

y-direction $pv_y|_y \Delta x\, \Delta z$ $pv_y|_{y+\Delta y} \Delta x\, \Delta z$ (4.5-5)

z-direction $pv_z|_z \Delta x\, \Delta y$ $pv_z|_{z+\Delta z} \Delta x\, \Delta y$

Viscous force $In(-)$	$\begin{bmatrix} \text{by } x \\ \text{momen-} \\ \text{tum} \end{bmatrix}$	$\begin{bmatrix} \text{by } y \\ \text{momen-} \\ \text{tum} \end{bmatrix}$	$\begin{bmatrix} \text{by } z \\ \text{momen-} \\ \text{tum} \end{bmatrix}$	
x-direction	$(\tau_{xx}v_x$	$+\tau_{xy}v_y$	$+\tau_{xz}v_z)$	$|_x \Delta y\, \Delta z$
y-direction	$(\tau_{yx}v_x$	$+\tau_{yy}v_y$	$+\tau_{yz}v_z)$	$|_y \Delta x\, \Delta z$
z-direction	$(\tau_{zx}v_x$	$+\tau_{zy}v_y$	$+\tau_{zz}v_z)$	$|_z \Delta x\, \Delta y$

Viscous force $Out(+)$	$\begin{bmatrix} \text{by } x \\ \text{momen-} \\ \text{tum} \end{bmatrix}$	$\begin{bmatrix} \text{by } y \\ \text{momen-} \\ \text{tum} \end{bmatrix}$	$\begin{bmatrix} \text{by } z \\ \text{momen-} \\ \text{tum} \end{bmatrix}$	
x-direction	$(\tau_{xx}v_x$	$+\tau_{xy}v_y$	$+\tau_{xz}v_z)$	$|_{x+\Delta x} \Delta y\, \Delta z$
y-direction	$(\tau_{yx}v_x$	$+\tau_{yy}v_y$	$+\tau_{yz}v_z)$	$|_{y+\Delta y} \Delta x\, \Delta z$,
z-direction	$(\tau_{zx}v_x$	$+\tau_{zy}v_y$	$+\tau_{zz}v_z)$	$|_{z+\Delta z} \Delta x\, \Delta y$

(4.5-6)

Body force

x-direction $-\rho\, \Delta x\, \Delta y\, \Delta z\, g_x v_x$

y-direction $-\rho\, \Delta x\, \Delta y\, \Delta z\, g_y v_y$ (4.5-7)

z-direction $-\rho\, \Delta x\, \Delta y\, \Delta z\, g_z v_z$

Substituting Eqs. 4.5-2 through 4.5-7 into Eq. 4.5-1, dividing the entire equation by $\Delta x \, \Delta y \, \Delta z$, and letting $\Delta x, \Delta y$, and Δz approach zero, we have

$$\frac{\partial}{\partial t}\rho\left(\hat{U}+\tfrac{1}{2}v^2\right) = -\left[\frac{\partial}{\partial x}\rho v_x\left(\hat{U}+\tfrac{1}{2}v^2\right) + \frac{\partial}{\partial y}\rho v_y\left(\hat{U}+\tfrac{1}{2}v^2\right) + \frac{\partial}{\partial z}\rho v_z\left(\hat{U}+\tfrac{1}{2}v^2\right)\right]$$

$$-\left(\frac{\partial q_x}{\partial x}+\frac{\partial q_y}{\partial y}+\frac{\partial q_z}{\partial z}\right)-\left(\frac{\partial}{\partial x}pv_x+\frac{\partial}{\partial y}pv_y+\frac{\partial}{\partial z}pv_z\right)$$

$$-\left[\frac{\partial}{\partial x}\left(\tau_{xx}v_x+\tau_{xy}v_y+\tau_{xz}v_z\right)+\frac{\partial}{\partial y}\left(\tau_{yx}v_x+\tau_{yy}v_y+\tau_{yz}v_z\right)\right.$$

$$\left.+\frac{\partial}{\partial z}\left(\tau_{zx}v_x+\tau_{zy}v_y+\tau_{zz}v_z\right)\right]+\rho\left(v_x g_x+v_y g_y+v_z g_z\right). \quad (4.5\text{-}8)$$

This is one form of the equation of energy. It can be expressed by the shorthand vector-tensor notation as follows:

$$\frac{\partial}{\partial t}\rho\left(\hat{U}+\tfrac{1}{2}v^2\right) = -\left(\nabla\cdot\rho v\left(\hat{U}+\tfrac{1}{2}v^2\right)\right)-(\nabla\cdot q)$$

$$-(\nabla\cdot pv)-(\nabla\cdot[\tau\cdot v])+\rho(v\cdot g). \quad (4.5\text{-}9)$$

This equation states that, for a fixed element in space,

$$\left\{\begin{array}{c}\text{Rate of energy}\\\text{increase}\\\text{(per unit volume)}\end{array}\right\} = \left\{\begin{array}{c}\text{Net rate of energy}\\\text{input by convection}\\\text{(per unit volume)}\end{array}\right\}$$

$$\frac{\partial}{\partial t}\rho\left(\hat{U}+\tfrac{1}{2}v^2\right) \qquad -\left(\nabla\cdot\rho v\left(\hat{U}+\tfrac{1}{2}v^2\right)\right)$$

$$+\left\{\begin{array}{c}\text{Net rate of energy}\\\text{input by conduction}\\\text{(per unit volume)}\end{array}\right\}+\left\{\begin{array}{c}\text{Rate of work done by}\\\text{body forces on the}\\\text{element (per unit volume)}\end{array}\right\}^*$$

$$-(\nabla\cdot q) \qquad\qquad \rho(v\cdot g)$$

*Note that in this term we express the work done on the system (the element), compared with the last term in Eq. 4.5-1 where the work is done on the surroundings. The sign here should be opposite that in Eq. 4.5-1.

$$- \left\{ \begin{array}{c} \text{Rate of work done} \\ \text{by element on fluid} \\ \text{against pressure forces} \\ \text{(per unit volume)} \end{array} \right\} - \left\{ \begin{array}{c} \text{Rate of work done} \\ \text{by element on fluid} \\ \text{against viscous forces} \\ \text{(per unit volume)} \end{array} \right\}.$$

$$- (\nabla \cdot p\mathbf{v}) \qquad\qquad\qquad - (\nabla \cdot [\tau \cdot \mathbf{v}])$$

4.6 THE EQUATION OF ENERGY (LAGRANGIAN DERIVATION)

In the Lagrangian derivation, we again consider an element of the fluid flowing along with the rest of it at a velocity $\mathbf{v}$. Since the position of the element changes, the *potential energy* Φ must be included in the total energy term. The first law of thermodynamics, expressed in terms of rates, states:

$$\left\{ \begin{array}{c} \text{Rate of change of} \\ \text{total energy} \\ (\text{KE} + \text{PE} + \text{IE}) \end{array} \right\}$$

$$= \left\{ \begin{array}{c} \text{Net rate of} \\ \text{heat input} \end{array} \right\} - \left\{ \begin{array}{c} \text{Net rate of work done by the system} \\ \text{via surface forces on surroundings} \end{array} \right\}. \qquad (4.6\text{-}1)$$

Here by surface forces we mean the pressure and viscous forces which act on the surfaces of the element. Since total energy is $\rho\,\Delta x\,\Delta y\,\Delta z\,\hat{E}$ (where $\hat{E} = \hat{U} + \hat{\Phi} + \frac{1}{2}v^2$ the total energy per unit mass),

$$\left(\begin{array}{c} \text{Rate of change of} \\ \text{total energy} \end{array} \right) = \rho\,\Delta x\,\Delta y\,\Delta z\,\frac{D\hat{E}}{Dt}. \qquad (4.6\text{-}2)$$

There is *no* heat in and out of the fluid element *by convection*, because an element following motion does *not* gain or lose material. The conduction and work terms are the same as in the Eulerian approach, except that the latter consists of only surface (pressure and viscous) forces as stated in Eq.

4.5-1. Following the same mathematical procedure as before, we obtain

$$\rho \frac{D\hat{E}}{Dt} = - \left(\frac{\partial q_x}{\partial x} + \frac{\partial q_y}{\partial y} + \frac{\partial q_z}{\partial z} \right) - \left(\frac{\partial}{\partial x} p v_x + \frac{\partial}{\partial y} p v_y + \frac{\partial}{\partial z} p v_z \right)$$

$$- \left[\frac{\partial}{\partial x} \left(\tau_{xx} v_x + \tau_{xy} v_y + \tau_{xz} v_z \right) + \frac{\partial}{\partial y} \left(\tau_{yx} v_x + \tau_{yy} v_y + \tau_{yz} v_z \right) \right.$$

$$\left. + \frac{\partial}{\partial z} \left(\tau_{zx} v_x + \tau_{zy} v_y + \tau_{zz} v_z \right) \right] \tag{4.6-3}$$

or

$$\rho \frac{D\hat{E}}{Dt} = - (\nabla \cdot \mathbf{q}) - (\nabla \cdot p\mathbf{v}) - (\nabla \cdot [\tau \cdot \mathbf{v}]). \tag{4.6-4}$$

The equation of energy in this form says that the rate of change of the total energy of the fluid is equal to the rate of heat conduction minus the work done on the surroundings by the fluid element via pressure and viscous forces. Or, for an element floating along with the bulk of the fluid,

$$\left\{ \begin{array}{c} \text{Rate of change of} \\ \text{total energy (per} \\ \text{unit volume)} \end{array} \right\} = \left\{ \begin{array}{c} \text{Net rate of energy} \\ \text{input by conduction} \\ \text{(per unit volume)} \end{array} \right\}$$

$$\rho \frac{D\hat{E}}{Dt} \qquad\qquad\qquad - (\nabla \cdot \mathbf{q})$$

$$- \left\{ \begin{array}{c} \text{Rate of work done by} \\ \text{element on surroundings} \\ \text{by pressure force} \\ \text{(per unit volume)} \end{array} \right\} - \left\{ \begin{array}{c} \text{Rate of work done by} \\ \text{element on surroundings} \\ \text{by viscous force} \\ \text{(per unit volume)} \end{array} \right\}$$

$$- (\nabla \cdot p\mathbf{v}) \qquad\qquad\qquad - (\nabla \cdot [\tau \cdot \mathbf{v}])$$

The equivalence of Eqs. 4.5-9 and 4.6-4 may not be very apparent until one considers the following. Let us write out

$$\rho \frac{D\hat{E}}{Dt} = \rho \frac{D}{Dt} \left(\hat{U} + \tfrac{1}{2} v^2 + \hat{\Phi} \right) \tag{4.6-5}$$

and expand the last term according to the definition of the substantial time derivative, that is,

$$\rho \frac{D\hat{\Phi}}{Dt} = \rho \frac{\partial \hat{\Phi}}{\partial t} + \rho(\mathbf{v} \cdot \boldsymbol{\nabla}\hat{\Phi}). \qquad (4.6\text{-}6)$$

We know that

$$\frac{\partial \Phi}{\partial t} = 0,$$

since the potential energy is dependent upon position only. Moreover, the only forces that give rise to potential energy are body forces, whether they be gravitational, electrostatic, or the like. According to the well-known relationship between force and potential

$$\mathbf{g} = -\boldsymbol{\nabla}\hat{\Phi}, \qquad (4.6\text{-}7)$$

we have

$$\rho \frac{D\hat{\Phi}}{Dt} = -\rho(\mathbf{v} \cdot \mathbf{g}). \qquad (4.6\text{-}8)$$

Furthermore, we can expand the first two terms on the right-hand side of Eq. 4.6-5:

$$\rho \frac{D}{Dt}\left(\hat{U} + \tfrac{1}{2}v^2\right) = \rho\left(\frac{\partial}{\partial t}\left(\hat{U} + \tfrac{1}{2}v^2\right) + \left(\mathbf{v} \cdot \boldsymbol{\nabla}\left(\hat{U} + \tfrac{1}{2}v^2\right)\right)\right) *$$

$$= \frac{\partial}{\partial t}\rho\left(\hat{U} + \tfrac{1}{2}v^2\right) - \left(\hat{U} + \tfrac{1}{2}v^2\right)\frac{\partial \rho}{\partial t}$$

$$+ \left(\boldsymbol{\nabla} \cdot \rho\mathbf{v}\left(\hat{U} + \tfrac{1}{2}v^2\right)\right) - \left(\hat{U} + \tfrac{1}{2}v^2\right)(\boldsymbol{\nabla} \cdot \rho\mathbf{v})$$

$$= \frac{\partial}{\partial t}\rho\left(\hat{U} + \tfrac{1}{2}v^2\right) + \left(\boldsymbol{\nabla} \cdot \rho\mathbf{v}\left(\hat{U} + \tfrac{1}{2}v^2\right)\right) - \left(\hat{U} + \tfrac{1}{2}v^2\right)\left(\frac{\partial \rho}{\partial t} + (\boldsymbol{\nabla} \cdot \rho\mathbf{v})\right)$$

$$= \frac{\partial}{\partial t}\rho\left(\hat{U} + \tfrac{1}{2}v^2\right) + \left(\boldsymbol{\nabla} \cdot \rho\mathbf{v}\left(\hat{U} + \tfrac{1}{2}v^2\right)\right), \qquad (4.6\text{-}9)$$

since according to the equation of continuity

$$\frac{\partial \rho}{\partial t} + (\boldsymbol{\nabla} \cdot \rho\mathbf{v}) = 0.$$

---

*The author apologizes for having to use triple parentheses here because brackets and braces are reserved for, respectively, vector and tensor quantities.

Substituting Eqs. 4.6-8 and 4.6-9 into Eq. 4.6-5 and then into Eq. 4.6-4 gives Eq. 4.5-9.

4.7 HEAT GENERATION BY SUBATOMIC REACTION

Both Eqs. 4.5-9 and 4.6-4 have a serious deficiency: that is, neither takes into account the effect of heat generation. For a multicomponent system with chemical reactions, once the mass *of a species* gained or lost due to chemical reactions is included in the continuity equation, as we will do in Chapter 5, the associated heat of reaction will be automatically included in the energy equation if proper derivational procedure is followed[1, 2]. However, even in the single-component nonreacting system under consideration, where there is no heat generated or absorbed by chemical reaction, some heat quantities are still missing because heat can be produced, for example, by nuclear fission, electrical dissipation,* and the like. These are *not* dealt with in the derivation because the principles of energy and mass conservation are valid only above the atomic level. Therefore, in cases where subatomic effects are present, these principles become "inoperative" and should be modified by adding the rate of heat generated per unit volume Q_G (cal/cm^3-sec). When the value of Q_G is negative, heat is being absorbed.

It should also be noted that anything above the atomic level is automatically taken care of in these equations—for example, heat generated by viscous dissipation ($\nabla \cdot [\tau \cdot v]$), which is at the molecular level. As a general rule, if the molecular theory is used to derive the equations of change, as is the usual case, then everything at or above the molecular level is included.

4.8 THE EQUATION OF MECHANICAL ENERGY

The equation of mechanical energy is not derived from any conservation principle regarding mechanical energy, since mechanical energy is *not* conserved (it can be transformed into and from thermal energy). However, if we take the dot product of velocity vector v with every term in the equation of motion (e.g., Eq. 4.4-12), we obtain

$$\rho \frac{D}{Dt}\left(\tfrac{1}{2}v^2\right) = -(v \cdot \nabla p) - (v \cdot [\nabla \cdot \tau]) + \rho(v \cdot g). \qquad (4.8-1)$$

*This occurs, for example, when electric current is passed through a wire. Heat is generated because of movement of electrons.

This is the equation of mechanical energy describing the kinetic energy change for an element of fluid following motion. Its Eulerian form can be obtained either mathematically from this or by taking the dot product of $\mathbf{v}$ and Eq. 4.4-19.

4.9 THE EQUATION OF THERMAL ENERGY

If we subtract Eq. 4.8-1 from Eq. 4.6-4 while introducing Eq. 4.6-8, we obtain*

$$\rho\left(\frac{D\hat{U}}{Dt}\right) = -(\nabla\cdot\mathbf{q}) - p(\nabla\cdot\mathbf{v}) - (\tau:\nabla\mathbf{v}). \tag{4.9-1}$$

This is called the equation of thermal energy, which states that the rate of increase of internal energy of fluid element following motion is equal to those of conduction, compression (reversible), and viscous dissipation (irreversible).

Detailed derivation of these relationships can be found in standard vectors and tensors textbooks. It suffices to explain here that, while $\nabla\mathbf{v}$ is the tensor

$$\left\{\begin{array}{ccc} \dfrac{\partial v_x}{\partial x} & \dfrac{\partial v_y}{\partial x} & \dfrac{\partial v_z}{\partial x} \\[2ex] \dfrac{\partial v_x}{\partial y} & \dfrac{\partial v_y}{\partial y} & \dfrac{\partial v_z}{\partial y} \\[2ex] \dfrac{\partial v_x}{\partial z} & \dfrac{\partial v_y}{\partial z} & \dfrac{\partial v_z}{\partial z} \end{array}\right\},$$

its double-dot product with another tensor (τ) is a scalar (see Eq. 2.16-18):

$$\tau_{xx}\frac{\partial v_x}{\partial x} + \tau_{xy}\frac{\partial v_x}{\partial y} + \tau_{xz}\frac{\partial v_x}{\partial z} + \tau_{yx}\frac{\partial v_y}{\partial x} + \cdots + \tau_{zz}\frac{\partial v_z}{\partial z}.$$

In terms of temperature T, if we consider internal energy as a function of volume and temperature, that is,

$$\hat{U} = \hat{U}(\hat{V}, T), \tag{4.9-2}$$

---

*In this derivation we have also used the vector and tensor identities proven in Problems 2.J and 2.L.

we can write (see Eq. 2.6-3)

$$d\hat{U} = \left(\frac{\partial \hat{U}}{\partial \hat{V}}\right)_T d\hat{V} + \left(\frac{\partial \hat{U}}{\partial T}\right)_{\hat{V}} dT. \tag{4.9-3}$$

Then, considering $\hat{V}$ and T functions of time, t, we can take the substantial time derivative as

$$\frac{D\hat{U}}{Dt} = \left(\frac{\partial \hat{U}}{\partial \hat{V}}\right)_T \frac{D\hat{V}}{Dt} + \left(\frac{\partial \hat{U}}{\partial T}\right)_{\hat{V}} \frac{DT}{Dt}. \tag{4.9-4}$$

Since

$$\left(\frac{\partial \hat{U}}{\partial T}\right)_{\hat{V}} = \hat{C}_V,$$

the heat capacity per unit mass at constant volume, and [3]

$$\left(\frac{\partial U}{\partial \hat{V}}\right)_T = T\left(\frac{\partial p}{\partial T}\right)_{\hat{V}} - p,$$

Eq. 4.9-4 becomes

$$\frac{D\hat{U}}{Dt} = \left[T\left(\frac{\partial p}{\partial T}\right)_{\hat{V}} - p\right]\frac{D\hat{V}}{Dt} + \hat{C}_V \frac{DT}{Dt}. \tag{4.9-5}$$

Since $\hat{V}$ is volume per unit mass, it is equal to the reciprocal of density. Then

$$\frac{D\hat{V}}{Dt} = \frac{D}{Dt}\left(\frac{1}{\rho}\right) = -\frac{1}{\rho^2}\frac{D\rho}{Dt} = \frac{1}{\rho}(\nabla \cdot \mathbf{v}). \tag{4.9-6}$$

The latter part of Eq. 4.9-6 is true by virtue of the equation of continuity (Eq. 4.3-8). Equation 4.9-5 now becomes

$$\rho\frac{D\hat{U}}{Dt} = \left[T\left(\frac{\partial p}{\partial T}\right)_{\hat{V}} - p\right](\nabla \cdot \mathbf{v}) + \rho\hat{C}_V \frac{DT}{Dt},$$

which, when substituted into Eq. 4.9-1, yields

$$\rho\hat{C}_V \frac{DT}{Dt} = -(\nabla \cdot \mathbf{q}) - T\left(\frac{\partial p}{\partial T}\right)_\rho (\nabla \cdot \mathbf{v}) - (\tau:\nabla\mathbf{v}), \tag{4.9-7}$$

which is the equation of energy in terms of fluid temperature, which changes because of heat conduction, expansion or compression, and viscous heating.

4.10 SPECIAL FORMS UNDER VARIOUS CONDITIONS

In this section we present several forms of the equation of energy under the conditions most frequently encountered. The assumption of constant thermal conductivity is also made, so that

$$\mathbf{q} = -k\nabla T. \tag{4.10-1}$$

1. *Newtonian fluid.* In this case the $-(\tau : \nabla \mathbf{v})$ in Eq. 4.9-7 can be replaced by the viscous dissipation function* Φ_v which in rectangular coordinates is

$$
\Phi_v = 2\left[\left(\frac{\partial v_x}{\partial x}\right)^2 + \left(\frac{\partial v_y}{\partial y}\right)^2 + \left(\frac{\partial v_z}{\partial z}\right)^2 \right] + \left(\frac{\partial v_y}{\partial x} + \frac{\partial v_x}{\partial y}\right)^2
$$

$$
+ \left(\frac{\partial v_z}{\partial y} + \frac{\partial v_y}{\partial z}\right)^2 + \left(\frac{\partial v_x}{\partial z} + \frac{\partial v_z}{\partial x}\right)^2
$$

$$
- \frac{2}{3}\left(\frac{\partial v_x}{\partial x} + \frac{\partial v_y}{\partial y} + \frac{\partial v_z}{\partial z}\right)^2, \tag{4.10-2}
$$

so that

$$\rho \hat{C}_v \frac{DT}{Dt} = k\nabla^2 T - T\left(\frac{\partial p}{\partial T}\right)_\rho (\nabla \cdot \mathbf{v}) + \mu \Phi_v. \tag{4.10-3}$$

2. *Ideal gas.* In this case $(\partial p/\partial T)_\rho = p/T$ and for gases the viscous heating effect can be neglected except under extreme conditions which are not likely to be encountered in biomedical work. Therefore,

$$\rho \hat{C}_v \frac{DT}{Dt} = k\nabla^2 T - p (\nabla \cdot \mathbf{v}). \tag{4.10-4}$$

3. *Constant pressure.* For a fluid at constant pressure $(\partial p/\partial T)_\rho = 0$ and $\hat{C}_v$ is replaced by $\hat{C}_p$, then Eq. 4.9-7 becomes

$$\rho \hat{C}_p \frac{DT}{Dt} = k\nabla^2 T - (\tau : \nabla \mathbf{v}). \tag{4.10-5}$$

4. *Incompressible fluid.* By virtue of the equation of continuity, $(\nabla \cdot v)$

*Caution against confusion here: Unfortunately the generally adopted symbols for potential energy (Φ) and viscous dissipation (Φ_v) differ only by a subscript. This is obtained by using Eq. 4.4-20.

$= 0$, plus the conditions that $\rho = $ constant, $\hat{C}_p = \hat{C}_v$, we have

$$\rho \hat{C}_p \frac{DT}{Dt} = k \nabla^2 T. \tag{4.10-6}$$

5. *Solids*. In the case of a solid, in addition to $\hat{C}_p = \hat{C}_v$ and ρ $=$ constant, $\mathbf{v} = 0$ so that $DT/Dt = \partial T/\partial t$. Therefore,

$$\rho \hat{C}_p \frac{\partial T}{\partial t} = k \nabla^2 T, \tag{4.10-7}$$

which is also called the heat-conduction equation since in a solid the only mode of heat transfer is conduction.

4.11 CYLINDRICAL AND SPHERICAL COORDINATES

The equations of change derived so far are in rectangular coordinates only. Those in cylindrical and spherical coordinates are also very important, since, for example, most body fluids flow in tubes or diffuse in cylindrical or spherical systems. There are two ways whereby equations of change can be derived in those coordinates. Both are straightforward.

In the first method, the basic principle is applied to an element the shape of which is characteristic of the coordinate system. This can be illustrated by the following example of deriving the equation of continuity in cylindrical coordinates, in which case we choose an element similar to a slice of pie, as depicted in Figure 4.11-1.In this example, the Eulerian approach is used and a mass balance is made on this fixed element of space. Since this element is taken to be small and eventually reduced to infinitesimal size, the volume can be approximated by that of a rectangular parallelepiped

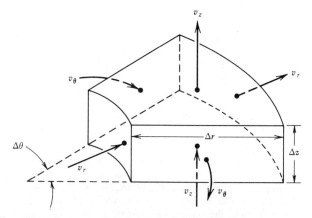

Figure 4.11-1. Mass balance on a fixed element in cylindrical space.

with sides $\Delta r, r\Delta\theta$, and Δz. Thus, during time period Δt

$$\text{Accumulation of mass} = \Delta(\rho r \Delta r \Delta\theta \Delta z) = r \Delta r \Delta\theta \Delta z \Delta\rho. \quad (4.11\text{-}1)$$

Note that although r is a variable, it can be taken out of the Δ, since the entire combination $r\Delta r\Delta\theta\Delta z$ (volume) is constant. For the same time period

$$\qquad\qquad r\text{-direction*} \qquad\qquad \theta\text{-direction} \qquad\qquad z\text{-direction}$$

Mass in:

$$\rho v_r|_r \Delta z \Delta\theta \Delta t \qquad \rho v_\theta|_\theta \Delta r \Delta z \Delta t \qquad \rho v_z|_z r \Delta r \Delta\theta \Delta t \quad (4.11\text{-}2)$$

Mass out:

$$\rho v_r|_{r+\Delta r} \Delta z \Delta\theta \Delta t \quad \rho v_\theta|_{\theta+\Delta\theta} \Delta r \Delta z \Delta t \quad \rho v_z|_{z+\Delta z} r \Delta r \Delta\theta \Delta t. \quad (4.11\text{-}3)$$

Substituting these quantities into Eq. 4.3-1 and dividing the entire equation by $r\Delta r\Delta\theta\Delta z \Delta t$, we have

$$\frac{\Delta\rho}{\Delta t} = -\frac{\rho v_r|_{r+\Delta r} - \rho v_r|_r}{r\Delta r} - \frac{\rho v_\theta|_{\theta+\Delta\theta} - \rho v_\theta|_\theta}{r\Delta\theta}$$

$$-\frac{\rho v_z|_{z+\Delta z} - \rho v_z|_z}{\Delta z}.$$

Reducing $\Delta r, \Delta\theta, \Delta z,$ and Δt to zero, we obtain

$$\frac{\partial\rho}{\partial t} = -\left(\frac{1}{r}\frac{\partial}{\partial r}\rho r v_r + \frac{1}{r}\frac{\partial}{\partial\theta}\rho v_\theta + \frac{\partial}{\partial z}\rho v_z\right) \quad (4.11\text{-}4)$$

or

$$\frac{\partial\rho}{\partial t} = -(\nabla\cdot\rho\mathbf{v}), \quad (4.11\text{-}5)$$

where

$$(\nabla\cdot\rho\mathbf{v}) = \frac{1}{r}\frac{\partial}{\partial r}(\rho v_r) + \frac{1}{r}\frac{\partial}{\partial\theta}(\rho v_\theta) + \frac{\partial}{\partial z}(\rho v_z), \quad (4.11\text{-}6)$$

which is really a slight extension of Eq. 2.17-40.

*Note that there is a slight difference between this and the case of rectangular coordinates Eqs. 4.3-3 and 4.3-4. In rectangular coordinates, the element is a cube; therefore, the surface areas at, for example, x and $x+\Delta x$ are equal. However, in the present case, the area $r\Delta\theta\Delta z$ at r is different from that at $r+\Delta r$, because of the difference in r, although $\Delta\theta$ and Δz remain unchanged. Therefore, r should be left to the left side, or "inside," of the vertical bar. However, for the z-direction, the vertical bar governs z and since r and z are mutually independent, r can be taken to the right of the z-direction vertical bar.

Similar developments can be made for equations of motion and energy, as well as using the Lagrangian approach, and also in spherical coordinates, in which case the element is shown in Figure 4.11-2. The resulting equations are tabulated in Section 4.12.

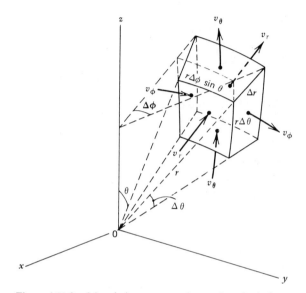

Figure 4.11-2. Mass balance on an element in spherical space.

The other method is to use the mathematical conversion between coordinate systems (see Section 2.17 and Problem 4.D).

4.12 SUMMARY AND TABLES

We have thus far shown the development of the basic equations used in solving problems in fluid flow and heat transfer for single-component systems. Complete sets of these equations, in various forms and coordinates systems are listed in Tables 4.12-1 through 4.12-3. It is impossible and useless to memorize them, but the reader is well advised to do the following:

1. Understand the basic principles behind these equations, their meaning, and their interrelationships.

2. Acquaint himself with these tables so that he knows what particular equations to use and where to find them, in any particular situation.

3. Acquaint himself with the manipulations of these equations, what terms to be neglected or modified, and so on.

Table 4.12-1 The Equation of Continuity in Various Coordinate Systems

A. Rectangular coordinates (x,y,z):

$$\frac{\partial \rho}{\partial t} + \frac{\partial}{\partial x}(\rho v_x) + \frac{\partial}{\partial y}(\rho v_y) + \frac{\partial}{\partial z}(\rho v_z) = 0 \qquad \nabla \cdot \rho \vec{v} = -\frac{\partial \rho}{\partial t}$$

B. Cylindrical coordinates (r,θ,z):

$$\frac{\partial \rho}{\partial t} + \frac{1}{r}\frac{\partial}{\partial r}(\rho r v_r) + \frac{1}{r}\frac{\partial}{\partial \theta}(\rho v_\theta) + \frac{\partial}{\partial z}(\rho v_z) = 0$$

C. Spherical coordinates (r,θ,ϕ):

$$\frac{\partial \rho}{\partial t} + \frac{1}{r^2}\frac{\partial}{\partial r}(\rho r^2 v_r) + \frac{1}{r\sin\theta}\frac{\partial}{\partial \theta}(\rho v_\theta \sin\theta) + \frac{1}{r\sin\theta}\frac{\partial}{\partial \phi}(\rho v_\phi) = 0$$

Table 4.12-2a The Equation of Motion in Rectangular Coordinates (x,y,z):

In terms of momentum fluxes:

A. x-component

$$\rho\left(\frac{\partial v_x}{\partial t} + v_x\frac{\partial v_x}{\partial x} + v_y\frac{\partial v_x}{\partial y} + v_z\frac{\partial v_x}{\partial z}\right) = -\frac{\partial p}{\partial x} - \left(\frac{\partial \tau_{xx}}{\partial x} + \frac{\partial \tau_{yx}}{\partial y} + \frac{\partial \tau_{zx}}{\partial z}\right) + \rho g_x$$

B. y-component

$$\rho\left(\frac{\partial v_y}{\partial t} + v_x\frac{\partial v_y}{\partial x} + v_y\frac{\partial v_y}{\partial y} + v_z\frac{\partial v_y}{\partial z}\right) = -\frac{\partial p}{\partial y} - \left(\frac{\partial \tau_{xy}}{\partial x} + \frac{\partial \tau_{yy}}{\partial y} + \frac{\partial \tau_{zy}}{\partial z}\right) + \rho g_y$$

C. z-component

$$\rho\left(\frac{\partial v_z}{\partial t} + v_x\frac{\partial v_z}{\partial x} + v_y\frac{\partial v_z}{\partial y} + v_z\frac{\partial v_z}{\partial z}\right) = -\frac{\partial p}{\partial z} - \left(\frac{\partial \tau_{xz}}{\partial x} + \frac{\partial \tau_{yz}}{\partial y} + \frac{\partial \tau_{zz}}{\partial z}\right) + \rho g_z$$

In terms of velocity gradients for a Newtonian fluid with constant ρ and μ:

D. x-component

$$\rho\left(\frac{\partial v_x}{\partial t} + v_x\frac{\partial v_x}{\partial x} + v_y\frac{\partial v_x}{\partial y} + v_z\frac{\partial v_x}{\partial z}\right) = -\frac{\partial p}{\partial x} + \mu\left(\frac{\partial^2 v_x}{\partial x^2} + \frac{\partial^2 v_x}{\partial y^2} + \frac{\partial^2 v_x}{\partial z^2}\right) + \rho g_x$$

E. y-component

$$\rho\left(\frac{\partial v_y}{\partial t} + v_x\frac{\partial v_y}{\partial x} + v_y\frac{\partial v_y}{\partial y} + v_z\frac{\partial v_y}{\partial z}\right) = -\frac{\partial p}{\partial y} + \mu\left(\frac{\partial^2 v_y}{\partial x^2} + \frac{\partial^2 v_y}{\partial y^2} + \frac{\partial^2 v_y}{\partial z^2}\right) + \rho g_y$$

F. z-component

$$\rho\left(\frac{\partial v_z}{\partial t} + v_x\frac{\partial v_z}{\partial x} + v_y\frac{\partial v_z}{\partial y} + v_z\frac{\partial v_z}{\partial z}\right) = -\frac{\partial p}{\partial z} + \mu\left(\frac{\partial^2 v_z}{\partial x^2} + \frac{\partial^2 v_z}{\partial y^2} + \frac{\partial^2 v_z}{\partial z^2}\right) + \rho g_z$$

Table 4.12-2b The Equation of Motion in Cylindrical Coordinates (r, θ, z)

In terms of momentum fluxes:

A. r-component

$$\rho\left(\frac{\partial v_r}{\partial t} + v_r \frac{\partial v_r}{\partial r} + \frac{v_\theta}{r}\frac{\partial v_r}{\partial \theta} - \frac{v_\theta^2}{r} + v_z \frac{\partial v_r}{\partial z} \right)$$

$$= -\frac{\partial p}{\partial r} - \left(\frac{1}{r}\frac{\partial}{\partial r}(r\tau_{rr}) + \frac{1}{r}\frac{\partial \tau_{r\theta}}{\partial \theta} - \frac{\tau_{\theta\theta}}{r} + \frac{\partial \tau_{rz}}{\partial z} \right) + \rho g_r$$

B. θ-component

$$\rho\left(\frac{\partial v_\theta}{\partial t} + v_r \frac{\partial v_\theta}{\partial r} + \frac{v_\theta}{r}\frac{\partial v_\theta}{\partial \theta} + \frac{v_r v_\theta}{r} + v_z \frac{\partial v_\theta}{\partial z} \right)$$

$$= -\frac{1}{r}\frac{\partial p}{\partial \theta} - \left(\frac{1}{r^2}\frac{\partial}{\partial r}(r^2\tau_{r\theta}) + \frac{1}{r}\frac{\partial \tau_{\theta\theta}}{\partial \theta} + \frac{\partial \tau_{\theta z}}{\partial z} \right) + \rho g_\theta$$

C. z-component

$$\rho\left(\frac{\partial v_z}{\partial t} + v_r \frac{\partial v_z}{\partial r} + \frac{v_\theta}{r}\frac{\partial v_z}{\partial \theta} + v_z \frac{\partial v_z}{\partial z} \right)$$

$$= -\frac{\partial p}{\partial z} - \left(\frac{1}{r}\frac{\partial}{\partial r}(r\tau_{rz}) + \frac{1}{r}\frac{\partial \tau_{\theta z}}{\partial \theta} + \frac{\partial \tau_{zz}}{\partial z} \right) + \rho g_z$$

In terms of velocity gradients for a Newtonian fluid with constant ρ and μ:

D. r-component

$$\rho\left(\frac{\partial v_r}{\partial t} + v_r \frac{\partial v_r}{\partial r} + \frac{v_\theta}{r}\frac{\partial v_r}{\partial \theta} - \frac{v_\theta^2}{r} + v_z \frac{\partial v_r}{\partial z} \right)$$

$$= -\frac{\partial p}{\partial r} + \mu\left[\frac{\partial}{\partial r}\left(\frac{1}{r}\frac{\partial}{\partial r}(r v_r) \right) + \frac{1}{r^2}\frac{\partial^2 v_r}{\partial \theta^2} - \frac{2}{r^2}\frac{\partial v_\theta}{\partial \theta} + \frac{\partial^2 v_r}{\partial z^2} \right] + \rho g_r$$

E. θ-component

$$\rho\left(\frac{\partial v_\theta}{\partial t} + v_r \frac{\partial v_\theta}{\partial r} + \frac{v_\theta}{r}\frac{\partial v_\theta}{\partial \theta} + \frac{v_r v_\theta}{r} + v_z \frac{\partial v_\theta}{\partial z} \right)$$

$$= -\frac{1}{r}\frac{\partial p}{\partial \theta} + \mu\left[\frac{\partial}{\partial r}\left(\frac{1}{r}\frac{\partial}{\partial r}(r v_\theta) \right) + \frac{1}{r^2}\frac{\partial^2 v_\theta}{\partial \theta^2} + \frac{2}{r^2}\frac{\partial v_r}{\partial \theta} + \frac{\partial^2 v_\theta}{\partial z^2} \right] + \rho g_\theta$$

F. z-component

$$\rho\left(\frac{\partial v_z}{\partial t} + v_r \frac{\partial v_z}{\partial r} + \frac{v_\theta}{r}\frac{\partial v_z}{\partial \theta} + v_z \frac{\partial v_z}{\partial z} \right)$$

$$= -\frac{\partial p}{\partial z} + \mu\left[\frac{1}{r}\frac{\partial}{\partial r}\left(r \frac{\partial v_z}{\partial r} \right) + \frac{1}{r^2}\frac{\partial^2 v_z}{\partial \theta^2} + \frac{\partial^2 v_z}{\partial z^2} \right] + \rho g_z$$

Table 4.12-2c The Equation of Motion in Spherical Coordinates (r,θ,ϕ)

In terms of momentum fluxes:

A. *r*-component

$$\rho\left(\frac{\partial v_r}{\partial t}+v_r\frac{\partial v_r}{\partial r}+\frac{v_\theta}{r}\frac{\partial v_r}{\partial\theta}+\frac{v_\phi}{r\sin\theta}\frac{\partial v_r}{\partial\phi}-\frac{v_\theta^2+v_\phi^2}{r}\right)=-\frac{\partial p}{\partial r}$$

$$-\left(\frac{1}{r^2}\frac{\partial}{\partial r}(r^2\tau_{rr})+\frac{1}{r\sin\theta}\frac{\partial}{\partial\theta}(\tau_{r\theta}\sin\theta)+\frac{1}{r\sin\theta}\frac{\partial\tau_{r\phi}}{\partial\phi}-\frac{\tau_{\theta\theta}+\tau_{\phi\phi}}{r}\right)+\rho g_r$$

B. *θ*-component

$$\rho\left(\frac{\partial v_\theta}{\partial t}+v_r\frac{\partial v_\theta}{\partial r}+\frac{v_\theta}{r}\frac{\partial v_\theta}{\partial\theta}+\frac{v_\phi}{r\sin\theta}\frac{\partial v_\theta}{\partial\phi}+\frac{v_r v_\theta}{r}-\frac{v_\phi^2\cot\theta}{r}\right)=-\frac{1}{r}\frac{\partial p}{\partial\theta}$$

$$-\left(\frac{1}{r^2}\frac{\partial}{\partial r}(r^2\tau_{r\theta})+\frac{1}{r\sin\theta}\frac{\partial}{\partial\theta}(\tau_{\theta\theta}\sin\theta)+\frac{1}{r\sin\theta}\frac{\partial\tau_{\theta\phi}}{\partial\phi}+\frac{\tau_{r\theta}}{r}-\frac{\cot\theta}{r}\tau_{\phi\phi}\right)+\rho g_\theta$$

C. *φ*-component

$$\rho\left(\frac{\partial v_\phi}{\partial t}+v_r\frac{\partial v_\phi}{\partial r}+\frac{v_\theta}{r}\frac{\partial v_\phi}{\partial\theta}+\frac{v_\phi}{r\sin\theta}\frac{\partial v_\phi}{\partial\phi}+\frac{v_\phi v_r}{r}+\frac{v_\theta v_\phi}{r}\cot\theta\right)=-\frac{1}{r\sin\theta}\frac{\partial p}{\partial\phi}$$

$$-\left(\frac{1}{r^2}\frac{\partial}{\partial r}(r^2\tau_{r\phi})+\frac{1}{r}\frac{\partial\tau_{\theta\phi}}{\partial\theta}+\frac{1}{r\sin\theta}\frac{\partial\tau_{\phi\phi}}{\partial\phi}+\frac{\tau_{r\phi}}{r}+\frac{2\cot\theta}{r}\tau_{\theta\phi}\right)+\rho g_\phi$$

In terms of velocity gradients for a Newtonian fluid with constant ρ and μ:[a]

D. *r*-component

$$\rho\left(\frac{\partial v_r}{\partial t}+v_r\frac{\partial v_r}{\partial r}+\frac{v_\theta}{r}\frac{\partial v_r}{\partial\theta}+\frac{v_\phi}{r\sin\theta}\frac{\partial v_r}{\partial\phi}-\frac{v_\theta^2+v_\phi^2}{r}\right)=-\frac{\partial p}{\partial r}$$

$$+\mu\left(\nabla^2 v_r-\frac{2}{r^2}v_r-\frac{2}{r^2}\frac{\partial v_\theta}{\partial\theta}-\frac{2}{r^2}v_\theta\cot\theta-\frac{2}{r^2\sin\theta}\frac{\partial v_\phi}{\partial\phi}\right)+\rho g_r$$

E. *θ*-component

$$\rho\left(\frac{\partial v_\theta}{\partial t}+v_r\frac{\partial v_\theta}{\partial r}+\frac{v_\theta}{r}\frac{\partial v_\theta}{\partial\theta}+\frac{v_\phi}{r\sin\theta}\frac{\partial v_\theta}{\partial\phi}+\frac{v_r v_\theta}{r}-\frac{v_\phi^2\cot\theta}{r}\right)=-\frac{1}{r}\frac{\partial p}{\partial\theta}$$

$$+\mu\left(\nabla^2 v_\theta+\frac{2}{r^2}\frac{\partial v_r}{\partial\theta}-\frac{v_\theta}{r^2\sin^2\theta}-\frac{2\cos\theta}{r^2\sin^2\theta}\frac{\partial v_\phi}{\partial\phi}\right)+\rho g_\theta$$

F. *φ*-component

$$\rho\left(\frac{\partial v_\phi}{\partial t}+v_r\frac{\partial v_\phi}{\partial r}+\frac{v_\theta}{r}\frac{\partial v_\phi}{\partial\theta}+\frac{v_\phi}{r\sin\theta}\frac{\partial v_\phi}{\partial\phi}+\frac{v_\phi v_r}{r}+\frac{v_\theta v_\phi}{r}\cot\theta\right)=-\frac{1}{r\sin\theta}\frac{\partial p}{\partial\phi}$$

$$+\mu\left(\nabla^2 v_\phi-\frac{v_\phi}{r^2\sin^2\theta}+\frac{2}{r^2\sin\theta}\frac{\partial v_r}{\partial\phi}+\frac{2\cos\theta}{r^2\sin^2\theta}\frac{\partial v_\theta}{\partial\phi}\right)+\rho g_\phi$$

[a]In these equations $\nabla^2=\dfrac{1}{r^2}\dfrac{\partial}{\partial r}\left(r^2\dfrac{\partial}{\partial r}\right)+\dfrac{1}{r^2\sin\theta}\dfrac{\partial}{\partial\theta}\left(\sin\theta\dfrac{\partial}{\partial\theta}\right)+\dfrac{1}{r^2\sin^2\theta}\left(\dfrac{\partial^2}{\partial\phi^2}\right).$

Table 4.12-3 The Equation of Energy

In terms of energy and momentum fluxes:

A. Rectangular coordinates

$$\rho \hat{C}_v \left(\frac{\partial T}{\partial t} + v_x \frac{\partial T}{\partial x} + v_y \frac{\partial T}{\partial y} + v_z \frac{\partial T}{\partial z} \right) = - \left[\frac{\partial q_x}{\partial x} + \frac{\partial q_y}{\partial y} + \frac{\partial q_z}{\partial z} \right]$$

$$- T \left(\frac{\partial p}{\partial T} \right)_\rho \left(\frac{\partial v_x}{\partial x} + \frac{\partial v_y}{\partial y} + \frac{\partial v_z}{\partial z} \right) - \left\{ \tau_{xx} \frac{\partial v_x}{\partial x} + \tau_{yy} \frac{\partial v_y}{\partial y} + \tau_{zz} \frac{\partial v_z}{\partial z} \right\}$$

$$- \left\{ \tau_{xy} \left(\frac{\partial v_x}{\partial y} + \frac{\partial v_y}{\partial x} \right) + \tau_{xz} \left(\frac{\partial v_x}{\partial z} + \frac{\partial v_z}{\partial x} \right) + \tau_{yz} \left(\frac{\partial v_y}{\partial z} + \frac{\partial v_z}{\partial y} \right) \right\}$$

B. Cylindrical coordinates

$$\rho \hat{C}_v \left(\frac{\partial T}{\partial t} + v_r \frac{\partial T}{\partial r} + \frac{v_\theta}{r} \frac{\partial T}{\partial \theta} + v_z \frac{\partial T}{\partial z} \right) = - \left[\frac{1}{r} \frac{\partial}{\partial r} (rq_r) + \frac{1}{r} \frac{\partial q_\theta}{\partial \theta} + \frac{\partial q_z}{\partial z} \right]$$

$$- T \left(\frac{\partial p}{\partial T} \right)_\rho \left(\frac{1}{r} \frac{\partial}{\partial r} (rv_r) + \frac{1}{r} \frac{\partial v_\theta}{\partial \theta} + \frac{\partial v_z}{\partial z} \right) - \left\{ \tau_{rr} \frac{\partial v_r}{\partial r} + \tau_{\theta\theta} \frac{1}{r} \left(\frac{\partial v_\theta}{\partial \theta} + v_r \right) \right.$$

$$\left. + \tau_{zz} \frac{\partial v_z}{\partial z} \right\} - \left\{ \tau_{r\theta} \left[r \frac{\partial}{\partial r} \left(\frac{v_\theta}{r} \right) + \frac{1}{r} \frac{\partial v_r}{\partial \theta} \right] + \tau_{rz} \left(\frac{\partial v_z}{\partial r} + \frac{\partial v_r}{\partial z} \right) \right.$$

$$\left. + \tau_{\theta z} \left(\frac{1}{r} \frac{\partial v_z}{\partial \theta} + \frac{\partial v_\theta}{\partial z} \right) \right\}$$

C. Spherical coordinates

$$\rho \hat{C}_v \left(\frac{\partial T}{\partial t} + v_r \frac{\partial T}{\partial r} + \frac{v_\theta}{r} \frac{\partial T}{\partial \theta} + \frac{v_\phi}{r \sin \theta} \frac{\partial T}{\partial \phi} \right) = - \left[\frac{1}{r^2} \frac{\partial}{\partial r} (r^2 q_r) \right.$$

$$\left. + \frac{1}{r \sin \theta} \frac{\partial}{\partial \theta} (q_\theta \sin \theta) + \frac{1}{r \sin \theta} \frac{\partial q_\phi}{\partial \phi} \right] - T \left(\frac{\partial p}{\partial T} \right)_\rho \left(\frac{1}{r^2} \frac{\partial}{\partial r} (r^2 v_r) \right.$$

$$\left. + \frac{1}{r \sin \theta} \frac{\partial}{\partial \theta} (v_\theta \sin \theta) + \frac{1}{r \sin \theta} \frac{\partial v_\phi}{\partial \phi} \right) - \left\{ \tau_{rr} \frac{\partial v_r}{\partial r} + \tau_{\theta\theta} \left(\frac{1}{r} \frac{\partial v_\theta}{\partial \theta} + \frac{v_r}{r} \right) \right.$$

$$\left. + \tau_{\phi\phi} \left(\frac{1}{r \sin \theta} \frac{\partial v_\phi}{\partial \phi} + \frac{v_r}{r} + \frac{v_\theta \cot \theta}{r} \right) \right\} - \left\{ \tau_{r\theta} \left(\frac{\partial v_\theta}{\partial r} + \frac{1}{r} \frac{\partial v_r}{\partial \theta} - \frac{v_\theta}{r} \right) \right.$$

$$\left. + \tau_{r\phi} \left(\frac{\partial v_\phi}{\partial r} + \frac{1}{r \sin \theta} \frac{\partial v_r}{\partial \phi} - \frac{v_\phi}{r} \right) + \tau_{\theta\phi} \left(\frac{1}{r} \frac{\partial v_\phi}{\partial \theta} + \frac{1}{r \sin \theta} \frac{\partial v_\theta}{\partial \phi} - \frac{\cot \theta}{r} v_\phi \right) \right\}$$

Table 4.12-3 (*Continued*)

In terms of the transport properties (for Newtonian fluids of constant ρ and k):

D. Rectangular coordinates

$$\rho\hat{C}_p\left(\frac{\partial T}{\partial t}+v_x\frac{\partial T}{\partial x}+v_y\frac{\partial T}{\partial y}+v_z\frac{\partial T}{\partial z}\right)=k\left[\frac{\partial^2 T}{\partial x^2}+\frac{\partial^2 T}{\partial y^2}+\frac{\partial^2 T}{\partial z^2}\right]$$

$$+2\mu\left\{\left(\frac{\partial v_x}{\partial x}\right)^2+\left(\frac{\partial v_y}{\partial y}\right)^2+\left(\frac{\partial v_z}{\partial z}\right)^2\right\}+\mu\left\{\left(\frac{\partial v_x}{\partial y}+\frac{\partial v_y}{\partial x}\right)^2\right.$$

$$\left.+\left(\frac{\partial v_x}{\partial z}+\frac{\partial v_z}{\partial x}\right)^2+\left(\frac{\partial v_y}{\partial z}+\frac{\partial v_z}{\partial y}\right)^2\right\}$$

E. Cylindrical coordinates

$$\rho\hat{C}_p\left(\frac{\partial T}{\partial t}+v_r\frac{\partial T}{\partial r}+\frac{v_\theta}{r}\frac{\partial T}{\partial \theta}+v_z\frac{\partial T}{\partial z}\right)=k\left[\frac{1}{r}\frac{\partial}{\partial r}\left(r\frac{\partial T}{\partial r}\right)+\frac{1}{r^2}\frac{\partial^2 T}{\partial \theta^2}+\frac{\partial^2 T}{\partial z^2}\right]$$

$$+2\mu\left\{\left(\frac{\partial v_r}{\partial r}\right)^2+\left[\frac{1}{r}\left(\frac{\partial v_\theta}{\partial \theta}+v_r\right)\right]^2+\left(\frac{\partial v_z}{\partial z}\right)^2\right\}+\mu\left\{\left(\frac{\partial v_\theta}{\partial z}+\frac{1}{r}\frac{\partial v_z}{\partial \theta}\right)^2\right.$$

$$\left.+\left(\frac{\partial v_z}{\partial r}+\frac{\partial v_r}{\partial z}\right)^2+\left[\frac{1}{r}\frac{\partial v_r}{\partial \theta}+r\frac{\partial}{\partial r}\left(\frac{v_\theta}{r}\right)\right]^2\right\}$$

F. Spherical coordinates

$$\rho\hat{C}_p\left(\frac{\partial T}{\partial t}+v_r\frac{\partial T}{\partial r}+\frac{v_\theta}{r}\frac{\partial T}{\partial \theta}+\frac{v_\phi}{r\sin\theta}\frac{\partial T}{\partial \phi}\right)=k\left[\frac{1}{r^2}\frac{\partial}{\partial r}\left(r^2\frac{\partial T}{\partial r}\right)\right.$$

$$\left.+\frac{1}{r^2\sin\theta}\frac{\partial}{\partial \theta}\left(\sin\theta\frac{\partial T}{\partial \theta}\right)+\frac{1}{r^2\sin^2\theta}\frac{\partial^2 T}{\partial \phi^2}\right]+2\mu\left\{\left(\frac{\partial v_r}{\partial r}\right)^2\right.$$

$$+\left(\frac{1}{r}\frac{\partial v_\theta}{\partial \theta}+\frac{v_r}{r}\right)^2+\left(\frac{1}{r\sin\theta}\frac{\partial v_\phi}{\partial \phi}+\frac{v_r}{r}+\frac{v_\theta\cot\theta}{r}\right)^2\right\}$$

$$+\mu\left\{\left[r\frac{\partial}{\partial r}\left(\frac{v_\theta}{r}\right)+\frac{1}{r}\frac{\partial v_r}{\partial \theta}\right]^2+\left[\frac{1}{r\sin\theta}\frac{\partial v_r}{\partial \phi}+r\frac{\partial}{\partial r}\left(\frac{v_\phi}{r}\right)\right]^2\right.$$

$$\left.+\left[\frac{\sin\theta}{r}\frac{\partial}{\partial \theta}\left(\frac{v_\phi}{\sin\theta}\right)+\frac{1}{r\sin\theta}\frac{\partial v_\theta}{\partial \phi}\right]^2\right\}$$

PROBLEMS

4.A Equation of Continuity by Lagrangian Approach

Derive the equation of continuity using the Lagrangian approach by formulating that a fluid element following motion neither gains nor loses any mass (even though its volume and/or shape might change) with respect to time; that is,

$$\frac{D}{Dt}(\rho \Delta V) = 0. \tag{4.A-1}$$

Using rectangular coordinates as an example—$\Delta V = \Delta x \, \Delta y \, \Delta z$—expanding Eq. 4.A-1, exchanging the D/Dt and Δ operations and reducing Δ to infinitesimal size, show that Eq. 4.3-8 results.

4.B Equation of Continuity for Incompressible Fluid

For an incompressible fluid, $\partial \rho / \partial t = 0$. Does this also mean $D\rho/Dt = 0$? Explain both physically and mathematically. Does this situation (i.e., both partial and substantial derivatives are simultaneously zero) necessarily occur in general for any variable, say, c?

4.C Equation of Motion by Eulerian Approach

Derive the equation of motion in the Eulerian form (Eq. 4.4-19 or any component thereof) by applying a momentum balance to an volumetric element *fixed in space*. The principle of momentum conservation states that, for this element,

$$\left\{ \begin{array}{c} \text{Rate of} \\ \text{momentum} \\ \text{accumulation} \end{array} \right\} = \left\{ \begin{array}{c} \text{Rate of} \\ \text{momentum} \\ \text{in} \end{array} \right\} - \left\{ \begin{array}{c} \text{Rate of} \\ \text{momentum} \\ \text{out} \end{array} \right\} + \left\{ \begin{array}{c} \text{Sum of forces} \\ \text{acting on sys-} \\ \text{tem} \end{array} \right\}.$$

$$\qquad (1) \qquad\qquad\qquad (2) \qquad\qquad\qquad (3) \qquad\qquad\qquad (4)$$

Under (2) and (3), the momentum can enter and leave the element by bulk flow (convection) and by molecular transport (shear and normal stresses). The forces now consist only of pressure and body forces. (Since the element is fixed, the viscous force we have in the Lagrangian approach appears as the rate of molecular momentum transport in this case—that is, the same quantity with a different interpretation because of the difference in viewpoint.) Also, give the physical significance of each term.

4.D Transformation of Equation of Continuity from Rectangular to Cylindrical Coordinate Systems

Using the appropriate formulas in Table 2.17-1 transform Eq. A to Eq. B in Table 4.12-1.

REFERENCES

1. Fulford, G. D. and Pei, D. C. T. *Ind. Eng. Chem.*, **61** (5), 47 (1969).
2. Hirschfelder, J. O.; Curtiss, C. F.; and Bird, R. B. *Molecular Theory of Gases and Liquids*, rev. ed. Wiley: New York, 1964.
3. Sage, B. H. *Thermodynamics of Multicomponent Systems*. Reinhold: New York, 1965.

CHAPTER FIVE

Multicomponent Systems

After developing the equations of change for pure substances, we are ready to do the same for multicomponent mixtures, which occur more frequently in biomedical problems. However, because the concentration of a mixture can be expressed in either mass or number of moles per unit volume and the composition in either mass or mole fraction, a variety of definitions and units for the mass flux results. These will be dealt with in the first three sections in this chapter. Then equations of change will be written for multicomponent systems.

5.1 MASS AND MOLAR CONCENTRATIONS

In a mixture containing n species, the concentration of the ith species can be expressed either as

$$\rho_i = \frac{\text{g of } i}{\text{cm}^3 \text{ of mixture}} \qquad (\text{mass concentration})*$$

or as

$$c_i = \frac{\text{g-moles of } i}{\text{cm}^3 \text{ of mixture}} \qquad (\text{molar concentration}).$$

When the mass concentrations of all species are summed, the following results:

$$\sum_{i=1}^{n} \rho_i = \frac{\text{g of mixture}}{\text{cm}^3 \text{ of mixture}} = \rho,$$

which by definition is the *total mass density*, or simply *density*, of the

---

*The mixture can be one of gases, liquids, and even solids—such as alloy. Also, this should be more generally written as mass of species i per unit volume but the more specific cgs units are employed here. The use of liter for volume may be preferable in many cases in biomedical work.

mixture. Similarly,

$$\sum_{i=1}^{n} c_i = \frac{\text{g-moles of mixture}}{\text{cm}^3 \text{ of mixture}} = c,$$

which may be called the *molar density* of the mixture.

Concentration can also be expressed as mass fraction or mole fraction:

$$\omega_i = \frac{\rho_i}{\sum\limits_{i=1}^{n} \rho_i} = \frac{\rho_i}{\rho}$$

$$x_i = \frac{c_i}{\sum\limits_{i=1}^{n} c_i} = \frac{c_i}{c}.$$

The reader may have discovered by now that we have used Greek letters to represent mass quantities, while English letters are used for molar quantities.

According to these definitions, the mass and molar concentrations of a species differ by a factor equal to its *molecular weight*; that is,

$$\rho_i = c_i M_i \qquad \text{or} \qquad c_i = \frac{\rho_i}{M_i}.$$

Then,

$$M = \frac{\rho}{c} = \sum_{i=1}^{n} x_i M_i,$$

which is the *number-mean molecular weight* of the mixture.

Table 5.1-1 Concentration Notations

	Mass	Relationship	Molar
Concentration	ρ_i	$\rho_i = c_i M_i$	c_i
Fraction	ω_i	$\omega_i = \dfrac{x_i M_i}{M}$	x_i
		where	
		$M = \sum\limits_{i=1}^{n} x_i M_i$	

A summary of these definitions and their relationships is shown in Table 5.1-1. Similar quantities for binary systems can be written usually with subscripts A and B representing the two species.

5.2 MASS AND MOLAR AVERAGE VELOCITIES

In a multicomponent system, each species moves at a different velocity characteristic of its chemical and physical nature. This is denoted by v_i, meaning the *velocity of species i*, which should not be mistaken as the velocity of the individual molecule. Since molecules are large in number and random in velocity distribution, it is impossible to consider them individually. The species velocity v_i is a statistical average of the velocities of all the molecules of species i in the mixture.

In a mixture of a species, the local *mass average velocity* is defined as

$$\mathbf{v} = \frac{\displaystyle\sum_{i=1}^{n} \rho_i \mathbf{v}_i}{\displaystyle\sum_{i=1}^{n} \rho_i} \tag{5.2-1}$$

or

$$\mathbf{v} = \frac{1}{\rho} \sum_{i=1}^{n} \rho_i \mathbf{v}_i$$

$$= \sum_{i=1}^{n} \omega_i \mathbf{v}_i. \tag{5.2-1a}$$

The quantity $\mathbf{v}_i - \mathbf{v}$ is called the *diffusion velocity of species i with respect to* $\mathbf{v}$. Since $\mathbf{v}_i$ is the velocity of species i with respect to a fixed coordinate system and $\mathbf{v}$ can be regarded as the velocity of the "mass center of gravity" of the system, $\mathbf{v}_i - \mathbf{v}$ is the velocity of i with respect to the mass center of gravity in movement.

Similarly, the local *molar average velocity* is defined as

$$\mathbf{v}^* = \frac{\displaystyle\sum_{i=1}^{n} c_i \mathbf{v}_i}{\displaystyle\sum_{i=1}^{n} c_i} \tag{5.2-2}$$

$$= \sum_{i=1}^{n} x_i \mathbf{v}_i, \tag{5.2-2a}$$

with $v_i - v^*$ called *diffusion velocity of species i with respect to the molar center of gravity*. The reader can readily see that v and v^* are different except when all species in the mixture have equal molecular weights, in which case the mass fraction would be equal to the mole fraction also.

5.3 MASS AND MOLAR FLUXES

As pointed out in Section 4.3, for a pure substance, ρv, or the density times the velocity, gives the mass flux. In a multicomponent system, if ρ is replaced by the concentration (either ρ_i or c_i), then its product with a velocity gives the mass or molar flux of the particular species i. Since there are various kinds of velocities and concentrations, a variety of fluxes results from all of the possible combinations. They can be best presented in a tabulated form (see Table 5.3-1).

Table 5.3-1 Mass and Molar Fluxes

		With respect to:		
		Stationary coordinates	Mass center of gravity	Molar center of gravity
	Velocity	v_i (cm/sec)	$v_i - v$ (cm/sec)	$v_i - v^*$ (cm/sec)
Flux	Concentration			
Mass	ρ_i (g/cm³)	$n_i = \rho_i v_i$	$j_i = \rho_i(v_i - v)$	$j_i^* = \rho_i(v_i - v^*)$
Molar	c_i (g-moles/cm³)	$N_i = c_i v_i$	$J_i = c_i(v_i - v)$	$J_i^* = c_i(v_i - v^*)$

The use of this table is easy and self-explanatory. For example, the molar concentration is c_i and the species velocity with respect to the mass center of gravity is $v_i - v$. Therefore, the molar flux with respect to mass center of gravity is $J_i = c_i(v_i - v)$.

Properties and interrelationships based on these definitions can be summarized in Tables 5.3-2 and 5.3-3.

Table 5.3-2 Sum of Mass and Molar Fluxes of Species

Flux	With respect to:		
	Stationary coordinates	Mass center of gravity	Molar center of gravity
Mass	$\sum\limits_{i=1}^{n} \mathbf{n}_i = \sum\limits_{i=1}^{n} \rho_i \mathbf{v}_i = \rho \mathbf{v}$	$\sum\limits_{i=1}^{n} \mathbf{j}_i = \sum\limits_{i=1}^{n} \rho_i (\mathbf{v}_i - \mathbf{v}) = 0$	$\sum\limits_{i=1}^{n} \mathbf{j}_i^* = \sum\limits_{i=1}^{n} \rho_i (\mathbf{v}_i - \mathbf{v}^*) = \rho(\mathbf{v} - \mathbf{v}^*)$
Molar	$\sum\limits_{i=1}^{n} \mathbf{N}_i = \sum\limits_{i=1}^{n} c_i \mathbf{v}_i = c \mathbf{v}^*$	$\sum\limits_{i=1}^{n} \mathbf{J}_i = \sum\limits_{i=1}^{n} c_i (\mathbf{v}_i - \mathbf{v}) = c(\mathbf{v}^* - \mathbf{v})$	$\sum\limits_{i=1}^{n} \mathbf{J}_i^* = \sum\limits_{i=1}^{n} c_i (\mathbf{v}_i - \mathbf{v}^*) = 0$

Table 5.3-3 Examples of Interrelationship of Various Mass and Molar Fluxes

	In terms of:					
	$\mathbf{n}_i$	N_i	$\mathbf{j}_i$	J_i	$\mathbf{j}_i^*$	J_i^*
$\mathbf{n}_i =$		$N_i M_i$	$\mathbf{j}_i + \rho_i \mathbf{v}$	$J_i M_i + \rho_i \mathbf{v}$	$\mathbf{j}_i^* + \rho_i \mathbf{v} + \omega_i \sum\limits_{k=1}^n \mathbf{j}_k^*$	$J_i^* M_i + \rho_i \mathbf{v}^*$
$N_i =$	$\dfrac{\mathbf{n}_i}{M_i}$		$\dfrac{\mathbf{j}_i}{M_i} + c_i \mathbf{v}$	$J_i + c_i \mathbf{v}$	$\dfrac{\mathbf{j}_i^*}{M_i} + c_i \mathbf{v} + \dfrac{\omega_i}{M_i} \sum\limits_{k=1}^n \mathbf{j}_k^*$	$J_i^* + c_i \mathbf{v}^*$
$\mathbf{j}_i =$	$\mathbf{n}_i - \omega_i \sum\limits_{k=1}^n \mathbf{n}_k$	$N_i M_i - \omega_i \sum\limits_{k=1}^n N_k M_k$	$\dfrac{\mathbf{j}_i}{M_i}$	$J_i M_i$	$\mathbf{j}_i^* + \omega_i \sum\limits_{k=1}^n \mathbf{j}_k^*$	$J_i^* M_i + \omega_i \sum\limits_{k=1}^n J_k^* M_k$
$J_i =$	$\dfrac{\mathbf{n}_i}{M_i} - \dfrac{\omega_i}{M_i} \sum\limits_{k=1}^n \mathbf{n}_k$	$N_i - \dfrac{\omega_i}{M_i} \sum\limits_{k=1}^n N_k M_k$	$\dfrac{\mathbf{j}_i}{M_i}$		$\dfrac{\mathbf{j}_i^*}{M_i} + \dfrac{\omega_i}{M_i} \sum\limits_{k=1}^n \mathbf{j}_k^*$	$J_i^* + \dfrac{\omega_i}{M_i} \sum\limits_{k=1}^n J_k^* M_k$
$\mathbf{j}_i^* =$	$\mathbf{n}_i - x_i M_i \sum\limits_{k=1}^n \dfrac{\mathbf{n}_k}{M_k}$	$N_i M_i - x_i M_i \sum\limits_{k=1}^n N_k$	$\mathbf{j}_i - x_i M_i \sum\limits_{k=1}^n \dfrac{\mathbf{j}_k}{M_k}$	$J_i M_i - x_i M_i \sum\limits_{k=1}^n J_k$		$J_i^* M_i$
$J_i^* =$	$\dfrac{\mathbf{n}_i}{M_i} - x_i \sum\limits_{k=1}^n \dfrac{\mathbf{n}_k}{M_k}$	$N_i - x_i \sum\limits_{k=1}^n N_k$	$\dfrac{\mathbf{j}_i}{M_i} - x_i \sum\limits_{k=1}^n \dfrac{\mathbf{j}_k}{M_k}$	$J_i - x_i \sum\limits_{k=1}^n J_k$	$\dfrac{\mathbf{j}_i^*}{M_i}$	

5.4 VARIOUS FORMS OF FICK'S LAW

Fick's law of diffusion, presented in Section 3.4, is based only on $\mathbf{j}$, the mass flux with respect to the mass center of gravity. Since there are five different other fluxes (Table 5.3-1), there are other forms of Fick's law, which we can obtain through the interrelationships in Table 5.3-3. These forms are given in Table 5.4-1. Note that these expressions are for binary (two-component) mixtures only (i.e., the index i takes on only the symbols A and B).

Table 5.4-1 Equivalent Forms of Fick's First Law of Binary Diffusion [1]

Flux	Gradient	Form of Fick's first law	
$\mathbf{n}_A$	$\nabla \omega_A$	$\mathbf{n}_A - \omega_A(\mathbf{n}_A + \mathbf{n}_B) = -\rho \mathcal{D}_{AB} \nabla \omega_A$	(A)
$\mathbf{N}_A$	∇x_A	$\mathbf{N}_A - x_A(\mathbf{N}_A + \mathbf{N}_B) = -c \mathcal{D}_{AB} \nabla x_A$	(B)
$\mathbf{j}_A$	$\nabla \omega_A$	$\mathbf{j}_A = -\rho \mathcal{D}_{AB} \nabla \omega_A$	(C)
$\mathbf{J}_A{}^*$	∇x_A	$\mathbf{J}_A{}^* = -c \mathcal{D}_{AB} \nabla x_A$	(D)
$\mathbf{j}_A$	∇x_A	$\mathbf{j}_A = -\left(\dfrac{c^2}{\rho}\right) M_A M_B \mathcal{D}_{AB} \nabla x_A$	(E)
$\mathbf{J}_A{}^*$	$\nabla \omega_A$	$\mathbf{J}_A{}^* = -\left(\dfrac{\rho^2}{c M_A M_B}\right) \mathcal{D}_{AB} \nabla \omega_A$	(F)
$c(\mathbf{v}_A - \mathbf{v}_B)$	∇x_A	$c(\mathbf{v}_A - \mathbf{v}_B) = -\dfrac{c \mathcal{D}_{AB}}{x_A x_B} \nabla x_A$	(G)

5.5 EQUATION OF CONTINUITY

The equation of continuity for the ith species in a mixture of n components can be derived following the same procedure as that for pure substances (Section 4.3) with the following changes:

1. The density ρ is replaced by mass concentration (or "partial density") ρ_i.
2. The mass flux $\rho \mathbf{v}$ is replaced by $\mathbf{n}_i$, since in this case the mass flux includes the diffusion flux in addition to the mass flux by virtue of bulk flow, or

$$\mathbf{n}_i = \rho_i \mathbf{v} + \mathbf{j}_i. \qquad (5.5-1)$$

3.   Although the total mass is conserved, that of an individual species is not, because of possible chemical reactions. Therefore, a term should be added to account for this, denoted by r_j (g/cm^3-sec), which is defined as the rate of production of species i. When r_j is negative, species i is being consumed or converted to other components.

With these modifications the resulting equation of continuity is

$$\frac{\partial \rho_i}{\partial t} + (\nabla \cdot \mathbf{n}_i) = r_i \qquad\qquad i = 1, 2, 3, \ldots, n. \qquad (5.5\text{-}2)$$

The reader may find it timely to review, by actually going through, the detailed procedure as we did in Section 4.3.

A similar equation of continuity can be obtained in terms of molar concentration, molar flux, and reaction rate in molar units. It is

$$\frac{\partial c_i}{\partial t} + (\nabla \cdot \mathbf{N}_i) = R_i \qquad\qquad i = 1, 2, 3, \ldots, n. \qquad (5.5\text{-}3)$$

Using identities relating mass fluxes and applying Fick's Law, Eqs. 5.5-2 and 5.5-3 can be rewritten

$$\frac{\partial \rho_i}{\partial t} + (\nabla \cdot \rho_i \mathbf{v}) = (\nabla \cdot \rho \, \mathcal{D}_{im} \, \nabla \omega_i) + r_i \qquad (5.5\text{-}4)$$

and

$$\frac{\partial c_i}{\partial t} + (\nabla \cdot c_i \mathbf{v}^*) = (\nabla \cdot c \, \mathcal{D}_{im} \, \nabla x_i) + R_i. \qquad (5.5\text{-}5)$$

In these equations, $\mathcal{D}_{im}$ is the *effective binary diffusivity* for the diffusion of i in a mixture, defined by

$$\mathbf{N}_i = -c \, \mathcal{D}_{im} \, \nabla x_i + x_i \sum_{j=1}^{n} \mathbf{N}_j \qquad (5.5\text{-}6)$$

or

$$\frac{1}{c \, \mathcal{D}_{im}} = \frac{\displaystyle\sum_{j=1}^{n} (1/c \, \mathcal{D}_{ij})(x_j \mathbf{N}_i - x_i \mathbf{N}_j)}{\mathbf{N}_i - x_i \displaystyle\sum_{j=1}^{n} \mathbf{N}_j}. \qquad (5.5\text{-}7)$$

For the mixture as a whole, if we sum the equations of continuity in

Table 5.5-1 The Equation of Continuity for Species A in a Binary Mixture

In terms of molar fluxes:

A. Rectangular coordinates

$$\frac{\partial c_A}{\partial t} + \left(\frac{\partial N_{Ax}}{\partial x} + \frac{\partial N_{Ay}}{\partial y} + \frac{\partial N_{Az}}{\partial z} \right) = R_A$$

B. Cylindrical coordinates

$$\frac{\partial c_A}{\partial t} + \left(\frac{1}{r} \frac{\partial}{\partial r} (rN_{Ar}) + \frac{1}{r} \frac{\partial N_{A\theta}}{\partial \theta} + \frac{\partial N_{Az}}{\partial z} \right) = R_A$$

C. Spherical coordinates

$$\frac{\partial c_A}{\partial t} + \left(\frac{1}{r^2} \frac{\partial}{\partial r} (r^2 N_{Ar}) + \frac{1}{r\sin\theta} \frac{\partial}{\partial \theta} (N_{A\theta} \sin\theta) + \frac{1}{r\sin\theta} \frac{\partial N_{A\phi}}{\partial \phi} \right) = R_A$$

In terms of molar concentration gradients, for constant ρ and $\mathscr{D}_{AB}$:

D. Rectangular coordinates

$$\frac{\partial c_A}{\partial t} + \left(v_x \frac{\partial c_A}{\partial x} + v_y \frac{\partial c_A}{\partial y} + v_z \frac{\partial c_A}{\partial z} \right) = \mathscr{D}_{AB} \left(\frac{\partial^2 c_A}{\partial x^2} + \frac{\partial^2 c_A}{\partial y^2} + \frac{\partial^2 c_A}{\partial z^2} \right) + R_A$$

E. Cylindrical coordinates

$$\frac{\partial c_A}{\partial t} + \left(v_r \frac{\partial c_A}{\partial r} + v_\theta \frac{1}{r} \frac{\partial c_A}{\partial \theta} + v_z \frac{\partial c_A}{\partial z} \right)$$

$$= \mathscr{D}_{AB} \left(\frac{1}{r} \frac{\partial}{\partial r} \left(r \frac{\partial c_A}{\partial r} \right) + \frac{1}{r^2} \frac{\partial^2 c_A}{\partial \theta^2} + \frac{\partial^2 c_A}{\partial z^2} \right) + R_A$$

F. Spherical coordinates

$$\frac{\partial c_A}{\partial t} + \left(v_r \frac{\partial c_A}{\partial r} + v_\theta \frac{1}{r} \frac{\partial c_A}{\partial \theta} + v_\phi \frac{1}{r\sin\theta} \frac{\partial c_A}{\partial \phi} \right)$$

$$= \mathscr{D}_{AB} \left(\frac{1}{r^2} \frac{\partial}{\partial r} \left(r^2 \frac{\partial c_A}{\partial r} \right) + \frac{1}{r^2\sin\theta} \frac{\partial}{\partial \theta} \left(\sin\theta \frac{\partial c_A}{\partial \theta} \right) + \frac{1}{r^2\sin^2\theta} \frac{\partial^2 c_A}{\partial \phi^2} \right) + R_A$$

mass units (Eq. 5.5-2) for all of the species $i = 1, \ldots, n$,

$$\frac{\partial \rho}{\partial t} + (\boldsymbol{\nabla} \cdot \rho \mathbf{v}) = 0, \tag{5.5-8}$$

since

$$\sum_{i=1}^{n} \rho_i = \rho, \qquad \sum_{i=1}^{n} \mathbf{n}_i = \rho \mathbf{v}, \qquad \text{and} \qquad \sum_{i=1}^{n} r_i = 0.$$

The last expression is true because the total mass of all species must be conserved, although individual species are not. The reader may have immediately recognized that Eq. 5.5-8 is exactly the same as Eq. 4.3-6, which means that *in mass units* the multicomponent system as a whole (that is, if we are not interested in the interdiffusion of species) can be treated as a pure substance as far as the equation of continuity is concerned.

However, if we sum the equations of continuity in molar units, we do not get a similar result:

$$\frac{\partial c}{\partial t} + (\boldsymbol{\nabla} \cdot c \mathbf{v}^*) = \sum_{i=1}^{n} R_i, \tag{5.5-9}$$

since although $c = \sum_{i=1}^{n} c_i$ and $\sum_{i=1}^{n} \mathbf{N}_i = c \mathbf{v}^*$, $\sum_{i=1}^{n} R_i \neq 0$ because the total number of moles before and after the reaction may be different, whereas the mass is conserved.

The equation of continuity in various forms for a binary mixture is presented in Table 5.5-1.

5.6 EQUATIONS OF MOTION AND ENERGY

Unlike the equation of continuity, for a multicomponent system, there is only one equation of motion and one equation of energy (i.e., for the mixture as a whole), although there are terms that deal with the various species individually. For example, their Lagrangian forms are as follows:

Motion:

$$\rho \frac{D \mathbf{v}}{D t} = -\boldsymbol{\nabla} p - [\boldsymbol{\nabla} \cdot \boldsymbol{\tau}] + \sum_{i=1}^{n} \rho_i \mathbf{g}_i \tag{5.6-1}$$

Energy:

$$\rho \frac{D}{Dt}\left\{ \hat{U} + \tfrac{1}{2}v^2 \right\} = -(\boldsymbol{\nabla} \cdot \mathbf{q}) - (\boldsymbol{\nabla} \cdot [\boldsymbol{\tau} \cdot \mathbf{v}])$$

$$-(\boldsymbol{\nabla} \cdot p\mathbf{v}) + \sum_{i=1}^{n} (\mathbf{n}_i \cdot \mathbf{g}_i). \qquad (5.6\text{-}2)$$

Comparison with Eqs. 4.4-12 and 4.6-4* shows that the only difference is that with a mixture, we allow that each species may have a different density and mass flux and that each may be influenced by a different body force. The latter does not usually happen, since gravitational acceleration is the same regardless of species. However, if electrostatic or magnetic forces are involved, for example, various species will very likely be subjected to different body forces, depending on their electrostatic or magnetic properties.

Equations 5.6-1 and 5.6-2 can be simplified by the introduction of a *pressure tensor*, $\boldsymbol{\pi}$,

$$\boldsymbol{\pi} = \boldsymbol{\tau} + p\boldsymbol{\delta},$$

where $\boldsymbol{\delta}$ is the *unit tensor*,

$$\left\{ \begin{array}{ccc} 1 & 0 & 0 \\ 0 & 1 & 0 \\ 0 & 0 & 1 \end{array} \right\}.$$

This means that for

$$\boldsymbol{\tau} = \left\{ \begin{array}{ccc} \tau_{xx} & \tau_{xy} & \tau_{xz} \\ \tau_{yx} & \tau_{yy} & \tau_{yz} \\ \tau_{zx} & \tau_{zy} & \tau_{zz} \end{array} \right\},$$

$\boldsymbol{\pi}$ is

$$\boldsymbol{\pi} = \left\{ \begin{array}{ccc} p + \tau_{xx} & \tau_{xy} & \tau_{xz} \\ \tau_{yx} & p + \tau_{yy} & \tau_{yz} \\ \tau_{zx} & \tau_{zy} & p + \tau_{zz} \end{array} \right\}.$$

That is, the pressure is combined with the normal stresses. With this, Eqs.

*Also noting that $\rho(D\hat{\Phi}/Dt) = -\rho(\mathbf{v} \cdot \mathbf{g})$; see Eq. 4.6-8 on p. 152.

5.6-1 and 5.6-2 can be rewritten, respectively:

$$\rho \frac{D\mathbf{v}}{Dt} = -[\nabla \cdot \boldsymbol{\pi}] + \sum_{i=1}^{n} \rho_i \mathbf{g}_i \qquad (5.6\text{-}5)$$

$$\rho \frac{D}{Dt}\{U + \tfrac{1}{2}v^2\} = -(\nabla \cdot \mathbf{q}) - (\nabla \cdot [\boldsymbol{\pi} \cdot \mathbf{v}]) + \sum_{i=1}^{n} (\mathbf{n}_i \cdot \mathbf{g}_i). \qquad (5.6\text{-}6)$$

The Eulerian forms can be written similarly and will be given in the following section.

5.7 SUMMARY

Eulerian Form

Continuity
   Species $i(i = 1, 2, \ldots, n)$:

$$\frac{\partial \rho_i}{\partial t} = -(\nabla \cdot [\rho_i \mathbf{v} + \mathbf{j}_i]) + r_i \qquad (5.7\text{-}1)$$

   Mixture:*

$$\frac{\partial \rho}{\partial t} = -(\nabla \cdot \rho \mathbf{v}) \qquad (5.7\text{-}2)$$

Motion

$$\frac{\partial}{\partial t}\rho \mathbf{v} = -[\nabla \cdot \{\rho \mathbf{v}\mathbf{v} + \boldsymbol{\pi}\}] + \sum_{i=1}^{n} \rho_i \mathbf{g}_i \qquad (5.7\text{-}3)$$

Energy

$$\frac{\partial}{\partial t}\rho\{\hat{U} + \tfrac{1}{2}v^2\} = -\left(\nabla \cdot \{\rho(U + \tfrac{1}{2}v^2)\mathbf{v} + \mathbf{q} + [\boldsymbol{\pi} \cdot \mathbf{v}]\}\right)$$

$$+ \sum_{i=1}^{n} (\mathbf{n}_i \cdot \mathbf{g}_i) \qquad (5.7\text{-}4)$$

*This is so because $\sum_{i=1}^{n} r_i = 0$, as explained in Section 5.5, and $\sum_{i=1}^{n} j_i = 0$, an identity derived from the definition of j_i; see Table 5.3-2, p. 171.

Lagrangian Form:

Continuity

Species $i(i=1,2,\ldots,n)$:

$$\rho\frac{D\omega_i}{Dt} = -(\nabla\cdot\mathbf{j}_i)+r_i \tag{5.7-5}$$

Mixture:

$$\frac{D\rho}{Dt} = -\rho(\nabla\cdot\mathbf{v}) \tag{5.7-6}$$

Motion

$$\rho\frac{D\mathbf{v}}{Dt} = -[\nabla\cdot\boldsymbol{\pi}]+\sum_{i=1}^{n}\rho_i\mathbf{g}_i \tag{5.6-5}$$

Energy

$$\rho\frac{D}{Dt}\left\{\hat{U}+\tfrac{1}{2}v^2\right\} = -(\nabla\cdot\mathbf{q})-(\nabla\cdot[\boldsymbol{\pi}\cdot\mathbf{v}])$$

$$+\sum_{i=1}^{n}(\mathbf{n}_i\cdot\mathbf{g}_i) \tag{5.6-6}$$

While there are many variations of these equations and interesting phenomena associated with them, we have to limit ourselves to this abbreviated presentation so that we will not be distracted from discussing the practical examples and surveying as many subject areas as possible.

PROBLEMS

5.A Diffusional Coefficient for Disrupted Bacterial Cells [2]

E. coli bacteria are extruded through a French pressure cell in various types of media. The bacteria concentration distribution, c_A, in the long plastic tube relative to the initial concentration, c_{A_0}, has been found to be

$$\frac{c_A}{c_{A_0}} = \frac{1}{2}\left\{1+\mathrm{erf}\left(\frac{x}{2\sqrt{\mathfrak{D}_{AB}t}}\right)\right\}, \tag{5.A-1}$$

where x is the distance down the tube; t, the time since the beginning of the diffusion; $\mathfrak{D}_{AB}$, the diffusional constant of *E. coli* in the medium; and

$\mathrm{erf}[x/(2\sqrt{\mathfrak{D}_{ABt}}\,)]$ is the error function as defined in Example 2.13-1—that is,

$$\mathrm{erf}\left(\frac{x}{2\sqrt{\mathfrak{D}_{AB}t}}\right)=\frac{2}{\sqrt{\pi}}\int_{0}^{x/(2\sqrt{\mathfrak{D}_{AB}t}\,)}e^{-y^2}dy, \qquad (5.\text{A-}2)$$

where y is the "dummy variable."

(a) Why would the c_A/c_{A0} versus x plot appear a straight line on probability graph paper? What does this fact mean to us in terms of data correlation?

(b) Show that, by obtaining the numerical values for the error function in Table 2.13-1, the diffusional coefficient can be calculated from the percent lengths at which the concentration ratios are 0.1 and 0.9, as

$$\mathfrak{D}_{AB}=\frac{x^2}{4(1.82)^2 t}. \qquad (5.\text{A-}3)$$

(c) From the data in the following graphs, estimate the diffusional coefficients of *E. coli* in sucrose, dextran, and beta galactosidase.

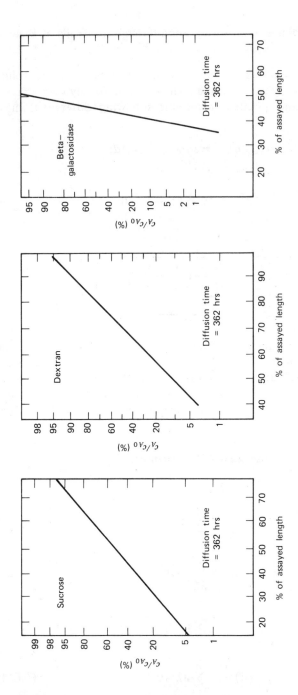

5.B Transfer of Solutes in a Moving Fluid through the Walls of Biological Tubes [3]

By modeling the renal tubule as such a thin tube that the solute concentration varies only axially while distributing uniformly over a cross-section, Bergmann and Dikstein made the following differential mass balance on the solute:

$$c_z v_z - c_{z+dz} v_{z+dz} = 2\pi r k_T (c_z - c_e) dz,$$

resulting in

$$c\frac{dv}{dz} + v\frac{dc}{dz} + 2\pi r k_T (c - c_e) = 0,$$

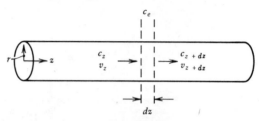

where k_T is the mass-transfer coefficient (mass or molar flux per unit concentration difference), v is the flow velocity, c_z and c_e are the solute concentrations at axial distance z and outside of the tubule, respectively. Analyze their derivation procedure, definitions, and results in view of the equation of continuity presented in this chapter. What assumptions have been implicitly made?

5.C Identities among Mass and Molar Quantities

Verify the following relationships:

$$\text{(a)} \qquad \sum_{i=1}^{n} N_i = cv \qquad\qquad (5.\text{C-}1)$$

$$\text{(b)} \qquad \sum_{i=1}^{n} j_i = 0 \qquad\qquad (5.\text{C-}2)$$

$$\text{(c)} \qquad \sum_{i=1}^{n} j_i^* = \rho(v - v^*) \qquad\qquad (5.\text{C-}3)$$

$$\text{(d)} \qquad \sum_{i=1}^{n} J_i^* = 0. \qquad\qquad (5.\text{C-}4)$$

What is the physical meaning of each of these relationships? Does it seem strange that two such similar quantities as $\mathbf{j}_i$ and $\mathbf{j}_i^*$, when summed, yield different results? Under what special condition would $\sum_{i=1}^{n} \mathbf{j}_i^*$ be equal to zero?

5.D Identities Between Mass and Molar Quantities

Verify that

$$\text{(a)} \qquad \mathbf{N}_i = \frac{\mathbf{j}_i}{M_i} + c_i \mathbf{v} \qquad\qquad \text{(5.D-1)}$$

$$\text{(b)} \qquad \mathbf{j}_i^* = \mathbf{J}_i M_i + x_i M_i \sum_{k=1}^{n} \mathbf{J}_k. \qquad\qquad \text{(5.D-2)}$$

What is the physical meaning of each of these relationships?

5.E Equation of Continuity for Multicomponent Systems

Derive the equation continuity in cylindrical coordinates for species i in a multicomponent mixture of n components by considering mass flux of i in and out of a cylindrical "pie" element as depicted in Figure 4.11-1. The procedure is suggested at the beginning of Section 5.5 and should result in something similar to Eq. B of Table 5.5-1, depending on whether mass or molar units are used.

REFERENCES

1. Bird, R. B.; Stewart, W. E.; and Lightfoot, E. N. *Transport Phenomena*. Wiley: New York, 1960, Chapter 16.
2. Lehman, R. C., and Pollard, E. *Biophys. J.*, **5**, 109 (1965).
3. Bergmann, F., and Dikstein, S. *J. Physiol.*, **145**, 14 (1959).

CHAPTER SIX
Biomedical Applications:
Equations of Change

Up to this point, we have briefly discussed the fundamental principles and derivations of the basic equations useful in solving an entire category of problems generally known as time-dependent distributed-parameter problems. Because the physical entities (concentration, velocity, and temperature) can generally vary from time to time and from place to place, these equations are necessarily differential in form. They are simply a collection of recipes from which working equations for specific problems can be obtained by reduction of terms. For instance, steady-state problems are those the time-derivative terms of which are zero. Another example: although Eq. F in Table 4.12-2(b) has a total of nine terms, only two of them will be left when applied to the Hagen–Poiseuille (H–P) flow (Example 6.1-1).

In future chapters, other aspects of transport phenomena—namely, interphase transport and macroscopic transport—and specific topics of physiological interest will be presented. However, it seems appropriate that at this point, where the differential aspect is concluded, we should take a few moments out to present some examples of its application.

It should be noted, however, that in this chapter we are attempting to cover as wide a range of problems as possible, and thus, we are necessarily arranging them in terms of physical phenomena (i.e., mass, momentum, or energy transport) instead of the physiological phenomena such as oxygen transport and hemodialysis. The latter type of coverage will be left to Chapters 9 and 10.

6.1 FLUID-FLOW PROBLEMS

The flow of fluids, being most easily envisioned and observed, will be used as our first group of examples. We will start with the simplest and most basic one, steady-state laminar Newtonian straight-tube flow—that is, H–P flow. Although this is highly idealized and, in some cases, not sufficiently realistic for direct physiological application, it does serve as the basis upon which more realistic problems are built. For instance, in Example 6.1-2 the time variation is added, while in Example 6.1-3 the time variation becomes periodic, which is quite relevant to arterial blood-flow problems. Then in Example 6.1-4, one more spatial variation is included that finds applications in seepage flow, such as the reabsorption process in renal tubules. As for the presence of cells in flow, we will leave it to Chapter 9, where the blood as a suspension is discussed.

Example 6.1-1   The Hagen–Poiseuille Flow.   In addition to the reason cited above, the use of the simple H–P flow as our first example carries special significance, because the analysis was first developed by, and thus named after, two scientists, one of whom was a physician [1]. In fact, there are ample cases in the literature that well-known engineering principles have been first developed from observing the various functions of the human body.

The H–P flow is the steady-state laminar flow of incompressible New-tonian fluid under a uniform force or pressure gradient. When mentioned without any other adjective, it is generally understood to be the flow in circular tube, which is the most common case and will be discussed in some detail here. The same type of flow occurring between two parallel plates is referred to as the *plane* H–P *flow*. Since derivation of the latter is very simple and similar to that of the former, we will merely present the result for the plane H–P flow case.

The H–P flow is graphically depicted in Figure 6.1-1, where the dimensions and coordinate system are also labeled. Since we are working with a cylindrical system we should start with Eq. B of Table 4.12-1. *p. 160* Furthermore, since the fluid is incompressible (constant ρ) and Newtonian (constant μ) and since the flow is in the z-direction, Eq. F in Table 4.12-2(b) should be used. The former gives us $\partial v_z / \partial z = 0$, since $\partial \rho / \partial t = 0$ — *steady state* (ρ constant) and $v_r = v_\theta = 0$ (flow in z-direction only). This is entirely reasonable and reinforces the fact that in steady-state incompressible flow in a straight tube of uniform cross-section, the velocity does not change in the direction of the flow, except near the entrance and exit regions.

In reducing the latter [Eq. F, Table 4.12-2(b)], which is the z-direction equation of motion, we note the following:

$$\frac{\partial v_z}{\partial t} = 0 \quad \text{(steady state)}$$

$$v_r = v_\theta = 0 \quad \text{and} \quad \frac{\partial v_z}{\partial z} = 0 \quad \text{(explained above)}$$

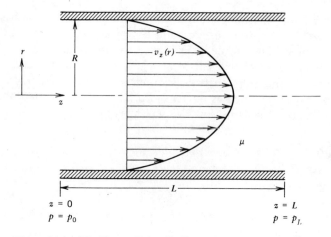

Figure 6.1-1.   The Hagen–Poiseuille flow.

$$\frac{\partial^2 v_z}{\partial \theta^2} = 0 \qquad (v_z \text{ not a function of } \theta \text{ because of radial symmetry})$$

$$\frac{\partial^2 v_z}{\partial z^2} = 0 \qquad \left(\because \frac{\partial v_z}{\partial z} = 0 \right)$$

$$g_z = 0 \qquad (\text{no body force in horizontal direction}).$$

Thus, we have

$$\frac{1}{r} \frac{d}{dr} \left(r \frac{dv_z}{dr} \right) = \frac{1}{\mu} \frac{dp}{dz}. \qquad (6.1\text{-}1)$$

The partial derivatives become total derivatives because v_z is now a function of r only and p that of z only. Since the tube is of uniform cross section, the pressure drop Δp ($= p_0 - p_L$) is evenly distributed over the tube length of L, which means

$$\frac{dp}{dz} = -\frac{\Delta p}{L} = \text{constant}. \qquad (6.1\text{-}2)$$

The minus sign is included because pressure (p_0) at $z = 0$ is higher than that (p_L) at $z = L$ (i.e., the pressure decreases with increasing axial distance).

Now Eq. 6.1-1 becomes

$$\frac{d}{dr}\left(r\frac{dv_z}{dr}\right) = -\frac{\Delta p}{\mu L}r,$$

which can be easily integrated to give

$$r\frac{dv_z}{dr} = -\frac{\Delta p}{2\mu L}r^2 + C_1. \qquad (6.1\text{-}3)$$

Using the boundary condition that dv_z/dr must be finite at $r=0$, we can easily conclude that C_1 must be equal to zero. Then dividing both sides of Eq. 6.1-3 by r and integrating again, we obtain

$$v_z = -\frac{\Delta p r^2}{4\mu L} + C_2. \qquad (6.1\text{-}4)$$

Applying the second boundary condition that $v_z=0$ at $r=R$ (which means no slipping of fluid at wall), we have

$$C_2 = \frac{\Delta p R^2}{4\mu L},$$

so that by substituting it back into Eq. 6.1-4, we have

$$v_z = \frac{\Delta p R^2}{4\mu L}\left[1 - \left(\frac{r}{R}\right)^2\right]. \qquad (6.1\text{-}5)$$

This is the parabolic velocity profile roughly sketched in Figure 6.1-1 for a certain value of $\Delta p R^2/4\mu L$. The implication here is that for each value of $\Delta p R^2/4\mu L$, we have a different parabola. (See also Figure 6.2-3.)

The parabolic nature can be more easily observed if we rewrite Eq. 6.1-5 in *dimensionless form* by letting

$$\phi = \frac{v_z}{\Delta p R^2/4\mu L} = \frac{v_z}{v_{z\,\text{max}}} \qquad \text{(dimensionless velocity)}$$

$$\xi = \frac{r}{R}, \qquad \text{(dimensionless radial coordinate)}$$

so that

$$\phi = 1 - \xi^2. \qquad (6.1\text{-}6)$$

Now we can see that in this form, a single parabola is sufficient to represent all cases because the effect of $\Delta p R^2/4\mu L$ has been incorporated in ϕ. (See also Figure 6.2-4.)

At the above we have indicated that the quantity $\Delta p R^2/4\mu L$ is the maximum velocity $v_{z_{max}}$ because if we let $dv_z/dr=0$ in conjunction with Eq. 6.1-5, we know that the maximum should occur at $r=0$ (according to Section 2.3;* actually this should be intuitively apparent). Using $r=0$ in Eq. 6.1-5 yields

$$v_{z_{max}} = \frac{\Delta p R^2}{4\mu L}.$$

It is of interest to note that according to the Newton's law of viscosity and Eq. 6.1-5,

$$\tau_{rz} = -\mu \frac{dv_z}{dr} = \frac{\Delta p r}{2L}$$

and the shear force at the wall is

$$\tau_{rz}|_{r=R} \cdot 2\pi R L = \Delta p \cdot \pi R^2,$$

which means that to maintain a steady flow a pressure force must be present that exactly overcomes the friction between the fluid and the tube.

Normally two other quantities are of major interest in a hydrodynamic problem such as this. The volumetric flow rate is simply the velocities at various points of the cross section summed over the entire cross section. Since the velocity profile is a continuous function, the process of integration, instead of summation, is used. The procedure has been explained in Section 2.7, so only the formulation will be presented here:

$$Q = \int_0^{2\pi} \int_0^R v_z(r)\, r\, dr\, d\theta$$

$$= 2\pi \left(\frac{\Delta p R^2}{4\mu L} \right) \int_0^R \left[1 - \left(\frac{r}{R} \right)^2 \right] r\, dr$$

$$= 2\pi \left(\frac{\Delta p R^2}{4\mu L} \right) R^2 \int_0^1 (1-\xi^2)\xi\, d\xi$$

$$\therefore Q = \frac{\Delta p \pi R^4}{8\mu L}. \tag{6.1-7}$$

*If we follow the principle strictly, we should take the second derivative as well. In this case, it is equal to $-\Delta p/2\mu L$ and inherently negative for all positive Δp. This means that there is only a maximum, no minimum, as the case should intuitively be.

This is the celebrated H–P equation relating the pressure drop to the flow rate, one of the basic equations in engineering work. It can also be used for any Newtonian biological fluid flowing laminarly in a circular tube of uniform diameter.

The average velocity is simply the volumetric flow rate divided by the cross-sectional area through which the fluid flows, or

$$\langle v_z \rangle = \frac{Q}{\pi R^2} = \frac{\Delta p R^2}{8 \mu L}. \tag{6.1-8}$$

Although the above pertains to horizontal flow under pressure gradient, the expressions can easily be modified to describe vertical and inclined flows under a combination of body forces and pressure gradient. For vertical flow, the gravitational force plays an important role; in this case we write

$$\Delta \mathscr{P} = \mathscr{P}_0 - \mathscr{P}_L, \tag{6.1-9}$$

where

$$\mathscr{P} = p - \rho g z. \tag{6.1-10}$$

Thus,

$$\Delta \mathscr{P} = \Delta p + \rho g L$$

or

$$\frac{\Delta \mathscr{P}}{L} = \frac{\Delta p}{L} + \rho g. \tag{6.1-11}$$

If the tube or the plane is on an incline which makes an angle β with the direction of gravity, Eq. 6.1-11 becomes

$$\frac{\Delta \mathscr{P}}{L} = \frac{\Delta p}{L} + \rho g \cos \beta, \tag{6.1-12}$$

which is the most general expression to be used for dp/dz in equations such as Eq. 6.1-1.

In the plane H–P flow, shown in Figure 6.1-2 the Newtonian fluid flows between two parallel plates that are so large that the fluid flow in the x-direction can be neglected (i.e., the flow is again one-dimensional in the z-direction). The thickness of the flow, or the distance between the plates, is $2B$. Other conditions remain the same as those in the tube flow. The result is listed in Table 6.1-1, where equivalent expression for the tube flow are also given as a comparison. The friction factor is included here merely to make the table more complete. Its meaning will be explained in Chapter

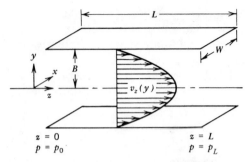

Figure 6.1-2.   The plane Hagen–Poiseuille flow.

7. It should also be noted that although in both the circular-tube and plane cases the velocity profiles are parabolic, their average velocities are different fractions of their respective maximum velocities, because of the different geometries involved.

Example 6.1-2   Unsteady-State Flow[2]. Although steady state is the ideal situation, almost all processes are unsteady-state, or time-dependent, in nature. Even a well-controlled chemical reactor or storage tank experiences slight fluctuations in temperature, tank level, product quality, and/or quantity, not to speak of the start-up and shut-down variations. This is even more true with physiological processes, where the unpredictability of biological beings makes the consideration of unsteady state necessary. We shall use the start-up of the laminar incompressible flow in circular tube as an example because it is an extension of the previous example of H–P flow. In this connection we might term the present case the unsteady-state H–P flow.

In this example an incompressible fluid initially rests in a circular tube. At time $t=0$ a pressure gradient of $\Delta p / L$ is applied to set the fluid in motion, so that eventually it becomes an H–P flow. Between the initial moment ($t=0$) and this eventuality (steady-state, $t=\infty$) there is a continuous and gradual development from a flat zero-velocity profile to the fully parabolic profile characteristic of the H–P flow. The example gives an expression for these intermediate profiles as they vary continuously with time.

Naturally, we again start with Eq. B of Table 4.12-1 and Eq. F of Table 4.12-2(b), except that in the present case the term $\partial v_z / \partial t$ does *not* vanish. However, $\partial \rho / \partial t$ is still equal to zero because of the constancy of ρ. Thus, in correspondence to Eq. 6.1-1, we have

$$\rho \frac{\partial v_z}{\partial t} = \frac{\Delta p}{L} + \frac{\mu}{r} \frac{\partial}{\partial r} \left(r \frac{\partial v_z}{\partial r} \right),$$

(6.1-13)

Table 6.1-1 Comparison of H–P Flows in Circular Tubes and Between Parallel Planes

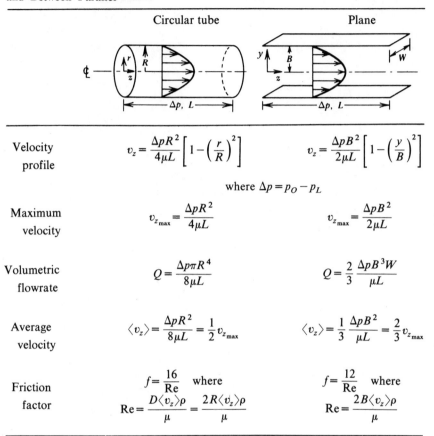

	Circular tube	Plane
Velocity profile	$v_z = \dfrac{\Delta p R^2}{4\mu L}\left[1-\left(\dfrac{r}{R}\right)^2\right]$	$v_z = \dfrac{\Delta p B^2}{2\mu L}\left[1-\left(\dfrac{y}{B}\right)^2\right]$

$$\text{where } \Delta p = p_O - p_L$$

	Circular tube	Plane
Maximum velocity	$v_{z\,max} = \dfrac{\Delta p R^2}{4\mu L}$	$v_{z\,max} = \dfrac{\Delta p B^2}{2\mu L}$
Volumetric flowrate	$Q = \dfrac{\Delta p \pi R^4}{8\mu L}$	$Q = \dfrac{2}{3}\dfrac{\Delta p B^3 W}{\mu L}$
Average velocity	$\langle v_z \rangle = \dfrac{\Delta p R^2}{8\mu L} = \dfrac{1}{2}v_{z\,max}$	$\langle v_z \rangle = \dfrac{1}{3}\dfrac{\Delta p B^2}{\mu L} = \dfrac{2}{3}v_{z\,max}$
Friction factor	$f = \dfrac{16}{\text{Re}}$ where $\text{Re} = \dfrac{D\langle v_z \rangle \rho}{\mu} = \dfrac{2R\langle v_z \rangle \rho}{\mu}$	$f = \dfrac{12}{\text{Re}}$ where $\text{Re} = \dfrac{2B\langle v_z \rangle \rho}{\mu}$

where the retention of the partial-derivative form should be carefully noted.

In addition to the two boundary conditions, an initial condition is also needed and given. They are

$$\text{I.C.:} \quad \text{at } t=0,\ v_z=0$$

$$\text{B.C.1:} \quad \text{at } r=0,\ v_z=\text{finite}$$

$$\text{B.C.2:} \quad \text{at } r=R,\ v_z=0.$$

The "classical" method of solving this problem is the separation of variable method briefly mentioned in Section 2.13, although the Laplace transform can also be used. We shall present the former here because this will be tied in with the use of the "unaccomplished" ratios in analogous heat- and mass-transfer problems. The use of the Laplace transform is left to the reader.

If we use the same dimensionless quantities as in the previous example, that is,

$$\phi = \frac{v_z}{\Delta p R^2/4\mu L} \quad \text{and} \quad \xi = \frac{r}{R}$$

plus a dimensionless time

$$\tau = \frac{\nu t}{R^2}, \quad \left(\nu = \frac{\mu}{\rho}, \text{kinematic viscosity}\right).$$

Eq. 6.1-13 and its associated initial and boundary conditions become

$$\frac{\partial \phi}{\partial \tau} = 4 + \frac{1}{\xi}\frac{\partial}{\partial \xi}\left(\xi \frac{\partial \phi}{\partial \xi}\right) \tag{6.1-14}$$

I.C.:   at $\tau = 0$, $\phi = 0$

B.C.1:   at $\xi = 0$, $\phi = $ finite

B.C.2:   at $\xi = 1$, $\phi = 0$.

The "cleanliness" of these expressions makes the advantage of using dimensionless quantities apparent. We can also recognize that the solution in the previous example, Eq. 6.1-6, must be a special one in the current problem, for $\partial \phi/\partial \tau = 0$ (i.e., steady state or $\tau = \infty$). Thus, we may write

$$\phi_\infty = \phi(\tau = \infty) = 1 - \xi^2, \tag{6.1-15}$$

but remembering that generally ϕ is a function of both ξ and τ.

We can now further write

$$\phi(\xi,\tau) = \phi_\infty(\xi) - \phi_t(\xi,\tau), \tag{6.1-16}$$

where ϕ_t is the transient velocity meaning "how far away are we from our goal—the eventual, steady state." It turns out that ϕ_t is a separable variable yielding a useful solution. Although ϕ may be theoretically separable, the subsequent process would run up to a dead end.

If now we insert Eq. 6.1-15 into Eq. 6.1-16 and then substitute into Eq.

6.1-14, we will obtain

$$\frac{\partial \phi_t}{\partial \tau} = \frac{1}{\xi} \frac{\partial}{\partial \xi} \left(\xi \frac{\partial \phi_t}{\partial \xi} \right), \tag{6.1-17}$$

for which the initial and boundary conditions are

    IC.:   at $\tau = 0$, $\phi_t = \phi_\infty$ (since $\phi = 0$ in Eq. 6.1-16)

    B.C.1:   at $\xi = 0$, $\phi_t = $ finite

    B.C.2:   at $\xi = 1$, $\phi_t = 0$ (since both ϕ and ϕ_∞ are zero in
                                             Eq. 6.1-16).

    Authors of most research papers and reference texts at this point would state that a separation of variable of the form

$$\phi_t(\xi, \tau) = f(\xi) \cdot g(\tau)$$

can be attempted. Actually, accumulated experience has told us that in most problems of this type, the definition of a transient quantity such as ϕ_t almost assures successful solution by the separation-of-variable method. The unaccomplished temperature and concentration ratios in analogous heat- and mass-transfer problems are also transient quantities (Examples 6.2-4, 6.2-5, 6.3-2(b) 6.3-3, 10.4-1, and 10.5-1).

    Using the above separation formula in Eq. 6.1-17 yields

$$\frac{1}{g} \frac{dg}{d\tau} = \frac{1}{f} \cdot \frac{1}{\xi} \frac{d}{d\xi} \left(\xi \frac{df}{d\xi} \right), \tag{6.1-18}$$

the left-hand side of which is a function of τ only and the right-hand side that of ξ only (note the change from partial to total derivatives in Eq. 6.1-18). For this equation to hold at all times, both sides must be equal to a constant, which we shall call $-k^2$. This means that this constant must be negative, and the reason for it will be apparent after we solve the first part of Eq. 6.1-18, which is now

$$\frac{1}{g} \frac{dg}{d\tau} = -k^2. \tag{6.1-19}$$

The solution to this simple first-order ordinary differential equation is (see Section 2.8)

$$g = C_1 e^{-k^2\tau}, \tag{6.1-20}$$

from which we can see that if $-k^2$ were exchanged for a positive quantity, g, and consequently ϕ_t, would *not* diminish to zero at steady state ($\tau = \infty$). In fact, it would infinitely increase with time, which is certainly unreasonable for this case.

We should also note that there is no initial condition associated with Eq. 6.1-19 because we do not know what g should be when $\phi_t = \phi_\infty$ at $\tau = 0$. Thus, we leave Eq. 6.1-20 with a "dangling" integration constant C_1 with the hope that we will be able to determine it later, which turns out to be the case.

After establishing the nature of this constant $-k^2$, we can proceed to solve the remainder of Eq. 6.1-18, which is

$$\frac{1}{\xi}\frac{d}{d\xi}\left(\xi\frac{df}{d\xi}\right) + k^2 f = 0 \qquad (6.1\text{-}21)$$

with the following boundary conditions:

B.C.1:   at $\xi = 0, f = $ finite

B.C.2:   at $\xi = 1, f = 0$ ($\therefore \phi_t = 0$ for all g).

According to Section 2.9, the solution of Eq. 6.1-21 is

$$f = C_2 J_0(k\xi) + C_3 Y_0(k\xi),$$

where J_0 and Y_0 are the Bessel functions of the first and second kinds, respectively. Using B.C.1, we find that C_3 must be equal to zero, for otherwise, the second term in this solution would not be finite as required, because $Y_0(0) = -\infty$, according to Figure 2.9-1(b). Subsequent application of B.C.2 to the remainder of this solution yields

$$C_2 J_0(k) = 0.$$

Since C_2 can no longer be zero (otherwise, there would be no solution, or, more correctly, there would be trivial solution $f = 0$), the only way for this equation to be fulfilled is

$$J_0(k) = 0.$$

According to Figure 2.9-1(a), this last equation is fulfilled when, and only when, the J_0 curve cuts across the horizontal axis, which occurs an infinite number of times. As briefly explained in Section 2.9 this means that we have the roots or "zeros" of the function $J_0(k)$ at $k_1 = 2.40483\ldots$, $k_2 = 5.52009\ldots$, $k_3 = 8.65373\ldots$, and so on. Still another way of putting it is that

k can take on only these values $k_1, k_2, k_3, \ldots$.

Thus, we have an infinite number of solutions for Eq. 6.1-21:

$$f_n = C_{2_n} J_0(k_n \xi), \qquad n = 1, 2, 3, \ldots, \infty$$

or for Eq. 6.1-17:

$$\phi_{t_n} = C_n' J_0(k_n \xi) e^{-k_n^2 \tau}, \qquad n = 1, 2, 3, \ldots, \infty \qquad (6.1\text{-}22)$$

where $C_n' = C_1 C_{2_n}$. We note here that B.C.2 did not give us the value of the integration constant C_2. Instead, it only specified the requirement for k. But we did not actually lose anything, because the two dangling constants, C_1 and C_{2_n}, are subsequently combined, to be solved by the initial condition.

However, Eq. 6.1-22 is not ready to be used with the initial condition as the latter is for ϕ_t, which, according to the principle of superimposition (Section 2.8), is a linear combination of all the ϕ_{t_n}, or

$$\phi_t = \sum_{n=1}^{\infty} L_n \phi_{t_n} \qquad \text{(where all the } L_n \text{ are constants)}$$

$$= \sum_{n=1}^{\infty} C_n'' J_0(k_n \xi) e^{-k_n^2 \tau}, \qquad (6.1\text{-}23)$$

where $C_n'' = L_n C_n'$. The application of the I.C. then yields

$$1 - \xi^2 = \sum_{n=1}^{\infty} C_n'' J_0(k_n \xi). \qquad (6.1\text{-}24)$$

It would be a grave mistake here simply to take a shortcut by declaring $C_n'' = (1 - \xi^2)/\Sigma J_0(k_n \xi)$ because C_n'' is inside the summation sign, which means

$$1 - \xi^2 = C_1'' J_0(k_1 \xi) + C_2'' J_0(k_2 \xi) + \cdots.$$

Use must then be made of the orthogonality relationships described in Section 2.11.

Referring to Table 2.9-2(D), we find that since all the k_n are the roots of $J_0(k_n) = 0$, formula (1) of the orthogonality relationships is applicable, whereupon we multiply both sides of Eq. 6.1-24 by $\xi J_0(k_m \xi)$ and integrate from 0 to 1. This yields

$$\int_0^1 \xi(1-\xi^2)J_0(k_m\xi)\,d\xi = \sum_{n=1}^{\infty} C_n'' \int_0^1 \xi J_0(k_m\xi)J_0(k_n\xi)\,d\xi$$

$$= \sum_{n=1}^{\infty} C_n'' \cdot \tfrac{1}{2}[J_0'(k_n)]^2 \delta_{mn}, \qquad (6.1\text{-}25)$$

in which the order of integration and summation is reversed, as they do not interfere with each other.

The Kronecker delta, δ_{mn}, is the mathematical symbol for saying what orthogonality means—that it vanishes unless $m=n$ for which $\delta_{mm}=1$. Eq. 6.1-25 says that of all the terms, only the term where $m=n$ survives. This means that Eq. 6.1-25 has become

$$\int_0^1 \xi(1-\xi^2)J_0(k_m\xi)\,d\xi = C_m'' \cdot \tfrac{1}{2}[J_0'(k_m)]^2. \qquad (6.1\text{-}26)$$

It involves only mechanical acts now, plus appropriate relationships in Table 2.9-2, to solve this last equation to obtain (see Appendix A)

$$C_m'' = \frac{8}{k_m^3 J_1(k_m)}. \qquad (6.1\text{-}27)$$

Realizing that m and n are merely dummy indices, we can finally combine Eqs. 6.1-15, 6.1-16, 6.1-23, and 6.1-27 to form the final solution:

$$\phi(\xi,\tau) = (1-\xi^2) - 8\sum_{n=1}^{\infty} \frac{J_0(k_n\xi)}{k_n^3 J_1(k_n)} e^{-k_n^2\tau} \qquad (6.1\text{-}28)$$

Figure 6.1-3 is a plot of Eq. 6.1-28 showing the various stages of the velocity profiles as they build up from zero to the fully parabolic one at $\tau = \infty$. The advantage of using dimensionless quantities should be evident here as the plotting of $\mu, \rho, t, R, \Delta p,$ and L individually has been spared. Dimensionless grouping allows their consolidation into only three variables that can be condensed on a single chart and with values that do *not* depend on the particular units used.

Example 6.1-3 Pulsatile Blood Flow. We are gradually taking up examples and problems more realistic in nature. The present one deals with pulsatile flow with a sinusoidal pressure gradient and is a slightly more generalized form of the analysis by Hershey and Song [3]. Although the

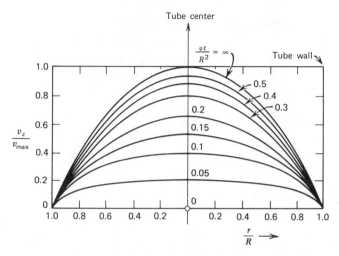

Figure 6.1-3. Unsteady-state velocity profiles for start-up flow [2].

time variation of blood pressure does not exactly follow a sine curve, it is a good illustration of the general nature of the flow and of the time lag between the pressure gradient and the flow rate. Fitting of the present model to more realistic data can be found elsewhere [4].

Starting again with the governing equations of continuity and motion, Eq. B of Table 4.12-1 and Eq. F of Table 4.12-2(b), we arrive at almost the same equation as Eq. 6.1-13 except that in the present case, the pressure gradient is no longer constant, but a sinusoidal function—that is,

$$-\frac{dp}{dz} = \frac{\Delta p}{L} + A \sin \omega t, \qquad (6.1\text{-}29)$$

where $\Delta p / L$ is the average pressure gradient level; A, the amplitude; and ω, the frequency of the periodic pressure fluctuation. This pulse function is shown in Figure 6.1-4. Note that the period, Δt, of fluctuation is related to frequency by $\Delta t = 2\pi/\omega$.

With Eq. 6.1-29 the equation of motion becomes

$$\frac{1}{\nu}\frac{\partial v_z}{\partial t} = \frac{1}{r}\frac{\partial}{\partial r}\left(r\frac{\partial v_z}{\partial r}\right) + \frac{\Delta p}{\mu L} + \frac{A}{\mu}\sin \omega t, \qquad (6.1\text{-}30)$$

where again $\nu = \mu/\rho$. It is quite obvious the solution for v_z consists of a steady-state portion and a periodic portion, corresponding to the average

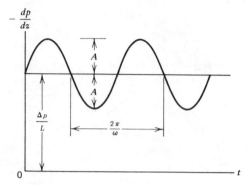

Figure 6.1-4. Idealized (sinusoidal) pressure pulse function.

and fluctuating pressure gradients, or

$$v_z(r,t) = v_{z_s}(r) + v_{z_p}(r,t). \tag{6.1-31}$$

Similar to the treatment in Example 6.1-2, it can be easily shown that when Eq. 6.1-31 is substituted into Eq. 6.1-30, two equations result:

$$0 = \frac{1}{r}\frac{d}{dr}\left(r\frac{dv_{z_s}}{dr}\right) + \frac{\Delta p}{\mu L}, \tag{6.1-32}$$

$$\frac{1}{\nu}\frac{\partial v_{z_p}}{\partial t} = \frac{1}{r}\frac{\partial}{\partial r}\left(r\frac{\partial v_{z_p}}{\partial r}\right) + \frac{A}{\mu}\sin\omega t. \tag{6.1-33}$$

The solution for Eq. 6.1-32 has been repeatedly given in Examples 6.1-1 and 6.1-2—that is,

$$v_{z_s}(r) = \frac{\Delta p R^2}{4\mu L}\left[1 - \left(\frac{r}{R}\right)^2\right], \tag{6.1-34}$$

while Eq. 6.1-33 can be solved by using the Laplace transform method. By noting the initial condition that at $t=0$, $v_{z_p}=0$, the Laplace transform of the entire Eq. 6.1-33 is

$$\frac{1}{\nu}p\bar{v}_{z_p} = \frac{1}{r}\frac{d}{dr}\left(r\frac{d\bar{v}_{z_p}}{dr}\right) + \frac{A}{\mu}\cdot\frac{\omega}{p^2+\omega^2}.* \tag{6.1-35}$$

*The author regrets any possible confusion here but the "p" in this Laplace transform development up to and including Eq. 6.1-44 refers to the image p-space, not pressure. Also the subscript "p" stands for "periodic."

Note here that v_{z_p} is originally a function of r and t. But after its transformation from t-space to the image p-space, the transformed variable $\bar{v}_{z_p}$ is now a function of r and p. Since $\bar{v}_{z_p}$ is no longer a function of t, the partial derivative with respect to r has now become a total derivative (see also Example 2.13-2).

The initial condition having been used, we are left with only the boundary conditions, which, after transformation, are

$$\text{B.C.1 at} \quad r=0, \quad \frac{d\bar{v}_{z_p}}{dr}=0, \quad \text{or } \bar{v}_{z_p}=\text{finite}$$

$$\text{B.C.2 at} \quad r=R, \quad \bar{v}_{z_p}=0.$$

According to Section 2.9, the *homogeneous* solution, the solution without the $(A/\mu)\cdot\omega/(p^2+\omega^2)$ term, to Eq. 6.1-35 is

$$\left(\bar{v}_{z_p}\right)_h = C_1 I_0\left(\sqrt{\frac{p}{\nu}}\,r\right) + C_2 K_0\left(\sqrt{\frac{p}{\nu}}\,r\right),$$

where I_0 and K_0 are the modified Bessel functions of the first and second kinds, respectively, and of zero order. For the *particular integral*, the trial solution is (see Table 2.8-2)

$$\left(\bar{v}_{z_p}\right)_p = B,^\dagger$$

where B is *not* a function of r. Substituting $\bar{v}_{z_p}=(\bar{v}_z)_h+(\bar{v}_{z_p})_p$ into Eq. 6.1-35 and equating coefficients of like power terms, we have

$$B = \frac{A}{\rho}\cdot\frac{\omega}{p(p^2+\omega^2)},$$

so that the *complete* solution is

$$\bar{v}_{z_p} = C_1 I_0\left(\sqrt{\frac{p}{\nu}}\,r\right) + C_2 K_0\left(\sqrt{\frac{p}{\nu}}\,r\right) + \frac{A}{\rho}\cdot\frac{\omega}{p(p^2+\omega^2)}. \quad (6.1\text{-}36)$$

From Figure 2.9-1(d) we note that $K_0(\sqrt{p/\nu}\,r)$ approaches infinity if $r\rightarrow0$, which would be in direct violation of B.C.1. This implies that C_2 must be

†Here the subscript p outside the parentheses stands for "particular integral."

equal to zero. Using B.C.2, we have

$$0 = C_1 I_0\left(\sqrt{\frac{p}{\nu}}\ R\right) + \frac{A}{\rho} \cdot \frac{\omega}{p(p^2+\omega^2)}\ .$$

Solving for C_1 and substituting it back into Eq. 6.1-36, while remembering that $C_2=0$, give us

$$\bar{v}_{z_p} = \frac{A}{\rho} \cdot \frac{\omega}{p(p^2+\omega^2)}\left[1 - \frac{I_0\left(\sqrt{p/\nu}\ r\right)}{I_0\left(\sqrt{p/\nu}\ R\right)}\right]. \qquad (6.1\text{-}37)$$

This means

$$v_{z_p}(r,t) = \mathcal{L}^{-1}\left\{\bar{v}_{z_p}(r,p)\right\}$$

$$= \frac{A}{\rho}\mathcal{L}^{-1}\left\{\frac{\omega}{p(p^2+\omega^2)}\left[1 - \frac{I_0\left(\sqrt{p/\nu}\ r\right)}{I_0\left(\sqrt{p/\nu}\ R\right)}\right]\right\}, \qquad (6.1\text{-}38)$$

which seems less horrifying if we realize that

$$\frac{\omega}{p(p^2+\omega^2)}\left[1 - \frac{I_0\left(\sqrt{p/\nu}\ r\right)}{I_0\left(\sqrt{p/\nu}\ R\right)}\right] = \frac{\omega}{p^2+\omega^2}\cdot\frac{I_0\left(\sqrt{p/\nu}\ R\right) - I_0\left(\sqrt{p/\nu}\ r\right)}{pI_0\left(\sqrt{p/\nu}\ R\right)}$$

$$= \bar{f}(p)\cdot\bar{g}(p), \qquad (6.1\text{-}39)$$

where

$$\bar{f}(p) = \frac{\omega}{p^2+\omega^2} \qquad (6.1\text{-}40)$$

$$\bar{g}(p) = \frac{I_0\left(\sqrt{p/\nu}\ R\right) - I_0\left(\sqrt{p/\nu}\ r\right)}{pI_0\left(\sqrt{p/\nu}\ R\right)}. \qquad (6.1\text{-}41)$$

The inverse of Eq. 6.1-40 is simply

$$f(t) = \mathcal{L}^{-1}\left\{\bar{f}(p)\right\} = \sin\omega t, \qquad (6.1\text{-}42)$$

while the ever-reliable Heaviside partial fraction expansion theorem gives us (see Appendix B)

$$g(t) = \mathcal{L}^{-1}\{\bar{g}(p)\} = 2 \sum_{k=1}^{\infty} \frac{J_0(\alpha_k r/R)}{\alpha_k J_1(\alpha_k)} \exp\left(\frac{-\alpha_k^2 \nu t}{R^2}\right), \quad (6.1\text{-}43)$$

where J_0 and J_1 are the Bessel functions of the first kind and, respectively, of the zero and first orders, with α_k being the roots of $J_0(\alpha_k)=0$; $\alpha_1 = 2.40483\ldots, \alpha_2 = 5.52009\ldots, \alpha_3 = 8.65373\ldots$, and so on. It is also of interest to note that the quantities ν, t, and R group automatically into the dimensionless time $\tau = \nu t/R^2$.

Using the *convolution integral*, formula A10 in Table 2.12-1, we can write

$$v_{z_p}(r,t) = \mathcal{L}^{-1}\{\bar{v}_{z_p}\} = \frac{A}{\rho}\mathcal{L}^{-1}\{\bar{f}(p)\bar{g}(p)\}$$

$$= \frac{A}{\rho}\int_0^t f(t-s)g(s)\,ds$$

$$= \frac{2A}{\rho}\sum_{k=1}^{\infty}\frac{J_0(\alpha_k r/R)}{\alpha_k J_1(\alpha_k)}F(\alpha_k,t), \quad (6.1\text{-}44)$$

where

$$F(\alpha_k,t) = \int_0^t \exp\left(\frac{-\alpha_k^2 \nu s}{R^2}\right)\sin\omega(t-s)\,ds. \quad (6.1\text{-}45)$$

This integration can be performed after expanding $\sin\omega(t-s)$ using standard trigonometry relationships. A more concise procedure is to let $t-s = u$, so that $ds = -du$ and that $u=t$ when $s=0$ and $u=0$ when $s=t$. Making these substitutions, we have (via Eq. 17, Table 2.4-1)

$$F(\alpha_k,t) = -\exp\left(\frac{-\alpha_k^2 \nu t}{R^2}\right)\int_t^0 \exp\left(\frac{\alpha_k^2 \nu u}{R^2}\right)\sin\omega u\,du$$

$$= \left[\exp\left(\frac{-\alpha_k^2 \nu t}{R^2}\right)\right]\left[\frac{\exp(\alpha_k^2 \nu u/R^2)((\alpha_k^2 \nu/R^2)\sin\omega u - \omega\cos\omega u)}{(\alpha_k^2 \nu/R^2)^2 + \omega^2}\right]_0^t$$

$$
= \frac{\omega \exp(-\alpha_k^2 \nu t/R^2) + (\alpha_k^2 \nu/R^2)\sin \omega t - \omega \cos \omega t}{(\alpha_k^4 \nu^2/R^4) + \omega^2}
$$

$$
= \frac{\omega \exp(-\alpha_k^2 \nu t/R^2)}{(\alpha_k^4 \nu^2/R^4) + \omega^2} + \frac{\sin(\omega t - \phi)}{\sqrt{(\alpha_k^4 \nu^2/R^4) + \omega^2}}, \tag{6.1-46}
$$

where

$$
\phi = \tan^{-1}\left(\frac{\omega R^2}{\alpha_k^2 \nu}\right). \tag{6.1-47}
$$

Therefore, the complete solution is

$$
v_z(r,t) = \frac{\Delta p R^2}{4\mu L}\left[1 - \left(\frac{r}{R}\right)^2\right] + \frac{2A}{\rho}\sum_{k=1}^{\infty} \frac{J_0(\alpha_k r/R)}{\alpha_k J_1(\alpha_k)}
$$

$$
\cdot \left\{\frac{\omega \exp(-\alpha_k^2 \nu t/R^2)}{(\alpha_k^4 \nu^2/R^4) + \omega^2} + \frac{\sin(\omega t - \phi)}{\sqrt{(\alpha_k^4 \nu^2/R^4) + \omega^2}}\right\}. \tag{6.1-48}
$$

The periodically changing velocity profiles are plotted in Figure 6.1-5. We can see that the first term on the right-hand side of Eq. 6.1-48 is the steady-state velocity distribution. The Bessel-function group multiplied by the exponential decay term is equivalent to the transient velocity due to initial start-up (see Eq. 6.1-28). In this case, the transient increases with increasing frequency but dies down rather quickly as time elapses.

The sine function contributes to the periodic nature of the solution. But the argument is $(\omega t - \phi)$ instead of ωt alone. This means that because of non-instant momentum transport, the velocity (response) lags the pressure gradient (driving potential) by a fractional period equal to $\phi/2\pi$ and ϕ is called the *phase angle*. In the ideal case of infinite viscosity, $\nu = \infty$ and $\phi = 0$ (i.e., there would be no time lag because of instant momentum transport). In another special case, $\omega = 0$ and the entire expression reverts to that of the steady-state H–P flow.

The volumetric flow rate can be obtained by integrating Eq. 6.1-48 over the circular cross-sectional area—that is,

$$
Q = \int_0^{2\pi}\int_0^R v_z(r,t)\, r\, dr\, d\theta
$$

$$
= 2\pi R^2 \int_0^1 \left[\frac{\Delta p R^2}{4\mu L}(1-\xi^2) - \frac{2A}{\rho}\sum_{k=1}^{\infty}\frac{J_0(\alpha_k \xi)}{\alpha_k J_1(\alpha_k)} F(\alpha_k, t)\right]\xi\, d\xi
$$

$$= \frac{\pi \Delta p R^4}{8\mu L} - \frac{4\pi R^2 A}{\rho} \sum_{k=1}^{\infty} \frac{F(\alpha_k, t)}{\alpha_k J_1(\alpha_k)} \int_0^1 \xi J_0(\alpha_k \xi) d\xi. \tag{6.1-49}$$

By setting $\eta = \alpha_k \xi$ and using Eq. 1 of Table 2.9-2 (C), we have

$$\int_0^1 \xi J_0(\alpha_k \xi) d\xi = \frac{1}{\alpha_k^2} \int_0^{\alpha_k} \eta J_0(\eta) d\eta$$

$$= \frac{1}{\alpha_k^2} [\eta J_1(\eta)]_0^{\alpha_k}$$

$$= \frac{1}{\alpha_k} J_1(\alpha_k). \tag{6.1-50}$$

Therefore,

$$Q = \frac{\pi \Delta p R^4}{8\mu L} + \frac{4\pi R^2 A}{\rho} \sum_{k=1}^{\infty} \frac{1}{\alpha_k^2} F(\alpha_k, t). \tag{6.1-51}$$

We note that the integration process is much easier than it appeared at first, because the complex expression for $F(\alpha_k, t)$ is not a function of r and is not involved in the procedure at all. Equation 6.1-50 is plotted in Figure 6.1-6 in which we can verify that there is indeed a time lag or phase angle ϕ between the flow and pressure. It should also be noted that the expressions derived here, including the phase-angle aspect, are essentially the same as those by Womersley [5], although they appear differently in form.

As large blood vessels such as arteries are elastic in nature, the next logical step would be to present an example for pulsatile flow in expandable and tapered ducts. However, this would involve the use of a stress–strain relationship and such quantities as Young's modulus and Poisson's ratio, which are of only peripheral interest as far as the present book is concerned. Also, the mathematical treatment would be quite complex. Therfore, to avoid such difficulties, we only refer the reader to pertinent references in this field [6, 7, 8].

Example 6.1-4   Two-Dimensional Flow in Renal Tubule [9]. In the previous three examples we dealt only with fluids flowing in one (the axial) direction. Even in the latter two situations, where there are two independent variables, only one variable pertains to space, the other concerning time variation. Thus, we have had only one-dimensional flows so far.

In biomedical work we often encounter flows in more than one direction. For example, in the rectangular-passage type of hemodialyzer, the

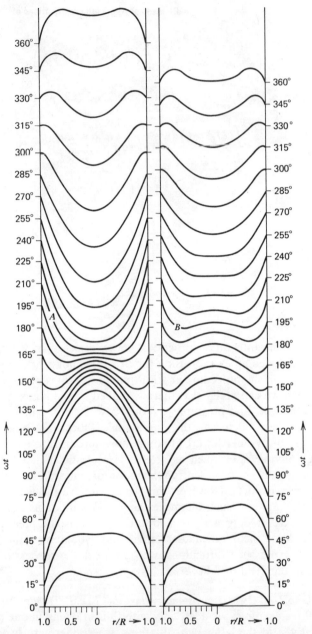

Figure 6.1-5. Time variation of velocity profile in pulsatile flow. (Adapted from [6]) Note that (a) this is only the periodic (fluctuating) portion of the velocity profile at various moments in a single cycle. The complete profile is formed by superimposing the periodic portion on top of the steady-state portion which is a parabolic function with a different maximum; and (b) $\omega_A < \omega_B$.

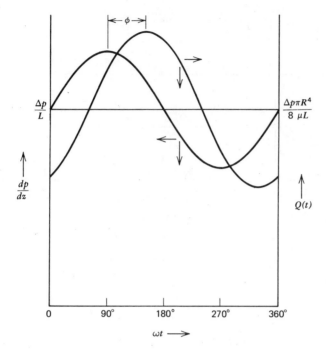

Figure 6.1-6. Pressure and flow rate cycles in pulsatile flow. Note that the flow trails the pressure by ϕ.

blood and the dialysate flow in alternate layers between flat membranes. The seepage across the membrane creates a velocity component in the fluid away from the centerplane of the flow channel in addition to that parallel to the centerplane in the direction of the pressure gradient.

In the present example we deal with the two-dimensional flow in the renal tubule in which a similar process occurs, but in cylindrical geometry. Here part of the water and some solutes (such as glucose) are reabsorbed through the walls while flowing along the tubule. This outward seepage creates the radial component of the velocity, which must be considered simultaneously with the axial component.

Figure 6.1-7 shows a schematic diagram and qualitative description of the system considered. The curved arrows show that part of the fluid in the tubule flows toward and through the walls, while the gradually shortened arrows in the axial (z-) direction show that the velocity and flow rate gradually decrease in the z-direction because of loss of fluid through the walls along the way. It should be noted that as the pressure decreases along the tubule, the radial flow decreases with it as well. As a simple

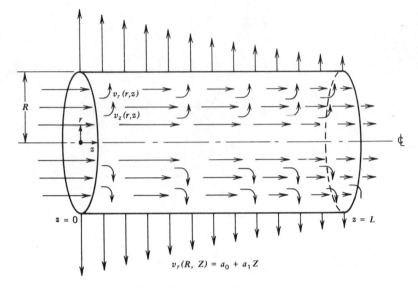

Figure 6.1-7.   Two-dimensional flow in renal tubule.

example, it is reasonable to assume that radial velocity at the wall decreases linearly in the direction of the flow, as represented by the expression $v_r(R,z) = a_0 + a_1 z$.

Starting with the equation of continuity (Eq. B, Table 4.12-1) and the radial (r-) and axial (z-) components of the equation of motion (Eqs. D and F, Table 4.12-2(b)), we have

$$\frac{1}{r}\frac{\partial}{\partial r}(r v_r) + \frac{\partial v_z}{\partial z} = 0 \tag{6.1-52}$$

$$\frac{1}{\mu}\frac{\partial p}{\partial r} = \frac{\partial}{\partial r}\left[\frac{1}{r}\frac{\partial}{\partial r}(r v_r)\right] + \frac{\partial^2 v_r}{\partial z^2} \tag{6.1-53}$$

$$\frac{1}{\mu}\frac{\partial p}{\partial z} = \frac{1}{r}\frac{\partial}{\partial r}\left(r\frac{\partial v_z}{\partial r}\right) + \frac{\partial^2 v_z}{\partial z^2} \tag{6.1-54}$$

by noting that all variables are time-independent (steady state) and θ-independent (radial symmetry); that $v_\theta = 0$; that ρ = constant (incompressible fluid); and that the small tubular diameter and slow flow make the viscous forces much larger than inertial forces, thus justifying the dropping of the velocity gradient terms on the left-hand sides of the equation of motion.

Now, instead of having a partial differential equation with two independent and one dependent variables, we have a set of three simultaneous partial differential equations with three dependent variables, p, v_r, and v_z each of which is a function of the same two independent variables (r and z). In the present case, the following boundary conditions govern:

B.C.1:   at $r = 0$, $\dfrac{\partial v_z}{\partial r}(0, z) = 0$,* $v_r(0, z) = 0$*

              and $v_z = $ finite

B.C.2:   at $r = R$, $v_z(R, z) = 0$ (no slip at walls)

              $v_r(R, z) = a_0 + a_1 z$

B.C.3:   at $z = 0$, $p = p_0$

B.C.4:   at $z = L$, $p = p_L$.

Equating the partial derivative with respect to z of the right-hand side of Eq. 6.1-53 to that with respect to r of the right-hand side of Eq. 6.1-54, and thus eliminating p for the time being, we have

$$\frac{\partial^2}{\partial r \partial z}\left[\frac{1}{r}\frac{\partial}{\partial r}(r v_r)\right] + \frac{\partial^3 v_r}{\partial z^3} = \frac{\partial}{\partial r}\left[\frac{1}{r}\frac{\partial}{\partial r}\left(r\frac{\partial v_z}{\partial r}\right)\right] + \frac{\partial^3 v_z}{\partial z^2 \partial r}. \quad (6.1\text{-}55)$$

Taking the partial derivative of this equation with respect to z and substituting Eq. 6.1-52 give

$$\left\{\frac{\partial}{\partial r}\left[\frac{1}{r}\frac{\partial}{\partial r}\left(r\frac{\partial}{\partial r}\left(\frac{1}{r}\frac{\partial}{\partial r}\right)\right)\right] + 2\frac{\partial}{\partial r}\left[\frac{1}{r}\frac{\partial}{\partial r}\left(\frac{\partial^2}{\partial z^2}\right)\right]\right.$$

$$\left. + \frac{1}{r}\frac{\partial^4}{\partial z^4}\right\}(r v_r) = 0. \quad (6.1\text{-}56)$$

We now postulate v_r to have the form

$$v_r(r, z) = g(r) \cdot f(z), \quad (6.1\text{-}57)$$

which does satisfy all of the differential equations and boundary conditions; for example, at $r = R$, Eq. 6.1-57 gives

$$v_r(R, z) = g(R) \cdot f(z). \quad (6.1\text{-}58)$$

*Because of radial symmetry.

Comparing this and B.C.2 leads to the following conclusion:

$$f(z) = a_0 + a_1 z \qquad (6.1\text{-}59)$$

and

$$g(R) = 1. \qquad (6.1\text{-}60)$$

Substitution of Eqs. 6.1-57 and 6.1-59 into Eq. 6.1-56 yields the following ordinary differential equation:

$$\frac{d}{dr}\left\{ \frac{1}{r}\frac{d}{dr}\left[r\frac{d}{dr}\left(\frac{1}{r}\frac{d}{dr}(rg) \right) \right] \right\} = 0, \qquad (6.1\text{-}61)$$

B.C.1(g):   At $r = 0$, $g(0) = 0$ (since $v_r = 0$, $f(z) \neq 0$)

B.C.2(g):   At $r = R$, $g(R) = 1$.

Successive integration of Eq. 6.1-61 and application of B.C.1(g) give

$$g(r) = C_1 r^3 + C_2 r \ln r + C_3 r \qquad (6.1\text{-}62)$$

so that

$$v_r(r, z) = f(z) \cdot g(r)$$
$$= (a_0 + a_1 z)(C_1 r^3 + C_2 r \ln r + C_3 r). \qquad (6.1\text{-}63)$$

Substituting Eq. 6.1-63 into Eq. 6.1-52, we have

$$\frac{\partial v_z}{\partial z} = -2(a_0 + a_1 z)[2C_1 r^2 + C_2(\ln r + \tfrac{1}{2}) + C_3]. \qquad (6.1\text{-}64)$$

Observing that

$$v_z = \int dv_z = \int \frac{\partial v_z}{\partial z} dz + G(r), \qquad (6.1\text{-}65)$$

where $G(r)$ is a function of r, yet to be determined, insertion of Eq. 6.1-64 gives

$$v_z = -[2C_1 r^2 + C_2(\ln r + \tfrac{1}{2}) + C_3] \cdot (2a_0 z + a_1 z^2) + G(r) \qquad (6.1\text{-}66)$$

Using B.C.1, we find that $C_2 = 0$ and the use of B.C.2 yields

$$G(R) - (2a_0 z + a_1 z^2)(2C_1 R^2 + C_3) = 0,$$

which should hold for all z. Thus, we have

$$G(R) = 0 \tag{6.1-67}$$

$$2C_1 R^2 + C_3 = 0. \tag{6.1-68}$$

Using B.C.2 on Eq. 6.1-63 gives

$$C_1 R^3 + C_3 R = 1. \tag{6.1-69}$$

Solving Eqs. 6.1-68 and 6.1-69 simultaneously, we obtain

$$C_1 = -\frac{1}{R^3}, \qquad C_3 = \frac{2}{R}.$$

Therefore,

$$v_r = (a_0 + a_1 z)\left[2\left(\frac{r}{R}\right) - \left(\frac{r}{R}\right)^3\right]. \tag{6.1-70}$$

Also, from Eq. 6.1-66 and the values of C_1 and C_3,

$$v_z = \frac{2}{R}\left[\left(\frac{r}{R}\right)^2 - 1\right](2a_0 z + a_1 z^2) + G(r). \tag{6.1-71}$$

Using this and Eq. 6.1-70 on Eq. 6.1-55 followed by successive integration gives

$$G(r) = -\frac{a_1 r^4}{2R^3} + D_1 r^2 + D_2 \ln r + D_3, \tag{6.1-72}$$

which is to be combined with Eq. 6.1-71. Using B.C.1, we find that $D_2 = 0$, and using B.C.2, we obtain

$$R^2 D_1 + D_3 = \frac{a_1 R}{2}. \tag{6.1-73}$$

We know that the volumetric flow rate at a certain axial distance is

$$Q(z) = 2\pi \int_0^R v_z(r,z) r \, dr \tag{6.1-74}$$

and that the flow rate at $z=0$ is

$$Q_0 = Q(0) = 2\pi \int_0^R v_z(r,0) r \, dr. \tag{6.1-75}$$

From Eqs. 6.1-71 and 6.1-72, we know that

$$v_z(r,0) = -\frac{a_1 r^4}{2R^3} + D_1 r^2 + D_3, \qquad (6.1\text{-}76)$$

using which on Eq. 6.1-75 gives

$$Q_0 = \pi R^2 \left(-\frac{a_1 R}{6} + \frac{D_1 R^2}{2} + D_3 \right)$$

or

$$\frac{D_1 R^2}{2} + D_3 = \frac{Q_0}{\pi R^2} + \frac{a_1 R}{6}. \qquad (6.1\text{-}77)$$

Solving Eqs. 6.1-73 and 6.1-77 simultaneously, we have

$$D_1 = \frac{2a_1}{3R} - \frac{2Q_0}{\pi R^4}$$

$$D_3 = \frac{2Q_0}{\pi R^2} - \frac{a_1 R}{6}.$$

Therefore,

$$v_z = \left[1 - \left(\frac{r}{R}\right)^2 \right] \left\{ \frac{2Q_0}{\pi R^2} - \frac{2}{R}(2a_0 z + a_1 z^2) - \frac{a_1 R}{2}\left[\frac{1}{3} - \left(\frac{r}{R}\right)^2 \right] \right\} \quad (6.1\text{-}78)$$

Using the Leibnitz formula for differentiating integrals (Eq. 2.6-15) on Eq. 6.1-74 with Eq. 6.1-78, we have

$$\frac{dQ}{dz} = 2\pi \int_0^R \frac{\partial}{\partial z}[rv_z(r,z)]\,dr + 2\pi \left\{ rv_z|_{r=R}\frac{dR}{dz} - rv_z|_{r=0}\frac{dO}{dz} \right\}^*$$

$$= 2\pi\left(-\frac{4}{R} \right)(a_0 + a_1 z)\int_0^R \left[1 - \left(\frac{r}{R}\right)^2 \right] r\,dr$$

$$= -2\pi R(a_0 + a_1 z). \qquad (6.1\text{-}79)$$

Note that this is the decrease in axial flow per unit tubular length and in magnitude should be exactly equal to the loss of fluid through radial flow;

*This means that since both limits are constant, we can move the differential operator inside the integral and differentiate the integrand partially with respect to z only.

that is, one should be readily able to prove that $v_r(R,z) \cdot 2\pi R = 2\pi R(a_0 + a_1 z)$.

From Eq. 6.1-79, we find

$$Q = \int \left(\frac{dQ}{dz} \right) dz + C$$

$$= \int -2\pi R(a_0 + a_1 z) dz + C$$

$$= -2\pi R \left(a_0 z + \frac{a_1 z^2}{2} \right) + C.$$

At $z = 0, Q = Q_0$; therefore, $C = Q_0$ and

$$\therefore Q(z) = Q_0 - \pi R(2a_0 z + a_1 z^2). \tag{6.1-80}$$

Substituting this into Eq. 6.1-78, we have, finally

$$v_z = \left[1 - \left(\frac{r}{R} \right)^2 \right] \left\{ \frac{2Q(z)}{\pi R^2} - \frac{a_1 R}{2} \left[\frac{1}{3} - \left(\frac{r}{R} \right)^2 \right] \right\}. \tag{6.1-81}$$

In this expression we can see that the axial velocity profile is a modified form of the parabolic profile of the H–P flow in the following two respects: (1) the volumetric flowrate, Q, instead of being a constant, decreases along the tube direction according to Eq. 6.1-80 because of loss of fluid radially through the walls; (2) additional distortion due to the varying nature of the radial flow, given by the term $(a_1 R/2) [\frac{1}{3} - (r/R)^2]$.

From Eq. 6.1-70 we can find that a maximum radial velocity

$$v_{r_{max}} = \frac{4}{9} \sqrt{6} \ (a_0 + a_1 z) \tag{6.1-82}$$

exists at $r = \sqrt{6}/3$ because v_r increases from zero at the centerline radially toward the walls as it picks up seeping material. But, it reaches a maximum and decreases further out as the increase in cylindrical area overshadows the increase in volumetric radial flow. The reader can find a similarly interesting behavior of the axial velocity maximum (see Problem 6.P).

Going back to Eqs. 6.1-53 and 6.1-54 and using Eqs. 6.1-70 and 6.1-81, we have

$$\frac{\partial p}{\partial r} = -\frac{8\mu r}{R^3} (a_0 + a_1 z) \tag{6.1-83}$$

$$\frac{\partial p}{\partial z} = -\frac{4a_1\mu}{R}\left[\left(\frac{r}{R}\right)^2 + \frac{2Q(z)}{a_1\pi R^3} + \frac{1}{3}\right], \qquad (6.1\text{-}84)$$

from which we can obtain (see Appendix C)

$$p(r,z)-p(0,0) = -\frac{4\mu}{R}(a_0+a_1 z)\left(\frac{r}{R}\right)^2 - \mu\left(\frac{4a_1}{3R} + \frac{8\overline{Q}}{\pi R^4}\right)z, \quad (6.1\text{-}85)$$

where

$$\overline{Q}(z) = \frac{1}{z}\int_0^z Q(z)\,dz \qquad (6.1\text{-}86)$$

is the average volumetric flow rate between $z=0$ and $z=z$.

If we want to consider the average pressure over a certain cross-section at $z=z$, we write

$$\langle p\rangle(z) = \frac{\displaystyle\int_0^{2\pi}\int_0^R p(r,z)r\,dr\,d\theta}{\displaystyle\int_0^{2\pi}\int_0^R r\,dr\,d\theta}$$

$$= \frac{2\pi}{\pi R^2}\int_0^R p(r,z)r\,dr \qquad (6.1\text{-}87)$$

With the substitution of Eq. 6.1-85, we have

$$\langle p\rangle(z)-p(0,0) = -\mu\left[\frac{2a_0}{R} + \left(\frac{8\overline{Q}(z)}{\pi R^4} + \frac{10a_1}{3R}\right)z\right], \qquad (6.1\text{-}88)$$

so that the pressure drop over tube length L is

$$\Delta\langle p\rangle = \langle p\rangle(0) - \langle p\rangle(L)$$

$$= \mu\left(\frac{8\overline{Q}(L)}{\pi R^4} + \frac{10a_1}{3R}\right)L. \qquad (6.1\text{-}89)$$

Note here that if $a_1 = 0$ (i.e., axially uniform reabsorption rate),

$$\Delta\langle p\rangle = \frac{8\mu L}{\pi R^4}\overline{Q}(L), \qquad (6.1\text{-}90)$$

which has the form of the H–P equation (Eq. 6.1-7) except that in the present case the flow rate is the average one from $z = 0$ to $z = L$ instead of a constant one. It will revert to the true H–P form when there is no seepage flow, that is, $a_0 = a_1 = 0$, in which case $Q(L) = Q_0$.

No data have been obtained showing the exact dependence of the reabsorption rate on the axial distance. However, in the simplified case of $a_1 = 0$ (axially uniform reabsorption rate), if we let

$$\alpha = \frac{Q_0 - Q(L)}{Q_0} \quad \text{or} \quad Q(L) = (1 - \alpha)Q_0 \qquad (6.1\text{-}91)$$

(i.e., the fraction of the original fluid that has been absorbed by the time it reaches $z = L$), it follows that, using Eq. 6.1-80 in Eq. 6.1-86,

$$a_0 = \frac{\alpha Q_0}{2\pi R L}$$

and*

$$\bar{Q}(L) = \left(1 - \frac{\alpha}{2}\right)Q_0,$$

which can be inserted into Eq. 6.1-90 to give

$$\Delta\langle p \rangle = \frac{8\mu L}{\pi R^4} Q_0 \left(1 - \frac{\alpha}{2}\right). \qquad (6.1\text{-}92)$$

Using this expression we can test order-of-magnitude agreement by observing that the upper and lower limits of Δp are at $\alpha = 0$ and $\alpha = 1$, respectively. Thus,

$$\frac{4\mu L}{\pi R^4} Q_0 \leqslant \Delta\langle p \rangle \leqslant \frac{8\mu L}{\pi R^4} Q_0.$$

Using the values reported by Gottschalk and Mylle [10],

$$R \approx 10^{-3} \text{cm}$$

$$Q_0 \approx 4 \times 10^{-7} \text{cm}^3/\text{sec}$$

$$\mu \approx 7 \times 10^{-3} \text{dynes sec/cm}^2$$

$$L \approx 1 \text{ cm}$$

$$\alpha \approx 0.8,$$

---

*This expression with the $(1 - \alpha/2)$ factor confirms the linear decrease of axial flow if the reabsorption rate is uniform.

the pressure drop is calculated to be 3.2 mmHg, according to Eq. 6.1-92, while it can vary between 2.7 and 5.4 mmHg.

It is quite possible that the radial reabsorption rate decreases exponentially in the direction of the flow. The reader will find it an interesting exercise following a similar procedure (see Problems 6.B and 6.C).

6.2   HEAT-TRANSFER PROBLEMS

In this Section we present examples dealing with the transport of thermal energy. Again our approach is to start from the simple cases of steady-state conduction in a single medium (Example 6.2-1) and in composite materials (Example 6.2-2). Then we proceed to the mathematically and physically more complex ones such as that with heat production (Example 6.2-3), multidimensional (Example 6.2-4), unsteady-state (Example 6.2-5), and convective (Example 6.2-6) heat transfer.

However, in order to discuss these examples knowledgeably, two additional concepts will be required. One, which we will discuss more formally in Chapter 7 in relation to mass and momentum transport, deals with heat transfer between a surface and the surroundings (Section 6.2a); the other is the heat addition and removal within the medium by fluid flow (Section 6.2b). Both of these are quite relevant to biomedical work.

a.   Sensible, Latent, and Radiative Heat Transfer at the Surface

Quite frequently in biomedical, as well as engineering, problems, the surface temperature of a system cannot be conveniently and/or accurately determined. In such cases, the ambient temperature is used and the surface temperature related to it by means of a simple phenomenological expression often amateurishly referred to as the Newton's "law" of cooling.*

In Figure 6.2-1, let us suppose that the surface temperature of the wall on the left is uniform at T_s, while the main body of the ambient environment is on the right at T_a. There is a region, which we call film, in which the temperature varies from T_s to T_a as shown.† Let us denote the thickness of this film by δ and imagine this as a case of pure conduction; we may then write, following Eq. 3.3-1,

$$q_x = k \frac{T_s - T_a}{\delta}. \qquad (6.2\text{-}1)$$

---

*Although Newton did originate this expression, it is not one of the fundamental laws of nature; rather, it is a convenient semiempirical way of describing a phenomenon.

†In the case of $T_s < T_a$ (heating instead of cooling), the curve would go the opposite way.

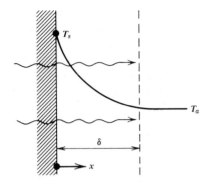

Figure 6.2-1. Surface convection.

Of course, we know that actually this is not a case of pure conduction, which, for rectangular geometry, would give a linear temperature distribution between T_s at the wall and T_a at the right-hand boundary of this film. We know that this clear-cut boundary does not exist in reality and that in addition to conduction, heat transfer in the film is aided by fluid flow. This is why the temperature at first drops quickly, but then levels off to T_a without a discontinuity at $x = \delta$. In order not to confuse this with pure conduction and because of the film thickness δ is difficult to ascertain, we rewrite Eq. 6.2-1 as

$$q_x = h(T_s - T_a), \tag{6.2-2}$$

where $h = k/\delta$ and is called the *heat-transfer coefficient* between the fluid and the wall.

Instead of being a "law" of cooling (or heating), Eq. 6.2-2 is actually the defining equation for the heat transfer coefficient (see Section 7.1). From it, we can see that h is the rate of heat transfer per unit temperature differential per unit surface area, while k, the thermal conductivity (with units of cal/cm-sec-°C), is the rate of heat transfer per unit temperature differential per unit thickness (or unit length in the heat-transfer path).

Flow-assisted heat transfer is termed *convection*, by which we often loosely mean forced convection, wherein the flow is generated by external means, such as pressure gradient through the use of a pump or blower. Strictly speaking, we should carefully distinguish it from *free convection*, wherein the flow, normally quite small, is generated by the density differential, which in turn is caused by the heating of the surrounding film by the warm surface. It is quite obvious, then, that heat-transfer coefficient is a function of velocity, among other factors. In fact, this is very important in the study of thermoregulation and gives rise to such quantities as the wind-chill index. We shall discuss this in more details in Chapter 7.

In considering surface heat transfer, we must not forget *thermal radiation*. According to the Stefan–Boltzmann law, the intensity of radiation is proportional to the fourth power of the absolute temperature of the emitting body. However, for most purposes in the study of radiative heat loss from the human body, it is sufficiently accurate to define a radiative heat-transfer coefficient, based on the first power of the temperatures of the emitting and receiving bodies (i.e., similar to Eq. 6.2-2):

$$q_x = h_r(T_s - T_w), \tag{6.2-3}$$

where T_w is the temperature of the surrounding walls, not of the ambient air. A typical and often-used value of h_r for nude human beings is 5 kcal/m^2–hr–°C. The relative importance of radiative to convective heat transfer is discussed elsewhere [11, 12].

Thus far, we have related only part of our story on surface heat transfer, the part that deals with sensible heat flow by virtue of a temperature difference. With physiological systems, especially on the skin of a human body and the tongue of a dog, the latent heat loss due to evaporation of water vapor, such as perspiration, is also very important. For this we can write

$$q_x = \hat{\lambda}_v \dot{w}, \tag{6.2-4}$$

where $\hat{\lambda}_v$ is the heat of vaporization per unit mass of water and $\dot{w}$ the mass rate of evaporation of water per unit total body surface area. If we further denote the humidity (partial density of water vapor) and the mass transfer coefficient of water vapor to the surrounding air by ρ_{H_2O} and $\mathcal{K}_{H_2O}$, respectively, and the fraction of skin actually covered by perspiration by β, we can write

$$\dot{w} = \mathcal{K}_{H_2O}\,\beta\,[\,(\rho_{H_2O})_{skin} - (\rho_{H_2O})_{air}\,], \tag{6.2-5}$$

so that Eq. 6.2-4 now becomes

$$q_x = \hat{\lambda}_v\,\mathcal{K}_{H_2O}\,\beta\,[\,(\rho_{H_2O})_{skin} - (\rho_{H_2O})_{air}\,]. \tag{6.2-6}$$

More will be said about the mass transfer coefficient $\mathcal{K}_{H_2O}$ in Chapter 7. It is important to understand here that (1) since both these sensible and insensible heat exchanges occur at the boundary surface of the medium, the use of Eqs. 6.2-2, 6.2-3, and 6.2-6 is in the boundary conditions and (2) since these three phenomena occur simultaneously in parallel, the heat fluxes in Eqs. 6.2-2, 6.2-3, and 6.2-6 are additive.

b. Heat Gain or Loss Within the Medium by Flow

In physiological systems heat is sometimes removed, and at other times added, by blood flowing through capillaries, which are dispersed fairly uniformly throughout the entire medium—the tissue. If the blood flows reasonably unidimensionally in relation to the macroscopic geometry, such as in the intermediate tissue adjacent to the brain [13] (see Figure 6.2-2), it can be accounted for by the convection terms in the equation of energy (see velocity terms on the left-hand sides of equations in Table 4.12-3).

However, in many other cases, such as the superficial tissue in the head (Figure 6.2-2) and both the deep and superficial tissues in the extremities [13] (Figure 6.2-9), the distribution of capillaries is such that direction of flow is random or prevails in both forward and reverse direction. It is a case of "instantaneous" local exchange of energy without a clear-cut bias or direction. As such, it cannot be treated in the convection terms because the net or overall velocity is zero. On the other hand, since it does not occur at boundaries, it should not appear in the boundary condition either, as discussed in Section 6.2a.

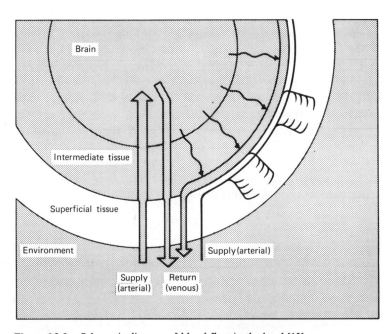

Figure 6.2-2.   Schematic diagram of blood flow in the head [13].

To place it in the differential equation of energy, one can see from the above discussion that this type of heat gain or loss is similar to the heat generation or consumption as described in Section 4.7 except that in the previous discussion a transformation of energy is involved, while in the present case the same form of heat is merely being brought in or carried away. Thus, we can write

$$Q_G = Q_G' + w_v \hat{C}_p (T - T_e), \qquad (6.2\text{-}7)$$

where Q_G (cal/cm^3-sec) is the net rate of heat gain per unit volume of medium, Q_G' is the rate of "true" heat generation or consumption, such as by subatomic (nuclear, electronic) reaction,* w_v is the perfusion rate per unit volume of medium (g/cm^3-sec), $\hat{C}_p$ is the heat capacity (cal/g-°C) of the heat-carrying fluid (e.g., blood), T is the local temperature, and T_e is the temperature of the entering fluid. In so writing, we are implying that the fluid assumes the local temperature T shortly after it enters and it leaves at this temperature T. If $T > T_e$, heat is being carried away, which is equivalent to heat consumption; if $T < T_e$, heat is being brought in, which is equivalent to generation. The use of Eq. 6.2-7 will be illustrated in Example 6.2-2.

Example 6.2-1 Temperature Profiles in Working Muscle. In this example, we consider the temperature profiles and its maximum rise in a muscle fiber, caused by the heat generated from exercising. We use this example not only because of its simplicity but also because of its analogy to the fluid-flow problem in Example 6.1-1, as we shall see by comparing the mathematical representations of the two.

Let us consider a cylindrical muscle of radius R, thermal conductivity k, and a uniform and constant rate of metabolic heat production of Q_G (cal/cm^3-sec). Also, let us assume that the surface of each muscle fiber is kept at a temperature T_s. By adding this heat generation term (see Section 4.7) to the conduction equation (Eq. 4.10-7),† we have, for steady state

$$k\nabla^2 T + Q_G = 0.$$

By noting that there is no temperature variation in the θ-direction (because of radial symmetry) or the z-direction (because of the long cylindrical rod),

---

*As for the chemical reaction in a multicomponent system, such as a biological one, the multicomponent equation of energy (Eq. 5.6-6 or 5.7-4) should be preferred, in which case the "true" heat generation by chemical reaction is implicitly included in $\hat{U}$ (see [16], pp. 562–563). Then there is no need to include this heat generation term "artificially".

†Or reducible from Eq. E of Table 4.12-3, by realizing that all velocity and shear-stress components are zero and $\hat{C}_v = \hat{C}_p$ for solids.

the above equation becomes

$$-\frac{k}{r}\frac{d}{dr}\left(r\frac{dT}{dr}\right)=Q_G, \qquad (6.2\text{-}8)$$

with the boundary conditions

B.C. 1: at $r=0$, $T=$ finite, or $\dfrac{dT}{dr}=0$

B.C. 2: at $r=R$, $T=T_s$.

Integrating Eq. 6.2-8 and applying B.C. 1, we have

$$\frac{dT}{dr}=-\frac{Q_G}{2k}r.$$

Integrating again and applying B.C. 2 yield

$$T=T_s+\frac{Q_G R^2}{4k}\left[1-\left(\frac{r}{R}\right)^2\right]. \qquad (6.2\text{-}9)$$

This expression, as shown in Figure 6.2-3, is parabolic in nature. It shows that the larger the heat generation and/or the poorer the heat-conduction capability, the higher the temperature rise in the medium.

Again, as in the case of H–P flow (Example 6.1-1), a better way to plot

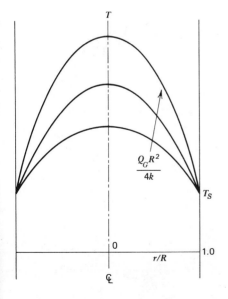

Figure 6.2-3. Temperature profiles in exercising muscle.

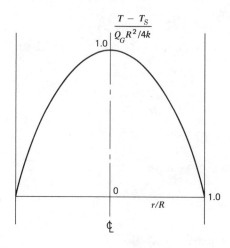

Figure 6.2-4. Dimensionless form of temperature profiles in exercising muscle.

this is to change Eq. 6.2-9 into dimensionless form:

$$\Theta = \frac{T - T_s}{Q_G R^2 / 4k} = 1 - \xi^2, \qquad (6.2\text{-}10)$$

where $\xi = r/R$, so that the various curves merge into one, as shown in Figure 6.2-4. Comparison of Eq. 6.2-10 with Eq. 6.1-6 and of Figure 6.2-4 with Figure 6.1-1 verifies the complete analogies between the two cases. Furthermore, we can make the following formal comparison:

	Example 6.1-1 *Fluid Flow*	*Example 6.2-1* *Heat Conduction*
Surface quantity	$v_z(R) = 0$ (velocity at tube wall is 0, B.C.2)	T_s
Maximum quantity at center	$\dfrac{\Delta P R^2}{4\mu L} = \dfrac{1}{4\nu}\left(\dfrac{\Delta P}{\rho L}\right) R^2$	$\dfrac{Q_G R^2}{4k} = \dfrac{1}{4\alpha}\left(\dfrac{Q_G}{\rho \hat{C}_p}\right) R^2$
Diffusivity	ν	α
Driving gradient	$\dfrac{\Delta P}{\rho L}$ ("momentum production" per unit mass)	$\dfrac{Q_G}{\rho \hat{C}_p}$ (energy production per unit heat capacity)

It is interesting to calculate the rate of heat dissipation at the cylindrical surface, for an arbitrary fiber length L by multiplying the surface area with the heat flux at the surface:

$$2\pi RL\cdot q_r|_{r=R} = 2\pi RL\left(-k\frac{dT}{dr}\right)\Bigg|_{r=R}$$

$$= \pi R^2 L\cdot Q_G, \qquad (6.2\text{-}11)$$

which is exactly equal to the total rate of heat production in the muscle fiber (volume times rate of heat production per unit volume). This verifies that we have a steady-state situation, since there will be no net accumulation of thermal energy. Thus, a stable temperature profile is maintained.

As a numerical illustration, let us take $R=1$ cm, $Q_G=5$ cal/cm³-hr, and $k=0.001$ cal/cm-sec-°C. The largest temperature rise in the fiber (at its center) can be calculated to be

$$T_{max} - T_s = \frac{Q_G R^2}{4k} = \frac{5\times 1^2}{4\times 60^2\times 10^{-3}} = 0.33°C,$$

so that if the fiber surface is at 37°C, the inside can be as high as 37.33°C.

Example 6.2-2   Heat Conduction Through Tissue, Fat, Skin, and Clothing or Fur. In the previous example, we dealt with the steady-state one-dimensional heat conduction in a single uniform medium, one of the simplest cases of all. In biomedical problems, quite often we encounter the conduction of heat through a series of media, each having its own thermal properties. An everyday phenomenon is that of the dissipation of body heat from the inner core of the body through the perfused tissue, subcutaneous fat layer, skin, and clothing or fur into the surrounding air. Except for the inner core and tissue, which involve convection due to blood flow, and for the ambient air, which involves convection due to air flow, the heat transfer through the three intermediate layers is pure conduction. But even these layers may have very different thermal properties and dimensions. Furthermore, some of these quantities, such as the thickness of the fat layer, may vary from person to person. Also, by varying the thickness and material of clothing, one can regulate the rate of body heat loss according to the prevailing ambient condition.

In this example we show how one can analyze the problem of heat conduction through such composite layers of different materials, in particular, with the incorporation of the concepts of surface heat exchange. However, we will only gloss over the heat gain or loss by flow within the

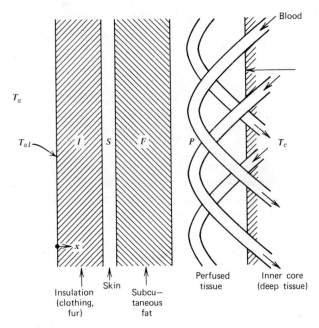

Figure 6.2-5.   Heat conduction through body layers.

tissue and the metabolic heat, and shall leave the detailed analysis to the next example.

Let us consider the schematic representation of the several layers as shown in Figure 6.2-5 and denote their temperatures by T, heat fluxes by q, thermal conductivities by k, and thicknesses by δ with appropriate superscripts representing the various layers as shown in the figure. The ambient and inner-core temperatures are assumed to be known at T_a and T_c (the latter being normally 37°C for human beings), respectively. The temperature of each boundary is denoted by the subscripts of the two layers sharing this boundary, such as T_{SF} and T_{FP}. Let us also assume that the layers are sufficiently thin compared with the size of the body so that curvatures as well as heat transfer in the y and z directions can be neglected. Thus, for steady state, we can reduce Eq. A of Table 4.12-3 to obtain

$$\frac{dq_x}{dx} = 0. \tag{6.2-12}$$

Integrating this simple expression for each layer, we have:

$$q_x^I = \text{constant} \qquad (6.2\text{-}13a)$$

$$q_x^S = \text{constant} \qquad (6.2\text{-}13b)$$

$$q_x^F = \text{constant} \qquad (6.2\text{-}13c)$$

$$q_x^P = \text{constant}. \qquad (6.2\text{-}13d)$$

Since the cross-sectional areas of the various layers are the same for rectangular geometry and at each boundary all of the heat fluxes from one layer must be passed on to the next, we can thus conclude that all of the constants in the above expressions must be the same,* or

$$q_x^I = q_x^S = q_x^F = q_x^P = -q_0. \qquad (6.2\text{-}14)$$

Here a negative sign is added because for the frequent case of $T_c > T_a$, the flow of heat is in the negative x-direction. However, this sign is not essential as q_0 could assume a negative value otherwise.

By virtue of the Fourier's law, Eq. 6.2-14 can be separated and rewritten as:

$$k^I \frac{dT^I}{dx} = q_0 \qquad (6.2\text{-}15a)$$

$$k^S \frac{dT^S}{dx} = q_0 \qquad (6.2\text{-}15b)$$

$$k^F \frac{dT^F}{dx} = q_0 \qquad (6.2\text{-}15c)$$

$$k_{\text{eff}}^P \frac{dT^P}{dx} = q_0 \qquad (6.2\text{-}15d)$$

where we have introduced the *effective thermal conductivity*, k_{eff}^P, to account for the fact that in perfused tissue there *is* heat brought in by blood flow as well as produced metabolically. Thus, the use of Eqs. 6.2-12 through 6.2-14 to include the perfused-tissue layer is not exact. Using such an "average" or "overall" quantity corrects the situation only partially since the heat flux in the perfused layer is not a constant (because of the heat input), but position-dependent. But for the construction of the overall heat transfer

---

*For cylindrical and spherical geometries this is no longer true because of the varying interfacial area between layers. Instead, the rate of heat transfer must be the same from layer to layer (see Problem 6.H).

coefficient, which we will do later in the present example, this is adequate. The boundary conditions accompanying Eqs. 6.2-15a, b, c, and d are

B.C. 1: at $x=0$, $q_0 = h_a(T^I - T_a)$ or $T^I = T_a + \dfrac{q_0}{h_a} = T_{aI}$

B.C. 2: at $x=\delta^I$, $T^I = T^S = T_{IS}$

B.C. 3: at $x=\delta^I + \delta^S$, $T^S = T^F = T_{SF}$

B.C. 4: at $x=\delta^I + \delta^S + \delta^F$, $T^F = T^P = T_{FP}$

B.C. 5: at $x=\delta^I + \delta^S + \delta^F + \delta^P$, $T^P = T_c$,

where h_a is the heat-transfer coefficient between the insulation and the ambient air. The inner-core temperature T_c is assumed to be uniform and known.

The solutions satisfying the above differential equations and boundary conditions are

$$T^I(x) = T_{aI} + \frac{q_0}{k^I}x \quad \left(T_{aI} = T_a + \frac{q_0}{h_a}\right) \tag{6.2-16a}$$

$$T^S(x) = T_{IS} + \frac{q_0}{k^S}(x - \delta^I) \tag{6.2-16b}$$

$$T^F(x) = T_{SF} + \frac{q_0}{k^F}[x - (\delta^I + \delta^S)] \tag{6.2-16c}$$

$$T^P(x) = T_{FP} + \frac{q_0}{k^P_{\text{eff}}}[x - (\delta^I + \delta^S + \delta^F)], \tag{6.2-16d}$$

which are plotted in Figure 6.2-6. From it, we note that for pure conduction in rectangular geometry, the temperature profiles are linear. The temperature gradient (i.e., change per unit thickness as represented by the slopes of these lines) is inversely proportional to the thermal conductivity of the medium. Thus, a good insulator (poor conductor, with a small k) would have a steeper line than a good conductor.

In all previous development it has been implicitly assumed that q_0 is known. Indeed, sometimes it is given, but at others it has to be computed in terms of the ambient and inner-core temperatures and the thermal properties of the various layers by first adopting B.C. 1 and then applying

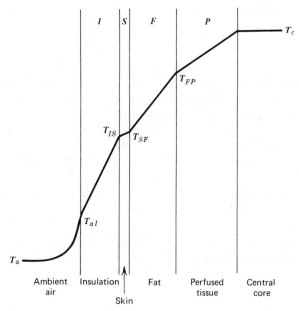

Figure 6.2-6. Temperature profiles in body layers.

B.C. 2 to Eq. 6.2-16a, B.C. 3 to Eq. 6.2-16b, B.C. 4 to Eq. 6.2-16c, and B.C. 5 to Eq. 6.2-16d, to obtain, respectively,

$$T_{aI} - T_a = \frac{q_0}{h_a} \tag{6.2-17a}$$

$$T_{IS} - T_{aI} = \frac{q_0}{k^I}\delta^I \tag{6.2-17b}$$

$$T_{SF} - T_{IS} = \frac{q_0}{k^S}\delta^S \tag{6.2-17c}$$

$$T_{FP} - T_{SF} = \frac{q_0}{k^F}\delta^F \tag{6.2-17d}$$

$$T_c - T_{FP} = \frac{q_0}{k^P_{\text{eff}}}\delta^P \tag{6.2-17e}$$

Now if we sum together Eqs. 6.2-17a–e, the intermediate boundary temperatures cancel out and we have

$$T_c - T_a = q_0\left(\frac{1}{h_a} + \frac{\delta^I}{k^I} + \frac{\delta^S}{k^S} + \frac{\delta^F}{k^F} + \frac{\delta^P}{k^P_{\text{eff}}}\right),\qquad (6.2\text{-}18)$$

so that we can calculate q_0 from only the two extreme temperatures T_c and T_a, and the thermal properties and thicknesses of the intermediate layers without knowing the boundary temperatures:

$$q_0 = \frac{T_c - T_a}{1/h_a + \delta^I/k^I + \delta^S/k^S + \delta^F/k^F + \delta^P/k^P_{\text{eff}}}.\qquad (6.2\text{-}19)$$

In fact, any of the intermediate boundary temperatures can be obtained by substituting the q_0 thus calculated into the appropriate equation from among Eqs. 6.2-17a–e.

In actual practice, it is sometimes expedient to combine thermal properties of the individual layers into a single overall heat-transfer coefficient, U, by defining

$$q_0 = U(T_c - T_a)\qquad (6.2\text{-}20)$$

or

$$\frac{1}{U} = \frac{1}{h_a} + \frac{\delta^I}{k^I} + \frac{\delta^S}{k^S} + \frac{\delta^F}{k^F} + \frac{\delta^P}{k^P_{\text{eff}}}.\qquad (6.2\text{-}21)$$

Here the analogy between heat (as well as mass and momentum) transfer and electrical network theory is quite apparent, as we can see that the resistances, being reciprocals of conductances, in series are additive.

Four further points must be noted. First, as footnoted earlier, for curvilinear coordinates, the rate of heat flow, not heat flux, is constant throughout the layers. Formally, instead of using Eq. A we should use Eqs. B and C in Table 4.12-3, which yields $(d/dr)(rq_r)=0$ and $(d/dr)(r^2q_r)=0$ for cylindrical and spherical geometries, respectively. Following through the entire analysis, we would find that the temperature profiles are no longer linear, but logarithmic and proportional to the reciprocal of the radial distance, respectively. Then the overall heat-transfer coefficient can be found to be

$$\frac{1}{U_{al}R_{al}} = \frac{1}{h_a R_{al}} + \frac{\ln(R_{al}/R_{IS})}{k^I} + \frac{\ln(R_{IS}/R_{SF})}{k^S}$$

$$+ \frac{\ln(R_{SF}/R_{FP})}{k^F} + \frac{\ln(R_{FP}/R_c)}{k^P_{\text{eff}}}\qquad (6.2\text{-}22)$$

for cylindrical geometry and

$$\frac{1}{U_{aI}R_{aI}^2} = \frac{1}{h_a R_{aI}^2} + \frac{1}{k^I}\left(\frac{1}{R_{IS}} - \frac{1}{R_{aI}}\right) + \frac{1}{k^S}\left(\frac{1}{R_{SF}} - \frac{1}{R_{IS}}\right)$$

$$+ \frac{1}{k^F}\left(\frac{1}{R_{FP}} - \frac{1}{R_{SF}}\right) + \frac{1}{k_{\text{eff}}^P}\left(\frac{1}{R_c} - \frac{1}{R_{FP}}\right) \quad (6.2\text{-}23)$$

for spherical geometry, where $R_{aI}, R_{IS}, \ldots$ are the radii of the intermediate boundaries as indicated by the subscripts. Here the subscript aI on U denotes that the definition of the overall heat-transfer coefficient is based on the surface area calculated from the outside radius of the insulation, obviously an easier quantity to measure than those of the tissue and fat layers, for example. Thus, when we use U_{aI}, the surface area of the insulation must be used with it. A similar U can be defined based on the skin surface area or radius, R_{IS}. This distinction does not occur in the rectangular geometry.

Second, when perspiration is involved, the term given in Eq. 6.2-6 must be included. To avoid complication, let us consider the perspiring nude body, which gives

$$\frac{Q}{A_{\text{skin}}} = U_{aS}(T_c - T_a) + \mathcal{H}_{\text{H}_2\text{O}}\,\beta\hat{\lambda}_v\left[(\rho_{\text{H}_2\text{O}})_{\text{skin}} - (\rho_{\text{H}_2\text{O}})_a\right]. \quad (6.2\text{-}24)$$

Third, when there is appreciable temperature difference between the body and the surrounding wall, the radiant energy loss must be included. Since in such cases, the surface temperature is usually known, we write

$$\frac{Q}{A_{aI}} = h_a(T_{aI} - T_a) + h_r(T_{aI} - T_w) \quad (6.2\text{-}25)$$

for a fully clothed body and

$$\frac{Q}{A_{\text{skin}}} = h_a(T_{\text{skin}} - T_a) + h_r(T_{\text{skin}} - T_w). \quad (6.2\text{-}26)$$

for a nude body, where T_w is the wall temperature.

Fourth, we must remember that we have "lumped" all of the temperature and heat flux variations due to metabolic production and gain from or loss to blood flow into this overall thermal conductivity, k_{eff}^P, which is only partially correct. To further correct and make a detailed analysis of the situation is the subject of the next example.

Example 6.2-3   Heat Gained by Tissue Through Blood Flow. In the previous example, we have glossed over the metabolic heat generation and

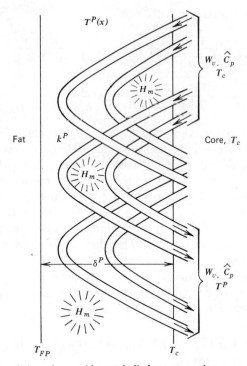

$T^P(x)$

$W_v, \hat{C}_p$
T_c

Fat k^P

Core, T_c

H_m

H_m

δ^P

$W_v, \hat{C}_p$
T^P

H_m

T_{FP} T_c

Figure 6.2-7. Perfused tissue layer with metabolic heat generation.

heat carried in and out by blood flow in the perfused tissue through the use of an overall, effective thermal conductivity. In the present example, we shall tackle this problem head-on and make a detailed analysis of it. Although the case of uniform metabolic heat generation has been thoroughly discussed in Example 6.2-1 and found to be quite simple and analogous to the H–P flow, the temperature-dependent heat carried by blood flow gives rise to some slight mathematical complications.

Let us consider only the perfused tissue layer from the previous example, as shown in Figure 6.2-7, with the same notations. The net rate of heat brought in by blood flow is $w_v \hat{C}_p (T_c - T^P)$ (see Section 6.2b for definitions), and that metabolically produced is H_m (cal/cm³-sec). Even with the blood flow in the capillaries dispersed throughout the tissue, we consider the entire layer *effectively* as a solid. The use of Eq. D in Table 4.12-3 with these two terms yields

$$k^P \frac{d^2 T^P}{dx^2} + w_v \hat{C}_p (T_c - T^P) + H_m = 0, \qquad (6.2\text{-}27)$$

with the following boundary conditions:

B.C.4: at $x = \delta^I + \delta^S + \delta^F$, $T^P = T^F = T_{FP}$

B.C.5: at $x = \delta^I + \delta^S + \delta^F + \delta^P$, $T^P = T_c$.

Solution of Eq. 6.2-27 can be greatly facilitated by the use of the following dimensionless quantities:

$$\Theta = \frac{T_c - T^P}{T_c - T_{FP}} \quad \text{(dimensionless temperature)}$$

$$\xi = \frac{x - (\delta^I + \delta^S + \delta^F)}{\delta^P} \quad \text{(dimensionless tissue depth)}$$

$$\gamma = \delta^P \sqrt{\frac{w_v \hat{C}_p}{k^P}} \quad \text{(perfusion–conduction ratio)}$$

$$\phi = \frac{H_m(\delta^P)^2}{k^P(T_c - T_{FP})}, \quad \text{(metabolism–conduction ratio)},$$

whereupon we have

$$\frac{d^2\Theta}{d\xi^2} - \gamma^2 \Theta = \phi \tag{6.2-28}$$

B.C.4: at $\xi = 0, \Theta = 1$

B.C.5: at $\xi = 1, \Theta = 0$.

We can see that in so doing we have not only made all quantities dimensionless but have also shifted the origin of the position coordinate from the insulation-air interface to the fat-tissue boundary.

The homogeneous solution to this equation (i.e., with ϕ temporarily removed) is

$$\Theta_h = C_1 \cosh \gamma\xi + C_2 \sinh \gamma\xi, \tag{6.2-29}$$

and the particular solution can be found by proposing a constant term corresponding to ϕ and solving for this undetermined constant (see Table 2.8-2). This yields

$$\Theta_p = -\frac{\phi}{\gamma^2}. \tag{6.2-30}$$

Combining Θ_h and Θ_p and applying the boundary conditions, we finally obtain*

$$\Theta = \left(1 + \frac{\phi}{\gamma^2}\right)\frac{\sinh\gamma(1-\xi)}{\sinh\gamma} + \frac{\phi}{\gamma^2}\left(\frac{\sinh\gamma\xi}{\sinh\gamma} - 1\right). \qquad (6.2\text{-}31)$$

This solution along with the temperature profiles of the other layers as represented by Eqs. 6.2-16a–c is plotted in Figure 6.2-8 for varying levels of γ and ϕ. It shows the phenomenon that for a nonexercising person exposed to varying degrees of cold environment, wherein the metabolism

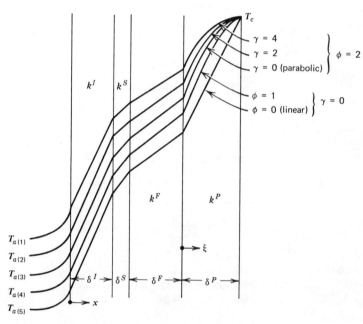

Figure 6.2-8. Temperature profiles in body layers with heat generation from perfused tissue layer.

*The same solution can also be obtained by defining still another modified variable

$$\Theta^* = \Theta + \frac{\phi}{\gamma^2}$$

to obtain a homogeneous differential equation in Θ^*.

stays fairly constant* at its basal rate, the rate of blood supply to tissue is regulated automatically and nonvoluntarily according to the ambient temperature. Several points are noted from the solution and this figure.

First, because of the heat gain, the temperature profile of the tissue layer is no longer linear. However, unlike the case of uniform heat pproduction in Example 6.2-1, the present distribution is not parabolic. The hyperbolic sine (sometimes sine, cosine, or hyperbolic cosine) functions are characteristic of transport processes in rectangular geometry,† wherein the rate of mass, momentum, or energy generation is a function of the driving potential itself.

Second, for nonlinear profiles, at the boundary between two layers, the slopes (temperature gradients) of the two layers differ by a factor equal to the inverse ratio of their thermal conductivities because of the inverse-proportionality relationship between the temperature gradient and thermal conductivity due to Fourier's law of heat conduction (Eq. 3.3-2) and the fact that there is no heat generation (i.e., continuous heat flux) at the boundary.

Third, in the extreme case of $w_v = 0$, no temperature-dependent heat is carried away by blood flow and we have only the uniform metabolically generated heat H_m. We can show, by solving Eq. 6.2-28 with $\gamma = 0$ or by setting $\gamma = 0$ in Eq. 6.2-31‡ that our solution becomes parabolic:

$$\Theta = (1 - \xi)\left(1 - \frac{\phi}{2}\xi\right). \tag{6.2-32}$$

The exact form of this equation differs from that of Eq. 6.2-10 because of the different geometries involved (similar to the difference between the solutions for the H–P flow and the plane H–P flow; see Example 6.1-1) and the shift in the spatial coordinate. Also the temperature coordinate is inverted in the present case ($\Theta = 0$ highest). If we further let $\phi = 0$, Eq. 6.2-32 becomes $\Theta = 1 - \xi$, a linear profile similar to those of the other layers, wherein there is no heat production.

Fourth, in the perfused tissue, although the dimensionless temperature Θ always numerically begins and terminates at 0 and 1, graphically it does not converge at the fat–tissue boundary because this end point ($\Theta = 1$)

---

*The temperature behavior for an exercising person with varying degrees of metabolic heat generation can be gleaned from Example 6.2-1, although the previous example is for a single cylindrical muscle fiber, while here we are discussing an entire tissue layer in rectangular geometry.

†Just as Bessel functions are for cylindrical geometry and the like; see Section 2.10.

‡With the application of the L'Hopital's rule three consecutive times.

varies with the value of T_{FP}, which in turn varies because of the different magnitudes of heat brought in through the tissue layer. The ambient temperature may change also. The determination of these boundary temperatures in terms of T_c, T_a, and the thermal properties of the various layers has been shown in the previous example, and its modification due to the change in T_{FP} will be shown later in the present example.

Last, the reason we have drawn Figure 6.2-8 with different ambient temperatures is that normally, as we mentioned earlier, for a nonexercising person the blood supply to the tissue changes mainly in reaction to a change in the ambient temperature. Here we can see that our physiological control system provides us with an excellent protection against cold environment in that as the ambient temperature drops the blood supply to tissue will decrease accordingly, so that the heat supply will also decrease. This will in turn decrease the intermediate temperatures somewhat, so that the slopes of the profiles, proportional to the rate of heat loss, will not increase to any appreciable extent. This protective mechanism works only to a limited degree, of course, in terms of time and coldness before a better extracorporeal insulation (clothing) is needed.

To obtain T_{FP} we have to use Eq. 6.2-31, the first derivative of which with respect to ξ at $\xi = 0$ can be found to be

$$\left. \frac{d\Theta}{d\xi} \right|_{\xi=0} = -\gamma \coth \gamma - \frac{\phi}{\gamma} \left(\frac{\cosh \gamma - 1}{\sinh \gamma} \right). \qquad (6.2\text{-}33)$$

Noting that

$$q_0 = k^P \frac{dT^P}{dx} \bigg|_{x = \delta^I + \delta^S + \delta^F} \qquad (6.2\text{-}34)$$

and the definition of Θ and ξ, we can write

$$q_0 = -\frac{k^P (T_c - T_{FP})}{\delta^P} \cdot \left. \frac{d\Theta}{d\xi} \right|_{\xi=0}, \qquad (6.2\text{-}35)$$

so that we have, with the substitution of Eq. 6.2-33,

$$T_c - T_{FP} = \frac{q_0}{k^P} \delta^P \frac{1}{\gamma \coth \gamma + (\phi/\gamma)[(\cosh \gamma - 1)/\sinh \gamma]}, \qquad (6.2\text{-}36)$$

in which we should remember that the temperature drop $(T_c - T_{FP})$ is also contained in ϕ (see definition of ϕ). Extracting this and rearranging, we finally have

$$T_c - T_{FP} = \left(\frac{\tanh\gamma}{\gamma}\right)\frac{q_0}{k^P}\delta^P + \frac{H_m}{w_v \hat{C}_p}\left(\frac{1-\cosh\gamma}{\cosh\gamma}\right). \quad (6.2\text{-}37)$$

Now if we sum together Eqs. 6.2-17a–d and 6.2-37, with the inclusion of the perspiration term from Eq. 6.2-6 (omitting the subscript H_2O), the intermediate boundary temperatures cancel out and we have

$$T_c - T_a = q_0\left(\frac{1}{h_a} + \frac{\delta^I}{k^I} + \frac{\delta^S}{k^S} + \frac{\delta^F}{k^F} + \frac{\tanh\gamma}{\gamma}\cdot\frac{\delta^P}{k^P}\right)$$

$$+ \frac{H_m}{w_v \hat{C}_p}\left(\frac{1-\cosh\gamma}{\cosh\gamma}\right) - \frac{\mathcal{K}\beta\hat{\lambda}_v}{h_a}(\rho_s - \rho_a) \quad (6.2\text{-}38)$$

or

$$q_0 = \frac{(T_c - T_a) - \left(H_m/w_v\hat{C}_p\right)\left[(1-\cosh\gamma)/\cosh\gamma\right] + \left(\mathcal{K}\beta\hat{\lambda}_v/h_a\right)(\rho_s - \rho_a)}{1/h_a + \delta^I/k^I + \delta^S/k^S + \delta^F/k^F + \left[(\tanh\gamma)/\gamma\right]\cdot(\delta^P/k^P)}.$$

$$(6.2\text{-}39)$$

Therefore, we can calculate q_0 when all other quantities are known. Then the intermediate boundary temperatures can be computed via Eqs. 6.2-17a–d and 6.2-37.

From Eq. 6.2-39 we can see that there are two alternate ways of dissipating excess body heat when ambient temperature is higher than normal: increasing the heat transfer coefficient h_a and increasing the evaporative rate. The former can be accomplished by increasing the air flow such as by using a fan or by immersing the body in a fluid with higher heat capacity,* such as water (see Chapter 7). The latter is achieved by fanning (increasing the mass transfer coefficient $\mathcal{K}$, see Chapter 7), wetting the body surface (increasing β), and/or increasing $\rho_s - \rho_a$ (such as via rubbing alcohol).[†] In fact, when the ambient temperature is equal to or higher than the body temperature, perspiration is the only means of dissipating body heat. Thus, on humid days in the summer, one feels

*This is also why in the winter one feels colder in damp air because it has a higher heat capacity than dry air.
†Because the vapor pressure of alcohol at the body surface temperature is higher than that of water, $\rho_s - \rho_a$ is larger for alcohol than for water.

especially hot because the higher ρ_a greatly reduces this mode of heat dissipation.

When H_m increases beyond normal values, such as when one exercises or is infected, the above means can also be used to prevent the body from reaching excessively high temperatures.

Example 6.2-4   Heat Transfer in Extremities [13].   In this example we shall consider steady-state heat conduction with heat generation in limbs. The analysis differs from the previous ones in several respects.

First, we are considering the entire arm or leg, instead of the layers near the surface. By comparison, the skin and fat layers are so thin that we now consider that each limb consists of only two tissue layers—the deep tissue (internal core)* and the superficial tissue. As such, we can no longer neglect the curvature, which means rectangular geometry is no longer applicable. We now view the limb as two concentric tissue cylinders.

Second, although in many cases cylinders with such a large length–diameter ratio (≈ 6) can be considered infinite (i.e., the heat loss from the ends is negligible compared with that through the cylindrical surface), in the present case there is a definite decreasing-temperature trend from the torso end of the limb to the other. Thus, we choose not to neglect the longitudinal heat flow and consider it as a finite cylinder. Hence, heat flow is two-dimensional (axial and radial).

Third, although in Example 6.2-2 we lumped all of the heat generation into the effective thermal conductivity and in Example 6.2-3 we made a detailed analysis of both the convective and generated heat with practically no simplifying assumptions, in the present case we are making a compromise by considering each heat term separately but assuming that the blood leaves the perfused tissue (either layer) not at the local tissue temperature but at the temperature of the venous blood. This means that the convective heat term is $w_v \hat{C}_p (T_A - T_V)$, where T_A and T_V are the arterial and venous blood temperatures, respectively, and assumed constant and known. Hence, the convective heat is not temperature- or position-dependent. Strictly speaking, this is not true, because the venous blood temperature is only the average temperature of all blood returned to the vein from various localities. That from near the skin surface is lower than T_V, and that from the inner core higher than T_V. However, since this assumption will greatly simplify our analysis and does not alter the basic features of the result, we will accept it.

Using the same nomenclature as the three previous examples, except I for the deep tissue and II for the superficial tissue, we can write, again

---

*The bones and blood vessels are also lumped into this core.

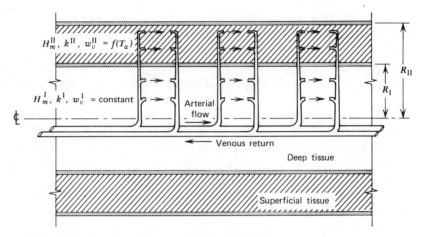

Figure 6.2-9. Two-tissue-layer model of the limb [13].

from Eq. E of Table 4.12-3:

$$k^{\mathrm{I}}\left[\frac{1}{r}\frac{\partial}{\partial r}\left(r\frac{\partial T^{\mathrm{I}}}{\partial r}\right)+\frac{\partial^2 T^{\mathrm{I}}}{\partial z^2}\right]+w_v^{\mathrm{I}}\hat{C}_p(T_A-T_V)+H_m^{\mathrm{I}}=0 \quad (6.2\text{-}40)$$

for deep tissue layer and

$$k^{\mathrm{II}}\left[\frac{1}{r}\frac{\partial}{\partial r}\left(r\frac{\partial T^{\mathrm{II}}}{\partial r}\right)+\frac{\partial^2 T^{\mathrm{II}}}{\partial z^2}\right]+w_v^{\mathrm{II}}\hat{C}_p(T_A-T_V)+H_m^{\mathrm{II}}=0 \quad (6.2\text{-}41)$$

for superficial tissue. The following boundary conditions govern:

B.C.1:   at $z=0$, $T^{\mathrm{I}}=T^{\mathrm{II}}=T_A$ for all $0\leqslant r\leqslant R_{\mathrm{II}}$

B.C.2:   at $z=L$, $\dfrac{\partial T^{\mathrm{I}}}{\partial z}=\dfrac{\partial T^{\mathrm{II}}}{\partial z}=0$ for all $0\leqslant r\leqslant R_{\mathrm{II}}$

B.C.3:   at $r=0$, $\dfrac{\partial T^{\mathrm{I}}}{\partial r}=0$ for all $0\leqslant z\leqslant L$

B.C.4:   at $r=R_{\mathrm{I}}$, $T^{\mathrm{I}}=T^{\mathrm{II}}$

$\qquad\qquad\qquad -k^{\mathrm{I}}\dfrac{\partial T^{\mathrm{I}}}{\partial r}=-k^{\mathrm{II}}\dfrac{\partial T^{\mathrm{II}}}{\partial r}$ for all $0\leqslant z\leqslant L$

B.C.5:   at $r=R_{\mathrm{II}}$, $-k^{\mathrm{II}}\dfrac{\partial T^{\mathrm{II}}}{\partial r}$

$\qquad\qquad\qquad =q_r=h_a(T^{\mathrm{II}}-T_a)+\dot{w}\hat{\lambda}_v$ for all $0\leqslant z\leqslant L,$

where L is the length of the limb.

Two additional assumptions are implicit in the above boundary conditions. One is that where the limb is connected to the torso, the temperature is the same as that of the arterial blood supply. This should not be too far from the truth, in view of the comparative large size of the torso. The other is that heat loss from the end of the limb (fingers or sole) is negligible. The latter assumption we make, despite the finite cylinder analysis, because we consider that heat supply from the torso from one end is gradually lost from the surface, so that toward the other end the gradient is very small. This tends to be true also because the blood at a lower temperature in the venous return in the hand or foot would partially absorb the heat in the supplying artery at a higher temperature.* This tends to "neutralize" the end temperature gradient to zero.

Again, we reduce Eqs. 6.2-40 and 6.2-41 and the associated boundary conditions into dimensionless forms for easy handling:

$$\frac{1}{\xi}\frac{\partial}{\partial \xi}\left(\xi\frac{\partial \Theta^{I}}{\partial \xi}\right) + \frac{\partial^{2}\Theta^{I}}{\partial \zeta^{2}} = \phi^{I} \tag{6.2-42}$$

$$\frac{1}{\xi}\frac{\partial}{\partial \xi}\left(\xi\frac{\partial \Theta^{II}}{\partial \xi}\right) + \frac{\partial^{2}\Theta^{II}}{\partial \zeta^{2}} = \phi^{II} \tag{6.2-43}$$

B.C.1:   at $\zeta = 0$, $\Theta^{I} = \Theta^{II} = 0$ for all $0 \leqslant \xi \leqslant 1$

B.C.2:   at $\zeta = \nu$, $\dfrac{\partial \Theta^{I}}{\partial \zeta} = \dfrac{\partial \Theta^{II}}{\partial \zeta} = 0$ for all $0 \leqslant \xi \leqslant 1$

B.C.3:   at $\xi = 0$, $\dfrac{\partial \Theta^{I}}{\partial \xi} = 0$ for all $0 \leqslant \zeta \leqslant 1$

B.C.4:   at $\xi = \kappa$, $\Theta^{I} = \Theta^{II}, \dfrac{\partial \Theta^{I}/\partial \xi}{\partial \Theta^{II}/\partial \xi} = \dfrac{k^{II}}{k^{I}}$ for all $0 \leqslant \zeta \leqslant 1$

B.C.5:   at $\xi = 1$, $\dfrac{\partial \Theta^{II}}{\partial \xi} = \text{Bi}(1 - \Theta^{II}) + \Lambda$ for all $0 \leqslant \zeta \leqslant 1$,

*See Example 8.8-4 for a discussion of the heat shunt mechanism.

where

$$\Theta^J = \frac{T^J - T_a}{T_A - T_a}, \ (J = \mathrm{I, II}) \qquad \text{(dimensionless temperature)}^*$$

$$\xi = \frac{r}{R_{\mathrm{II}}} \qquad \text{(dimensionless radial coordinate)}$$

$$\zeta = \frac{z}{R_{\mathrm{II}}} \qquad \text{(dimensionless axial coordinate)}$$

$$\phi^J = \frac{w_v^J \hat{C}_p (T_A - T_V) + H_m^J}{k^J (T_A - T_a)} R_{\mathrm{II}}^2, \qquad (J = \mathrm{I, II})$$

$$\kappa = \frac{R_{\mathrm{I}}}{R_{\mathrm{II}}}$$

$$\mathrm{Bi} = \frac{h_a R_{\mathrm{II}}}{k^{\mathrm{II}}} \qquad \left(\begin{array}{l} \text{the Biot number, sequential} \\ \text{convection- conduction ratio} \end{array} \right)^{\dagger}$$

$$\nu = \frac{L}{R_{\mathrm{II}}} \qquad \text{(length-radius ratio of limb)}$$

$$\Lambda = \frac{\dot{w} \hat{\lambda}_v}{k^{\mathrm{II}} (T_A - T_a)} \qquad \text{(perspiration-conduction ratio).}$$

Equations 6.2-42 and 6.2-43 can be solved at least three different ways. For the sake of brevity here and since all of them are methods previously employed, we will leave the details to Appendix D and only describe the essense of the procedures here. One method is to reduce them to ordinary

---

*Here we have a temptation to use the "unaccomplished temperature ratio" $(T_A - T^J)/(T_A - T_a)$. But in problems involving convection (mass or energy carried by flow), it is more common to obtain the "steady-state" solution, Θ_∞, first and then define a "transient" solution as the difference between this steady-state and the current quantities, such as we have done in Examples 6.1-2 and 6.1-3. This transient quantity is in effect an "unaccomplished" one. Since we will use this transient-solution concept, as to be explained later, we use the "accomplished" temperature ratio here so that we will not negate the transient concept by doing it twice.

†We add the word *sequential* to distinguish it from the Nusselt number, to be discussed in Chapter 7. The Nusselt number compares the convective and conductive capabilities of the same fluid, while the Biot number compares the convective capability of a fluid with the conductive capability of a solid with which it is in a sequential heat-transfer path.

differential equations by taking its Laplace transform with respect to ζ. The second one is similar to Example 6.1-2, in which the unsteady-state solution is considered to consist of a steady-state part ($\partial\phi/\partial\tau = 0$ for $\tau \to \infty$) that becomes a function of ξ only and a transient part that is a function of both ξ and τ. This transient part is then expressed as the product of two separate functions, one of ξ and the other of τ.

Applied to the present example, we could imagine that if the limb were long enough, we would have an infinite cylinder for which the differential equations are ordinary ones:

$$\frac{1}{\xi}\frac{d}{d\xi}\left(\xi\frac{d\Theta_\infty^I}{d\xi}\right) = \phi^I, \ldots, \tag{6.2-44}$$

where Θ_∞^I now stands for the solution for Θ^I when $\zeta = \infty$, and so on. This solution is relatively simple; using it we can define a "transient"* solution that is in turn separated into functions of ξ and ζ in product fashion.

The third alternative is a variation of this in that we first consider the radial dimension to be infinite in extent, which in fact gives us an infinite slab of finite thickness (L). This means we first solve

$$\frac{d^2\Theta_\infty^I}{d\zeta^2} = \phi^I, \ldots, \tag{6.2-45}$$

where Θ_∞^I now stands for the solution when $\xi = \infty$, and so on. Then we "carve out"† the cylinder by imposing the cylindrical derivative,

$$\frac{1}{\xi}\frac{\partial}{\partial\xi}\left(\xi\frac{\partial\Theta^I}{\partial\xi}\right), \text{ etc.}$$

(i.e., substitute back in Eq. 6.2-42 $\Theta^I = \Theta_\infty^I - \Theta_{\cdots r''}^I$, etc.) and separate $\Theta_{\cdots r''}^I(\xi,\zeta)$ as the product of two functions, one of ξ and the other of ζ.

In any case, the solutions obtained are (details in Appendix D):

$$\Theta^I = 1 + \frac{\phi^I\zeta}{2}(2\nu - \zeta) - \sum_{n=0}^{\infty} C_n I_0(\beta_n\xi)\sin\beta_n\zeta \tag{6.2-46}$$

$$\Theta^{II} = 1 + \frac{\phi^{II}\zeta}{2}(2\nu - \zeta) - \sum_{n=0}^{\infty}\{B_{1n}I_0(\beta_n\xi) + B_{2n}K_0(\beta_n\xi)\}\sin\beta_n\zeta, \tag{6.2-47}$$

---

*Quotation marks are used here because the problem is not really a time-dependent one. Actually, we are saying "away from the infinite cylinder"

†See the discussion of Example 6.2-5.

where

$$\beta_n = \frac{(n+\frac{1}{2})\pi}{v}, \qquad n = 0,1,2,3,\ldots,$$

and the constants B_{1n}, B_{2n}, and C_n can be obtained from solving the simultaneous algebraic equations

$$(B_{1n} - C_n)I_0(\beta_n\kappa) + B_{2n}K_0(\beta_n\kappa) = \frac{2(\phi^{II}-\phi^{I})}{v\beta_n^3}$$

$$B_{1n}F_n + B_{2n}G_n = \frac{2(\text{Bi}+\Lambda)}{v\beta_n} + \frac{2\text{Bi}\phi^{II}}{v\beta_n^3}$$

$$C_n = \frac{k^{II}}{k^{I}}\left[B_{1n} - B_{2n}\frac{K_1(\beta_n\kappa)}{I_1(\beta_n\kappa)}\right],$$

where

$$F_n = \text{Bi}I_0(\beta_n) + \beta_n I_1(\beta_n)$$

$$G_n = \text{Bi}K_0(\beta_n) - \beta_n K_1(\beta_n).$$

These temperature profiles are plotted in Figure 6.2-10 in both the longitudinal and radial directions, for a value of $\kappa = 0.8$. Note the suddenly increasing drop of temperature starting at this point because we artificially "install" a larger heat loss in the superficial region.

A similar problem for the head (spherical geometry) has also been worked out [13], but it is even more complex because it is divided into three regions in one of which the blood flows and carries heat undimensionally in the r-direction. This means the convective (v_r) term in the equation of energy is retained. Without presenting this problem in detail, it should be mentioned that there is potential importance in such analysis because it has been found in an industrial or aerospace enviroment if it is difficult to keep the entire surroundings at a comfortable temperature, at least the head must be surrounded by a reasonable temperature for the neural system to function properly. This analysis is thus quite useful in the design of water-circulated and cooled headgears.

A similar study connecting the various body parts (head, neck, torso, limbs, etc.) with arterial blood supply and venous return has also been made [14]. Rate of heat transfer through perfused tissue can also be used to determine the rate of blood flow [15].

Example 6.2-5 Sterilization of Canned Goods. One problem frequently encountered in the food and medical-supply industries or even at home is the thermal sterilization of canned or bottled goods ranging from pharma-

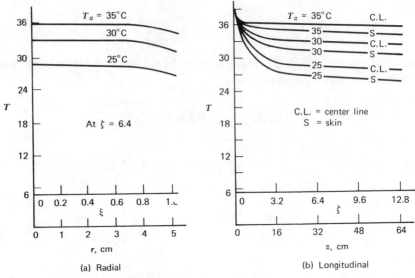

Figure 6.2-10. Temperature profiles in the arm at various environmental temperatures [13]. The following parameters are used:

τ_a	ϕ_w^{I}	ϕ_w^{II}	ϕ_H^{I}	ϕ_H^{II}	βi	Λ
25	0.1015	0.1758	0.0500	0.03535	1.321	0.0208
30	0.0803	0.3205	0.0436	0.03080	1.321	0.0398
35	0.0502	0.3780	0.0436	0.03080	1.321	0.0945

$$\phi^J = \phi_w^J + \phi_H^J; \quad \phi_w^J = \frac{w_v^J \hat{C}_p (T_A - T_V)}{k^J (T_A - T_a)} R_{\mathrm{II}}^2, \quad \phi_H^J = \frac{H_m^J R_{\mathrm{II}}^2}{k^J (T_A - T_a)}$$

ceutical products to baby formula. In such cases as drugs that are thermally degradable, precise temperature and time controls are of critical importance. In this example, we will show how heat-transfer analysis helps predict the time required and maintain such controls.

We shall consider the simple case of a cylindrical can with solid content initially at T_i being immersed in a steambath such that the surface temperature of the can is always at T_s after $t=0$. Thermal energy "diffuses" into the can, so to speak, from all directions. We assume that even with liquid content, the convection inside is negligible. Thus, our problem is simplified to that of pure conduction. The situation is schematically depicted in Figure 6.2-11 where the origin of the cylindrical coordinate system is purposely located at the point where the axis of the

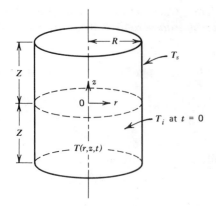

Figure 6.2-11. Thermal sterilization of canned goods.

cylinder penetrates the centerplane between the top and bottom surfaces of the can. The advantage of doing this will be apparent toward the end of this example.

Normally, the height and diameter of the can are of similar magnitude, so that it can be considered neither an infinite cylinder nor a thin slab. We must consider it a finite cylinder so that the temperature of the content is a function of the radial and axial coordinates as well as time; that is, $T = T(r,z,t)$. In short, we have an unsteady-state problem and a two-dimensional problem wrapped into one.

With the convective velocities neglected, the conduction equation as reduced from Eq. E of Table 4.12-3 is

$$\rho \hat{C}_p \frac{\partial T}{\partial t} = k \left[\frac{1}{r} \frac{\partial}{\partial r} \left(r \frac{\partial T}{\partial r} \right) + \frac{\partial^2 T}{\partial z^2} \right], \tag{6.2-48}$$

where ρ, $\hat{C}_p$, and k are, respectively, the density, heat capacity (specific heat), and thermal conductivity of the content of the can. The initial and boundary conditions accompanying Eq. 6.2-48 are

I.C.:	at $t=0$,	$T = T_i$	
B.C.1:	at $r=0$,	$\dfrac{\partial T}{\partial r} = 0$	for all $-Z \leqslant z \leqslant +Z$
B.C.2:	at $r=R$,	$T = T_s$	for all $-Z \leqslant z \leqslant +Z$
B.C.3:	at $z=0$,	$\dfrac{\partial T}{\partial z} = 0$	for all $0 \leqslant r \leqslant R$
B.C.4:	at $z = \pm Z$,	$T = T_s$	for all $0 \leqslant r \leqslant R$.

Hoping to be able to solve this problem by the method of separation of variables, we define the *unaccomplished temperature ratio* as the dimensionless temperature

$$\Theta = \frac{T_s - T}{T_s - T_i},$$

along with

$$\xi = \frac{r}{R} \qquad \text{(dimensionless radial coordinate)}$$

$$\zeta = \frac{z}{Z} \qquad \text{(dimensionless axial coordinate)}$$

$$\Lambda = \frac{Z}{R} \qquad \text{(height : diameter ratio)}$$

$$\tau = \frac{kt}{\rho \hat{C}_p R^2} \qquad \text{(dimensionless time).}^*$$

Equation 6.2-48 and the initial and boundary conditions now become

$$\frac{\partial \Theta}{\partial \tau} = \frac{1}{\xi}\frac{\partial}{\partial \xi}\left(\xi \frac{\partial \Theta}{\partial \xi}\right) + \frac{1}{\Lambda^2}\frac{\partial^2 \Theta}{\partial \zeta^2} \qquad (6.2\text{-}49)$$

I.C.:   at $\tau = 0, \Theta = 1$

B.C.1:   at $\xi = 0, \dfrac{\partial \Theta}{\partial \xi} = 0$ for all $-1 \leqslant \zeta \leqslant +1$

B.C.2:   at $\xi = 1, \Theta = 0$ for all $-1 \leqslant \zeta \leqslant +1$

B.C.3:   at $\zeta = 0, \dfrac{\partial \Theta}{\partial \zeta} = 0$ for all $0 \leqslant \xi \leqslant 1$

B.C.4:   at $\zeta = \pm 1, \Theta = 0$ for all $0 \leqslant \xi \leqslant 1$.

Postulating the solution to be of the form

$$\Theta(\xi, \zeta, \tau) = F(\xi) \cdot G(\zeta) \cdot H(\tau), \qquad (6.2\text{-}50)$$

*Remember that the group $k/\rho\hat{C}_p$ is thermal diffusivity.

we can rewrite Eq. 6.2-49 as

$$\frac{1}{H}\frac{dH}{d\tau} = \frac{1}{F}\cdot\frac{1}{\xi}\frac{d}{d\xi}\left(\xi\frac{dF}{d\xi}\right) + \frac{1}{\Lambda^2}\cdot\frac{1}{G}\frac{d^2G}{d\zeta^2}, \tag{6.2-51}$$

the left-hand side of which is a function of τ only, and the right-hand side, of ξ and ζ only. For this to be always true, both sides must be equal to a constant—the same one, of course. Moreover, as shown in Example 6.1-2, since we expect Θ, and hence H, to decrease with time, this constant must be negative, say, $-\beta^2$. Thus, from the left-hand side, we have

$$\frac{1}{H}\frac{dH}{d\tau} = -\beta^2, \tag{6.2-52}$$

which readily gives

$$H = C_1 e^{-\beta^2\tau}, \tag{6.2-53}$$

where C_1 is the integration constant to be evaluated in combination with others later. The right-hand side of Eq. 6.2-51, being also equal to $-\beta^2$, can be rearranged to

$$-\frac{1}{F}\cdot\frac{1}{\xi}\frac{d}{d\xi}\left(\xi\frac{dF}{d\xi}\right) = \frac{1}{\Lambda^2}\frac{1}{G}\frac{d^2G}{d\zeta^2} + \beta^2. \tag{6.2-54}$$

Using a similar argument that the left-hand side is a function of ξ only and the right-hand side of ζ only, we can separate Eq. 6.2-54 again, via the constant* α^2, into

$$\frac{1}{\xi}\frac{d}{d\xi}\left(\xi\frac{dF}{d\xi}\right) + \alpha^2 F = 0 \tag{6.2-55}$$

and

$$\frac{d^2G}{d\zeta^2} + (\beta^2 - \alpha^2)\Lambda^2 G = 0 \tag{6.2-56}$$

with

B.C.1:   at $\xi = 0$, $\dfrac{dF}{d\xi} = 0$

B.C.2:   at $\xi = 1$, $F = 0$

B.C.3:   at $\zeta = 0$, $\dfrac{dG}{d\zeta} = 0$

B.C.4:   at $\zeta = \pm 1$, $G = 0$.

---

*Whether we write it as $+\alpha^2$ or as $-\alpha^2$ does *not* matter, because the solution will "correct itself," as we have pointed out in Example 6.1-2. In fact, the same applies to β^2.

The solution of Eq. 6.2-55 with B.C. 1 is

$$F = C_2 J_0(\alpha \xi),$$

(6.2-57)

while that of Eq. 6.2-56 with B.C. 3 is

$$G = C_3 \cos\left(\Lambda\sqrt{\beta^2 - \alpha^2}\ \zeta\right),$$

(6.2-58)

where it appears that the integration constants C_2 and C_3 can be evaluated by application of B.C. 2 and B.C. 4 to Eqs. 6.2-57 and 6.2-58, respectively. However, instead of achieving this, this procedure yields the restrictions on α and β. That is, α must be the roots of

$$J_0(\alpha) = 0,$$

(6.2-59)

with $\alpha_1 = 2.40483\ldots$, $\alpha_2 = 5.52009\ldots$, $\ldots$, while β must be such that

$$\cos\left(\Lambda\sqrt{\beta^2 - \alpha_n^2}\ \right) = 0$$

or*

$$\Lambda\sqrt{\beta_{mn}^2 - \alpha_n^2}\ = (m - \tfrac{1}{2})\pi$$

or

$$\beta_{mn}^2 = \alpha_n^2 + \frac{(m - \tfrac{1}{2})^2 \pi^2}{\Lambda^2}, \qquad m = 0, \pm 1, \pm 2, \ldots.$$

(6.2-60)

Putting together the three solutions, Eqs. 6.2-53, 6.2-57, and 6.2-58, with the principle of superimposition, as we have done in Example 6.1-2, we have†

$$\Theta = \sum_{m=1}^{\infty} \sum_{n=1}^{\infty} A_{mn} J_0(\alpha_n \xi) \cos{(m - \tfrac{1}{2})\pi\zeta} \cdot \exp\left\{ -\left[\alpha_n^2 + \frac{(m - \tfrac{1}{2})^2 \pi^2}{\Lambda^2} \right] \tau \right\},$$

(6.2-61)

where

$$A_{mn} = C_1 C_{2n} [C_{3(+m)} + C_{3(-m+1)}].$$

(6.2-6)

*Note that we must use a different index m so as not to confuse it with the index on the argument of the Bessel function.
†Note here that we have "folded" the cosine series at between $m=0$ and $m=1$. Since $\cos(1 - \tfrac{1}{2})\pi\zeta = \cos(0 - \tfrac{1}{2})\pi\zeta$, $\cos(2 - \tfrac{1}{2})\pi\zeta = \cos(-1 - \tfrac{1}{2})\pi\zeta$, and so on, each two cosine terms can actually be combined into one via the combined constant $C_{3(m)} + C_{3(-m+1)}$.

Applying the initial condition to Eq. 6.2-61, we obtain

$$1 = \sum_{m=1}^{\infty} \sum_{n=1}^{\infty} A_{mn} J_0(\alpha_n \xi) \cos\left(m - \tfrac{1}{2}\right)\pi\zeta, \qquad (6.2\text{-}63)$$

from which A_{mn} can be "liberated" by using the orthogonality relationships of Bessel and cosine functions (see Appendix E). Thus, we finally have

$$\Theta(\xi, \zeta, \tau) = 4 \sum_{m=1}^{\infty} \sum_{n=1}^{\infty} \frac{(-1)^{m-1}}{(m - \tfrac{1}{2})\pi} \cdot \frac{J_0(\alpha_n \xi)}{\alpha_n J_1(\alpha_n)}$$

$$\cdot \cos\left(m - \tfrac{1}{2}\right)\pi\zeta \cdot \exp\left\{-\left[\alpha_n^2 + \frac{(m - \tfrac{1}{2})^2 \pi^2}{\Lambda^2}\right]\tau\right\}. \qquad (6.2\text{-}64)$$

If we separate the double series in Eq. 6.2-64, the result becomes extremely interesting and revealing:

$$\Theta(\xi, \zeta, \tau) = 2 \sum_{m=1}^{\infty} \frac{(-1)^{m-1}}{(m - \tfrac{1}{2})\pi} \cos\left(m - \tfrac{1}{2}\right)\pi\zeta \cdot \exp\left[-\frac{(m - \tfrac{1}{2})^2 \pi^2 \tau}{\Lambda^2}\right]$$

$$\cdot 2 \sum_{n=1}^{\infty} \frac{J_0(\alpha_n \xi)}{\alpha_n J_1(\alpha_n)} e^{-\alpha_n^2 \tau}. \qquad (6.2\text{-}65)$$

We can see that the second part is the solution for the heating of an infinite cylinder, similar to that for the unsteady-state flow in a circular tube (Eq. 6.1-28)* and identical to the transient diffusion in cylindrical tissues (Eq. 6.3-12), while the first part is similar to that for finite slab (Eq. 11.1-31, Ref. 16, p. 356).[†] This means the solution for a finite cylinder is the combination of those for the infinite cylinder and finite slab. The interesting thing is that geometrically the finite cylinder can actually be considered as the "cross" between an infinite cylinder and the parallel surfaces of a finite slab as shown in Figure 6.2-12. In fact, it is theoretically permissible to multiply solutions for simple geometries together to obtain that for a more complex geometry, as long as the directions of their axes are perpendicular to each other.

*The slight difference is created by the extra term $(1 - \xi^2)$ in the fluid-flow problem.
[†]Thus, we have located the origin of the coordinate system as described at the beginning because in most problems of flat-plate geometry, the centerplane is used as the zero plane.

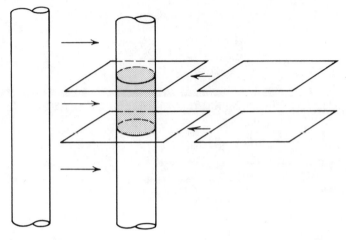

Figure 6.2-12. The finite cylinder as the intersection between an infinite cylinder and the parallel surfaces of a finite slab.

For those who do not wish to wade through all of the derivations and numerical calculations, the solutions for the three basic geometries are available graphically in Figure 6.2-13[‡] so that the temperature anywhere inside of a composite body of any dimension can be obtained by simply multiplying the appropriate unaccomplished temperature ratios of the constituent geometries together.

Such composition of solutions applies to diffusion and fluid flow as well but only for the "unaccomplished" concentration ratio or the transient velocity, such as ϕ_t in Example 6.1-2, whichever the case may be.

In cases where the surface temperature is not specified, the boundary condition at $r = R$ may be replaced by

$$-k\frac{\partial T}{\partial r}\bigg|_{r=R} = h_a(T - T_a)\bigg|_{r=R} \tag{6.2-66}$$

(where h_a is the heat-transfer coefficient and T_a is the ambient temperature) similar to those of Examples 6.2-2 through 6.2-4. The resulting temperature profile can be shown to be a function of the Biot number, $\mathrm{Bi} = h_a R/k$, in addition to the dimensionless time and coordinates—that is, $T(r/R, z/Z, \alpha t/R^2, h_a R/k)$.[‡] The graphical representation would be greatly expanded. It is customary to make a plot of the temperature ratio, $(T_a - T)/(T_a - T_i)$ versus $\alpha t/R^2$ at different $h_a R/k$ values for each coordinate position (see, for example, Ref. 17, p. 149).

[‡]Where α is now the thermal diffusivity, $k/\rho\hat{C}_p$.

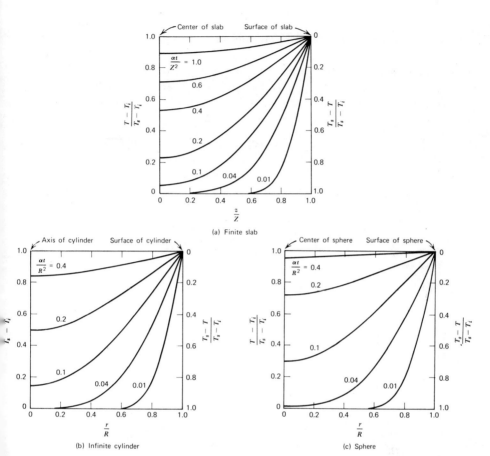

Figure 6.2-13. Temperature profiles inside solids of the three basic geometrical shapes. In each chart the left-hand ordinate is the accomplished temperature ratio $(T - T_i)/(T_s - T_i)$, while the right-hand is the unaccomplished one, $(T_s - T)/(T_s - T_i)$. It is the unaccomplished ratio that has the multiplicative property. These charts are applicable to analogous mass transfer and fluid flow problems. [From Carslaw, H.S.; and Jaeger, J.C. *Conduction of Heat in Solids.* Oxford Univ. Press: 1959, pp. 101, 200 & 234].

From a solution such as Eq. 6.2-65, the lowest, as well as the average, temperature inside the can at any time can be calculated. The time required for any spot to reach a certain minimum temperature can also be determined. If the content of the can is a liquid, vigorous shaking would produce the average temperature. However, after mixing, the heat-transfer rate would be altered because the temperature gradient is disturbed; in fact, it becomes a lumped-parameter system.

Exercise.   Write down (do not derive) the unsteady-state temperature distribution in a rectangular brick being heated from all sides, by considering it as the intersection of the parallel surface of three mutually perpendicular finite slabs as follows:

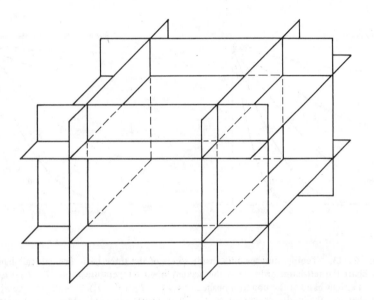

Example 6.2-6   Forced Convection.   In this example, we consider a fluid at temperature T_0 flowing through a circular tube and entering a heat zone at $z = 0$. The heating is such that the heat flux is always constant (q_1) at the wall. This system is depicted in Figure 6.2-14. We include this problem because it introduces the concept of flow-averaged (or cup-mixing) temperature, as distinguished from the volume-averaged temperature. The difference between the two is analogous to that between the large-vessel and capillary hematocrits, to be discussed in Chapter 9. A secondary reason for using this example is that it involves a different kind of

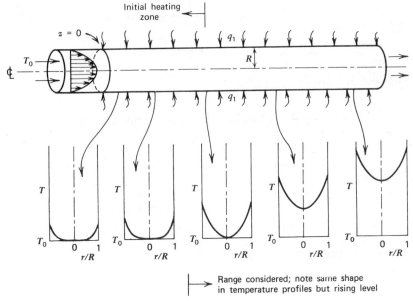

Figure 6.2-14. Heating of fluid with constant wall flux.

separation of variables, potentially useful in physiological transport problems, such as in Example 10.4-1.

The complete solution of this problem [18] can be obtained by the same separation-of-variables method as in the last example and in Example 6.1-2. Here we only solve for the limiting case of large downstream distance from the entrance of the heating zone.

Assuming laminar flow, the parabolic velocity profile is substituted into Eq. E of Table 4.12-3, which upon simplification yields

$$\rho \hat{C}_p v_{max}\left[1-\left(\frac{r}{R}\right)^2\right]\frac{\partial T}{\partial z}=k\frac{1}{r}\frac{\partial}{\partial r}\left(r\frac{\partial T}{\partial r}\right). \tag{6.2-67}$$

Here we have also assumed that in the axial direction, heat transported by flow is much larger than that by conduction, so that the term $k(\partial^2 T/\partial z^2)$ is negligible compared with the forced-convection term. The boundary conditions are

$$\text{B.C. 1:} \qquad \text{at } r=0, \frac{\partial T}{\partial r}=0$$

$$\text{B.C. 2:} \qquad \text{at } r = R, \; q_r = -k\frac{\partial T}{\partial r} = -q_1^*$$

$$\text{B.C. 3:} \qquad \text{at } z = 0, \; T = T_0$$

Using the dimensionless quantities

$$\Theta = \frac{T - T_0}{q_1 R / k} \qquad \text{(dimensionless temperature)}^\dagger$$

$$\xi = \frac{r}{R} \qquad \text{(dimensionless radial coordinate)}$$

$$\zeta = \frac{zk}{\rho \hat{C}_p v_{\max} R^2} \qquad \text{(dimensionless axial coordinate)},$$

Eq. 6.2-67 and its boundary conditions become

$$(1 - \xi^2)\frac{\partial \Theta}{\partial \zeta} = \frac{1}{\xi}\frac{\partial}{\partial \xi}\left(\xi \frac{\partial \Theta}{\partial \xi}\right) \tag{6.2-68}$$

$$\text{B.C.1:} \qquad \text{at } \xi = 0, \; \frac{\partial \Theta}{\partial \xi} = 0$$

$$\text{B.C.2:} \qquad \text{at } \xi = 1, \; \frac{\partial \Theta}{\partial \xi} = 1$$

$$\text{B.C.3:} \qquad \text{at } \zeta = 0, \; \Theta = 0.$$

Here a little physical insight helps a great deal. We know that initially the temperature is uniform at T_0. With heating from outside, a certain curved temperature profile will be gradually developed. We can see that after a certain stage, the curvature will be established and will no longer change with distance, but the heat supply per unit tube length will be consumed mainly in raising the level of the entire temperature profile proportionally (see Fig. 6.2-14). In other words, the temperature profile at a large distance from the heating-zone entrance can be considered as

---

*We use $-q_1$ here since we know that in such a heating problem, q_1 is in the direction opposite to r. Actually this sign is immaterial because otherwise we could define Θ later with $-q_1$ in the denominator, or let the dimensionless solution have an opposite sign, as in Section 9.8 of Ref. 16. With this in mind, we can see that the solution here can be applied to both heating and cooling problems.

†Note that we have no upper temperature limit, and hence, no unaccomplished or accomplished ratio can be defined.

composing of a nonchanging curve, $\phi(\xi)$, on top of a linearly increasing level, $K\zeta$, or

$$\Theta = K\zeta + \phi(\xi), \tag{6.2-69}$$

which means that we have separated the dependent variable, Θ, into two additive, rather than multiplicative, parts.

Note here, too, that we have lost one of the three boundary conditions, since $\zeta = 0$ is outside of the range considered. We have to create artificially a fourth one based on the fact that the heat "absorbed" by the fluid through the walls over a certain distance z from the entrance should be exactly equal to that carried out by the fluid flow at z, or

$$2\pi R z q_1 = \int_0^{2\pi} \int_0^R \rho \hat{C}_p (T - T_0) v_z r \, dr \, d\theta \tag{6.2-70}$$

or

$$\text{B.C. 4:} \qquad \zeta = \int_0^1 \Theta(\xi, \zeta)(1 - \xi^2)\xi \, d\xi. \tag{6.2-71}$$

Substituting Eq. 6.2-69 into Eq. 6.2-68, we have

$$K(1 - \xi^2) = \frac{1}{\xi} \frac{d}{d\xi}\left(\xi \frac{d\phi}{d\xi}\right), \tag{6.2-72}$$

which upon integration and application of B.C. 1, B.C. 2, and B.C. 4 yields

$$\Theta(\xi, \zeta) = 4\zeta + \xi^2 - \tfrac{1}{4}\xi^4 - \tfrac{7}{24}. \tag{6.2-73}$$

The area-averaged or volume-averaged temperature, as usual, is

$$\langle T \rangle(z) = \frac{\displaystyle\int_0^{2\pi} \int_0^R T(r,z) r \, dr \, d\theta}{\displaystyle\int_0^{2\pi} \int_0^R r \, dr \, d\theta} \tag{6.2-74}$$

or

$$\langle \Theta \rangle(\zeta) = 2\int_0^1 \Theta(\xi, \zeta)\xi \, d\xi, \tag{6.2-75}$$

while another average temperature can be defined by using $v_z(r)$ as the weighting factor:

$$\frac{\langle v_z T \rangle (z)}{\langle v_z \rangle} = \frac{\int_0^{2\pi} \int_0^R T(r,z) v_z(r) r \, dr \, d\theta}{\int_0^{2\pi} \int_0^R v_z(r) r \, dr \, d\theta} \qquad (6.2\text{-}76)$$

or

$$\frac{\langle v_z \Theta \rangle (\zeta)}{\langle v_z \rangle} = 4 \int_0^1 \Theta(\xi,\zeta)(1-\xi^2) \xi \, d\xi. \qquad (6.2\text{-}77)$$

The latter is the temperature obtained by cutting off the tube at z, bleeding the fluid into a container, and thoroughly mixing it before measuring its temperature. Thus, it is called the flow-averaged, or cup-mixing, temperature. It is more representative of the temperature of the region in which the fluid flows faster. One can readily see that in an analogous situation of nonuniform substrate concentration or cell distribution in flowing blood, we have flow-averaged concentration or large-vessel hematocrit as well.

Exercise.   Use the temperature distribution Eq. 6.2-73 to evaluate the volume-averaged and flow-averaged temperature for the present case. Which is larger? Is this true always? If not, how can we determine when one is larger or smaller?

6.3 MASS TRANSFER PROBLEMS

As in the previous two sections, our presentation of biomedical mass transfer examples will be from simple to complex. It is rather appropriate that the first example chosen is the classical model proposed by a distinguished scientist who won the Nobel prize for his contribution in the medical field. While we plan to take the reader through the usual "ascension" to more exotic analyses almost in parallel to the previous momentum and energy transfer examples, we must caution here that because of the various types and definitions of mass and molar fluxes involved, the analogy, even in mathematical formality, is sometimes subject to modification.

Example 6.3-1   Krogh's Tissue Cylinder.   Some of the earliest, indeed classical, work in oxygen transport was done by August Krogh [19], who proposed that there was a tissue cylinder surrounding each capillary blood vessel that fed oxygen to, and accepted metabolically produced CO_2 from, tissue. Optimal geometry suggests that these cylinders be arranged in hexagonal form, as shown in Figure 6.3-1; a single vessel-tissue cylinder is shown enlarged in Figure 6.3-2.

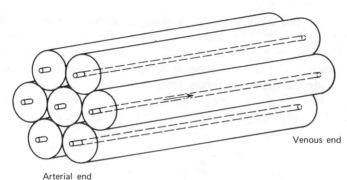

Figure 6.3-1. Krogh tissue-cylinder arrangement [19].

In assuming steady-state diffusion of oxygen through the tissue in which the oxygen is also consumed according to a zero-order metabolic reaction and that the longitudinal oxygen concentration gradient is negligible, the following ordinary differential equation is obtained from the binary-system (oxygen and tissue) continuity equation in cylindrical coordinates (Eq. E, Table 5.5-1):*

$$\mathcal{D}_{AB}\frac{1}{r}\frac{d}{dr}\left(r\frac{dp_A}{dr}\right)=m, \qquad (6.3\text{-}1)$$

where we use the oxygen tension or partial pressure, p_A, as a measure of concentration, as is the usual practice. The boundary conditions for this problem are

$$\text{B.C.1 at } r=R_1, p_A=p_{A_i}$$

$$\text{B.C.2 at } r=R_2, \frac{dp_A}{dr}=0,$$

where R_1 and R_2 are, respectively, the outer radius of the capillary vessel (or the inner radius of the tissue cylinder) and the (outer) radius of the tissue cylinder. B.C.1 implies that the oxygen tension is known at the

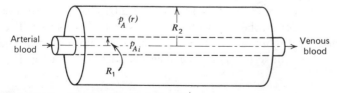

Figure 6.3-2. Krogh's tissue cylinder [19].

*Since oxygen is consumed, $R_A = -m$.

capillary wall and taken to be the average value between the arterial and
venous ends, which is a reasonable assumption as a starting point,
although recent work by Reneau et al. [20] has modified it and improved
the accuracy. B.C.2 is true, because at the cylinder–cylinder boundary,
there are identical oxygen fluxes from both sides; that is, geometrical
symmetry dictates a zero oxygen-tension gradient at the surface of each
cylinder. By integrating Eq. 6.3-1 and applying the boundary conditions,
we have

$$p_A(r)=p_{A_i}+\frac{m}{\mathfrak{D}_{AB}}\left(\frac{r^2-R_1^2}{4}\right)-\frac{a^2}{2}\ln\frac{r}{R_1},\qquad(6.3\text{-}2)$$

where $a=nR_1\sqrt{m/\mathfrak{D}_{AB}}$, the metabolism–diffusion ratio. Note that since
we assume no concentration gradient in the longitudinal (z-) direction,
both at the boundary and in the tissue, the oxygen partial pressure is a
function of the radial coordinate only.

Equation 6.3-2 can be rewritten in dimensionless form:

$$\pi(\xi)=\frac{p_{A_i}-p_A(r)}{mR_1^2/4\mathfrak{D}_{AB}}$$

$$=(1-\xi^2)+2n^2\ln\xi,\qquad(6.3\text{-}3)$$

where $\xi=r/R_1$ and $n=R_2/R_1$. This equation can be conveniently plotted,
as shown in Figure 6.3-3. Note, again, that by using dimensionless quantities we do not have to concern ourselves with the individual values of p_{A_i},
m, R_1, R_2, $\mathfrak{D}_{AB}$, and so on. Only their combinations count. This makes the
graph much more compact and useful. More oxygen transport problems
will be discussed in Chapter 10.

Example 6.3-2   Absorption and Diffusion of γ-Globulin by Lung Tissues.
In calculating the rate of diffusion and absorption of γ-globulin by lung
tissues, Colquhoun [21] treated the latter as long solid cylinders surrounded
by a solution of constant concentration. The difference between this type
of tissue cylinder and Krogh's should be noted, since in Krogh's the tissue
part is an annulus and the oxygen diffuses radially but outwardly in the
annular tissue space. In the present case, the diffusion is toward the axis of
the tissue cylinder all the way to the center with no void space. The system
is depicted in Figure 6.3-4. The equation of continuity for γ-globulin
(represented by subscript A) in one such cylinder is, from Eq. E of Table
5.5-1,

$$\frac{\partial c_A}{\partial t}=\mathfrak{D}_{AB}\frac{1}{r}\frac{\partial}{\partial r}\left(r\frac{\partial c_A}{\partial r}\right)+R_A.\qquad(6.3\text{-}4)$$

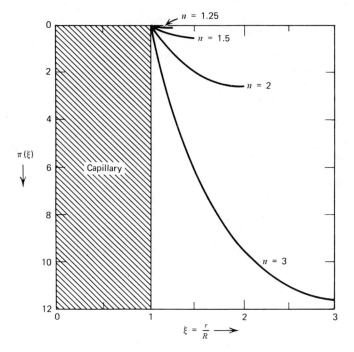

Figure 6.3-3. Radial oxygen-concentration profiles in Krogh's tissue cylinder.

Two different cases were considered. The first was the steady state with first-order reaction. In this case $\partial c_A / \partial t = 0$ and $R_A = -kc_A$ (since γ-globulin is consumed by the reaction). Equation 6.3-4 then becomes

$$\mathscr{D}_{AB} \frac{1}{r} \frac{d}{dr}\left(r\frac{dc_A}{dr}\right) - kc_A = 0, \qquad (6.3\text{-}5)$$

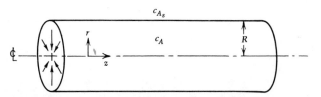

Figure 6.3-4. Diffusion of γ-globulin in the lung-tissue cylinder.

with the following boundary conditions:

B.C.1:   at $r=0$, $\dfrac{dc_A}{dr}=0$ (radial symmetry)

B.C.2:   at $r=R$, $c_A=c_{A_s}$ (known solution concentration)

The general solution to Eq. 6.3-5 is

$$c_A = C_1 I_0(br) + C_2 K_0(br),$$

which I_i and K_i are the modified Bessel functions of the first and second kind, respectively, and of the ith order, of the argument in the parentheses, and $b = \sqrt{k/\mathscr{D}_{AB}}$. Using B.C.1 yields $C_2 = 0$, since $dK_0(br)/dr = -bK_1(br)$ [Eq. B(16) of Table 2.9-2 with $p=0$ but the r there is the br here] and $K_1(0) = \infty$ [see Figure 2.9-1(d)]. Using B.C.2 yields the solution

$$\frac{c_A}{c_{A_s}} = \frac{I_0(br)}{I_0(bR)} . \tag{6.3-6}$$

The amount of γ-globulin in the cylinder (constant, because of steady state) is [Eq. C(5) of Table 2.9-2 with $p=1$ but the r there is the br here]

$$M_s = \int\!\!\int_V\!\!\int c_A\, dV = \frac{2\pi L c_{A_s}}{I_0(bR)} \int_0^R I_0(br) r\, dr$$

$$= \frac{2\pi R L c_{A_s}}{b} \cdot \frac{I_1(bR)}{I_0(bR)} . \tag{6.3-7}$$

The amount of γ-globulin in the cylinder if there were no chemical reaction ($k=0$) would be*

$$M_e = M_s|_{b=0}$$

$$= \pi R^2 L c_{A_s} \tag{6.3-8}$$

which should naturally be true since with no comsumption the tissue cylinder would simply be soaked with γ-globulin at the same concentration as outside at infinite time.

*Since $I_0(0)=1, I_1(0)=0$, then $M_e=0/0$ and the use of L'Hopital's rule is necessitated in order to obtain the result as shown in Eq. 6.3-8.

The ratio

$$\frac{M_s}{M_e} = \frac{2}{bR} \cdot \frac{I_1(bR)}{I_0(bR)} \tag{6.3-9}$$

is the fraction of γ-globulin remaining in the tissue at steady state and $(1 - M_s/M_e)$ is that chemically reacted. The fraction M_s/M_e approaches zero as b approaches infinity since with rapid chemical reaction ($k \to \infty$) and/or extremely slow diffusion ($\mathcal{D}_{AB} \to 0$), the γ-globulin tends to be completely reacted with the tissue before it has any opportunity to diffuse into the cylinder. Thus, the tissue is devoid of γ-globulin. This is the other extreme case from that of $b = 0$.

The steady-state rate of uptake of γ-globulin per unit length of tissue cylinder is

$$F(\infty) = \frac{1}{L} \cdot 2\pi R L \, \mathcal{D}_{AB} \frac{dc_A}{dr}\bigg|_{r=R}$$

$$= 2\pi R \sqrt{k \, \mathcal{D}_{AB}} \, c_{A_s} \frac{I_1(bR)}{I_0(bR)}. \tag{6.3-10}$$

The second case considered was the transient state, with no chemical reaction. In this case a partial differential equation results from Eq. 6.3-4:

$$\frac{\partial c_A}{\partial t} = \mathcal{D}_{AB} \frac{1}{r} \frac{\partial}{\partial r} \left(r \frac{\partial c_A}{\partial r} \right), \tag{6.3-11}$$

with the same boundary conditions as in the previous case plus the initial condition that at $t = 0, c_A = c_{A_0}$ (the initial γ-globulin concentration in the tissue). By properly defining dimensionless variables, the solution to Eq. 6.3-11 can be obtained by either separation of variables or Laplace transform. The procedure is similar to those of Examples 6.1-2 and 6.2-5, with the solution being

$$\Gamma(\xi, \tau) = 2 \sum_{n=1}^{\infty} \frac{J_0(\alpha_n \xi)}{\alpha_n J_1(\alpha_n)} e^{-\alpha_n^2 \tau}, \tag{6.3-12}$$

where

$$\Gamma = \frac{c_{A_s} - c_A}{c_{A_s} - c_{A_0}} \qquad \text{(unaccomplished concentration ratio)}$$

$$\xi = \frac{r}{R} \qquad \text{(dimensionless radial coordinate)}$$

$$\tau = \frac{\mathcal{D}_{AB} t}{R^2} \qquad \text{(dimensionless time)}$$

and α_n $(n = 1, 2, \ldots, \infty)$ are the roots of $J_0(\alpha_n \xi) = 0$. Note that in solving for the integration constant, the orthogonality relationship of the Bessel function has to be used.

The amount of γ-globulin that has diffused into the ith cylinder (with radius R_i and length L_i) at time t is

$$m_i(t) = \int\int\int_{V_i} \left[c_{A_i}(r_i, t) - c_{A_0} \right] dV_i$$

$$m_i(\tau_i) = 2\pi R_i^2 L_i (c_{A_s} - c_{A_0}) \int_0^1 \Gamma(\xi, \tau_i) \xi \, d\xi$$

$$= (c_{A_s} - c_{A_0}) \pi R_i^2 L_i \left\{ 1 - 4 \sum_{n=1}^{\infty} \frac{\exp(-\alpha_n^2 \tau_i)}{\alpha_n^2} \right\}. \qquad (6.3\text{-}13)^*$$

At steady-state ($\tau = \infty$),

$$m_i(\infty) = (c_{A_s} - c_{A_0}) \pi R_i^2 L_i. \qquad (6.3\text{-}14)$$

The difference between Eqs. 6.3-8 and 6.3-14 is that the former gives the γ-globulin content of the tissue cylinder at steady state, including that initially present, while the latter is the amount that has diffused into the cylinder at steady state since time $t = 0$, for which the initially present amount, namely, $c_{A_0} \pi R_i^2 L_i$ must be deducted.

The transient γ-globulin content of the ith cylinder as a fraction of the

*There is an index i on τ because $\tau_i = \mathcal{D}_{AB} t / R_i^2$, but none on ξ (even though $\xi = r / R_i$) since it is a dummy variable in the integration.

steady-state content is

$$f_i(t) = \frac{m_i(t)}{m_i(\infty)}$$

$$= 1 - 4 \sum_{n=1}^{\infty} \frac{1}{\alpha_n^2} \exp(-\alpha_n^2 \tau_i). \qquad (6.3\text{-}15)$$

Colquhoun also considered a population of cylinder size. Let P_i be the fraction of cylinders with radius R_i and length L_i; the overall γ-globulin uptake, is

$$M(t) = \sum_i P_i m_i(t)$$

$$= (c_{A_s} - c_{A_0}) \sum_i f_i(t) P_i \pi R_i^2 L_i \qquad (6.3\text{-}16)$$

or

$$F(t) = \frac{M(t)}{M(\infty)} = \frac{\sum_i f_i(t) P_i R_i^2 L_i}{\sum_i P_i R_i^2 L_i}. \qquad (6.3\text{-}17)$$

Example 6.3-3   Convective Oxygen Transport in Flowing Blood Film [22]. In this example we shall discuss the mass transfer, not in a stationary phase, but in a moving fluid. The particular systems considered is that of a blood film flowing downward adjacent to a solid wall while oxygen diffuses across this film. Thus, oxygen is transported in two directions, aided by the gravity flow of blood. This theoretical analysis finds application in the falling-film type (such as rotating disk) of blood oxygenator and is the same used by engineers in studying the mass transfer in the so-called wetted-wall columns.

Figure 6.3-5 is a schematic diagram of such a system. Under normal conditions in the vertical (z-) direction, the oxygen diffusion rate is much smaller than its rate of transport by flow. This means that the term $\mathcal{D}_{\text{eff}}(\partial^2 c_A / \partial^2 z)$ is negligible compared with $v_z(\partial c_A / \partial z)$.* The equation of continuity thus becomes

$$v_z \frac{\partial c_A}{\partial z} = \mathcal{D}_{\text{eff}} \frac{\partial^2 c_A}{\partial x^2}, \qquad (6.3\text{-}18)$$

*But this does not mean $\mathcal{D}_{\text{eff}}(\partial^2 c_A / \partial x^2)$ is negligible as well, since it is the only transport in the x-direction.

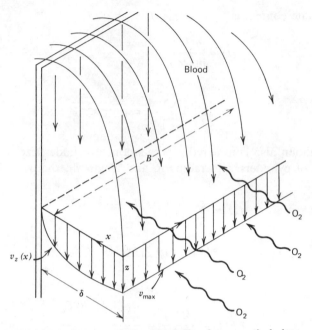

Figure 6.3-5. Oxygen diffusion across falling blood film along a vertical plate.

with the boundary condition

$$\text{B.C. 1:} \qquad \text{at } x = 0, c_A = c_{A_i}$$

$$\text{B.C. 2:} \qquad \text{at } x = \delta, \frac{\partial c_A}{\partial x} = 0$$

$$\text{B.C. 3:} \qquad \text{at } z = 0, c_A = c_{A_0},$$

where c_{A_i} is the oxygen concentration at the gas–liquid interphase and c_{A_0} that of the deoxygenated blood entering from the top of the plate, both supposedly known. The second boundary condition is true because oxygen cannot penetrate the wall. Note that we have implicitly assumed that there is no oxygen transport either by flow or by diffusion in the y-direction. Furthermore, in Eq. 6.3-18 the chemical reaction term representing the oxygen–hemoglobin association is noticeably absent because it has been incorporated into the "effective diffusivity" $\mathcal{D}_{\text{eff}}$.

Via a procedure similar to that of Example 6.1-1, the velocity distribution has been found also to be parabolic [Section 2.2 of Ref. 16.). However, to simplify our problem and because the fluid film is usually very thin, it is taken to be uniform*—that is, $v_z = v_0$ (constant)—for our example.

Using the dimensionless variables

$$\Gamma(\xi,\zeta) = \frac{c_{A_i} - c_A(r,z)}{c_{A_i} - c_{A_0}},\qquad \begin{array}{l}\text{(the unaccomplished}\\ \text{concentration ratio)}\end{array}$$

$$\xi = \frac{x}{\delta},\qquad \text{(dimensionless depth)}$$

$$\zeta = \frac{z\,\mathcal{D}_{AB}}{v_0\delta^2},\qquad \begin{array}{l}\text{(dimensionless}\\ \text{longitudinal distance)}\end{array}$$

Eq. 6.3-18 can be neatly rewritten

$$\frac{\partial \Gamma}{\partial \zeta} = \frac{\partial^2 \Gamma}{\partial \xi^2}, \tag{6.3-19}$$

and the boundary conditions become

$$\begin{array}{lll}\text{B.C.1:} & \text{at } \xi=0, & \Gamma=0\\[4pt] \text{B.C.2:} & \text{at } \xi=1, & \dfrac{\partial \Gamma}{\partial \xi}=0\\[8pt] \text{B.C.3:} & \text{at } \zeta=0, & \Gamma=1.\end{array}$$

By assuming a solution of the form†

$$\Gamma(\xi,\zeta) = \Xi(\xi)Z(\zeta) \tag{6.3-20}$$

and following a procedure similar to those used in Examples 6.1-2, and 6.2-5, we have

$$Z = Ae^{-b^2\zeta} \tag{6.3-21}$$

and

$$\Xi = B_1 \sin b\xi + B_2 \cos b\xi. \tag{6.3-22}$$

Applying B.C. 1 on Eq. 6.3-22 eliminates the cosine term, and applying B.C. 2 leads to the conclusion that b must be equal to $(n-\frac{1}{2})\pi$, where $n=0, \pm 1, \pm 2,\ldots$, for B_1 to be nonzero; otherwise, the solution would be meaningless. Recombining this part with Eq. 6.3-21 and using the ortho-

---

*This is why laminar jets are used for many gas adsorption studies because they do have uniform velocity profiles [23].

†The "separation of variables" method. This problem can also be solved by the Laplace transform method.

gonal relationship of the sine function (Eq. A of Table 2.11-1), we have*

$$\Gamma(\xi,\zeta) = \frac{4}{\pi} \sum_{n=1}^{\infty} \frac{1}{2n-1} \sin(n-\tfrac{1}{2})\pi\xi \cdot \exp\left[-(n-\tfrac{1}{2})^2 \pi^2 \zeta\right]. \quad (6.3\text{-}23)$$

"Translating" this back to dimensional form and using the definition[†]

$$W_A = \int_0^B \int_0^H J_{A_x}^* \bigg|_{x=0} dz\, dy \qquad (6.3\text{-}24)$$

for the rate of oxygen absorbed over the longitudinal distance H, where B is the width of the wall and

$$J_{A_x}^* = -\mathcal{D}_{AB} \frac{\partial c_A}{\partial x},$$

we have

$$W_A = \frac{8 B L v_z \left(c_{A_i} - c_{A_0}\right)}{\pi^2} \sum_{n=1}^{\infty} \frac{1}{(2n-1)^2} \left\{1 - \exp\left[-(n-\tfrac{1}{2})^2 \frac{\pi^2 H \, \mathcal{D}_{AB}}{v_0 \delta^2}\right]\right\}.$$

$$(6.3\text{-}25)$$

The average oxygen concentration is

$$\langle c_A \rangle = c_{A_i} - \left(c_{A_i} - c_{A_0}\right) \frac{8}{\pi^2} \sum_{n=1}^{\infty} \frac{1}{(2n-1)^2} \exp\left[-(n-\tfrac{1}{2})^2 \frac{\pi^2 H \, \mathcal{D}_{AB}}{v_0 \delta^2}\right].$$

$$(6.3\text{-}26)$$

Experimental work by Hershey et al. [22] showed that there was disagreement between the model and the data, indicating that incorporation of chemical reaction into the diffusivity makes it no longer valid to be considered constant. In fact, Buckles et al. [24], by taking into account the oxygen–hemoglobin reaction and considering blood to be a non-Newtonian fluid according to Casson's model (see Section 9.1), developed a model that yielded closer agreement. However, the mathematics was such that the solution could not be expressed in closed form and that a

---

*Note that the sine series is also folded at between $n=0$ and $n=1$, similar to the cosine series in Example 6.2-5.
[†]By this, we imply that since x_A is very small, $N_{A_x} = J_{A_x}$.

numerical solution was obtained.

   Exercise.   Obtain the same expression as Eq. 6.3-25 by using

$$W_A = \int_0^B \int_0^\delta \left[c_A(x, H) - c_{A_0} \right] v_0 \, dx \, dy.$$

Why can one do this?

PROBLEMS

6.A Wall Slippage of a Newtonian Fluid

   Certain biological fluids exhibit slippage phenomena at the tube wall; that is, instead of $v_z = 0$, the boundary condition at $r = R$ is $v_z = v_R$. Show that
(a)   the velocity profile of the laminar flow of a Newtonian fluid in a straight circular tube with wall slippage is

$$v_z(r) = v_R + \frac{\Delta P R^2}{4\mu L} \left[1 - \left(\frac{r}{R} \right)^2 \right]. \tag{6.A-1}$$

(b)   If the *apparent* viscosity, η_a, of such a flow situation is defined in the H–P fashion by

$$Q = \frac{\pi R^4 \Delta P}{8 \eta_a L},$$

it is related to the inherent viscosity of the fluid, μ, by

$$\frac{1}{\eta_a} = \frac{1}{\mu} + \frac{4 v_R}{\tau_R R}, \tag{6.A-2}$$

where $\tau_R = \Delta P R / 2L$ and is the shear stress at the wall.
(c)   Sometimes the wall slippage velocity v_R is a function of τ_R, Eq. 6.A-2 can then be rewritten as

$$\frac{1}{\eta_a} = \frac{1}{\mu} + \frac{4 \zeta(\tau_R)}{R}. \tag{6.A-3}$$

What is this function $\zeta(\tau_R)$?

6.B Two-Dimensional Flow in Renal Tubule with Exponentially Decreasing Bulk Flow [25, 26]

Subsequent to his work cited in Example 6.1-4, Macey considered the case of exponentially decreasing (with downstream distance) bulk flow at the wall—that is,

$$Q(z) = Q_0 e^{-\alpha z} \qquad (6.B-1)$$

where α is the "decline constant" for wall seepage rate.

(a)   Show that the same set of differential equations as in Example 6.1-4 govern the radial and axial velocity profiles of the flow.

(b)   Write down the boundary conditions for the present case. In particular, by differentiating

$$Q(z) = 2\pi \int_0^R r v_z(r, z) \, dr, \qquad (6.B-2)$$

with respect to z (using the Leibnitz formula; see Section 2.6), show that the boundary condition for the bulk flow at the wall becomes one of exponential decline of seepage rate; that is,

$$v_r(R, z) = v_0 e^{-\alpha z} \qquad (6.B-3)$$

where $v_0 = Q_0 \alpha / 2\pi R$, the wall permeation velocity at tubule inlet. An alternate way of doing this is to differentiate Eq. (6.B-1) and compare the result with that of a differential-macroscopic mass balance over an incremental tubule length dz.

(c)   By defining the *stream function** ψ such that

$$\frac{1}{r} \frac{\partial \psi}{\partial z} \equiv v_r \qquad (6.B-4)$$

$$-\frac{1}{r} \frac{\partial \psi}{\partial r} \equiv v_z. \qquad (6.B-5)$$

*Also called the *potential function*, its meaning is that the lines represented by the equation $\psi = constant$ are lines of constant potentials, or streamlines traced out by the particles of the fluid in a steady-state flow. In the present case they would be obtained by plotting

$$[ArJ_1(\lambda r) + Br^2 J_2(\lambda r)]e^{-\lambda z} = C$$

on a r-versus-z graph after the constants have been evaluated. For more on stream functions, see Section 4.2 of Ref. 16.

We have already shown in Problem 2.B that

$$E^2(E^2\psi)=0, \tag{6.B-6}$$

where the operator E^2 is

$$E^2=\left(\frac{\partial^2}{\partial r^2}-\frac{1}{r}\frac{\partial}{\partial r}+\frac{\partial^2}{\partial z^2}\right). \tag{6.B-7}$$

What are the boundary conditions on ψ now?
(d)   Eqs. 6.B-3, 6.B-4, and 6.B-5 suggests a solution of the form

$$\psi = f(r)e^{-\lambda z} + F(r).$$

Show that the solution of $f(r)$ is such that

$$\psi=[ArJ_1(\lambda r)+ Br^2J_2(\lambda r)]e^{-\lambda z} + F(r), \tag{6.B-8}$$

where A, B, and λ are constants.
(e)   Combining Eqs. 6.B-4, 6.B-5, and 6.B-8 and the appropriate boundary conditions, show that

$$v_r=\frac{J_1(\alpha R)J_1(\alpha r)-(r/R)J_0(\alpha R)J_2(\alpha r)}{J_1^2(\alpha R)-J_2(\alpha R)J_0(\alpha R)}v_0e^{-\alpha z} \tag{6.B-9}$$

$$v_z=\frac{J_1(\alpha R)J_0(\alpha r)-(r/R)J_0(\alpha R)J_1(\alpha r)}{J_1^2(\alpha R)-J_2(\alpha R)J_0(\alpha R)}v_0e^{-\alpha z}$$

$$+\left(\frac{2Q_0}{\pi R^2}-\frac{4v_0}{\alpha R}\right)\left[1-\left(\frac{r}{R}\right)^2\right]. \tag{6.B-10}$$

(f)   Kozinski et al. [26] obtained the intratubular pressure distribution as (proven in Problem 2.F)

$$p(r,z)-p(0,0)=-\left(\frac{8Q_0\mu}{\pi R^4}-\frac{16v_0\mu}{\alpha R^3}\right)z$$

$$\tag{6.B-11}$$

$$-\frac{2v_0\mu}{R}(1-e^{-\alpha z})\frac{J_0(\alpha R)J_0(\alpha r)}{J_1^2(\alpha R)-J_2(\alpha R)J_0(\alpha R)},$$

while Macey's original solution [25] does not include the linear z-term. Which is correct? Why? Can you make this conclusion without checking the detailed mathematical steps—that is, simply from the appearance of the final equations?

(g)   Consider the limiting case for a very small tube and/or slow decline rate—that is, $\alpha R \ll 1$. Show that in this case

$$v_z = v_z(0,z)\left[1-\left(\frac{r}{R}\right)^2\right],$$ (6.B-12)

$$v_r = v_r(R,z)\left[2\frac{r}{R}-\left(\frac{r}{R}\right)^3\right],$$ (6.B-13)

where

$$v_z(0,z) = \frac{2Q_0}{\pi R^2} - \frac{4v_0}{\alpha R}(1-e^{-\alpha z}),$$ (6.B-14)

$$v_r(R,z) = v_0 e^{-\alpha z}.$$ (6.B-15)

Compare this with those in Example 6.1-4.

6.C   Seepage Rate in Hemodialyzer Slit [26]

Consider the blood compartment in a rectangular hemodialyzer as a narrow slit of half-thickness b. Using a coordinate system such that the blood flows in the z-direction while the seepage is outward toward the porous membranes at $x = \pm b$ and assuming again exponentially decreasing bulk flow in the downstream direction, show that the equations corresponding to those in Problem 6.B are

$$Q(z) = Q_0 e^{-\alpha z},$$ (6.C-1)

$$Q(z) = 2W\int_0^b v_z(x,z)dx,$$ (6.C-2)

$$v_x(b,z) = v_0 e^{-\alpha z},$$ (6.C-3)

where

$$v_0 = \frac{Q_0 \alpha}{2W},$$

W = lateral width of compartment (y-direction),

$$v_x \equiv \frac{\partial \psi}{\partial z}, \tag{6.C-4}$$

$$v_z \equiv -\frac{\partial \psi}{\partial x}, \tag{6.C-5}$$

$$\nabla^2(\nabla^2 \psi) = 0, \tag{6.C-6}$$

in which

$$\nabla^2 = \frac{\partial^2}{\partial x^2} + \frac{\partial^2}{\partial z^2} \tag{6.C-7}$$

$$\psi = (A'x \cos \lambda x + B' \sin \lambda x)e^{-\lambda z} + G(x) \tag{6.C-8}$$

$$v_x = \frac{(\alpha \cos \alpha b)(x \cos \alpha x) - \cos \alpha b \sin \alpha x + \alpha b \sin \alpha b \sin \alpha x}{\alpha b - \cos \alpha b \sin \alpha b} v_0 e^{-\alpha z} \tag{6.C-9}$$

$$v_z = \frac{\alpha b \sin \alpha b \cos \alpha x - \alpha x \cos \alpha b \sin \alpha x}{\alpha b - \cos \alpha b \sin \alpha b} v_0 e^{-\alpha z}$$

$$+ \left(\frac{3Q_0}{4Wb} - \frac{3v_0}{2\alpha b} \right) \left[1 - \left(\frac{x}{b} \right)^2 \right] \tag{6.C-10}$$

$$p(x,z) - p(0,0) = -\left(\frac{3Q_0 \mu}{2Wb^3} - \frac{3v_0 \mu}{\alpha b^3} \right) z$$

$$\tag{6.C-11}$$

$$- 2\mu \alpha v_0 (1 - e^{-\alpha z}) \frac{\cos \alpha b \cos \alpha x}{\alpha b - \sin \alpha b \cos \alpha b}$$

$$v_z = v_z(0,z) \left[1 - \left(\frac{x}{b} \right)^2 \right] \tag{6.C-12}$$

$$v_x = \frac{v_x(b,z)}{2} \left[3\frac{x}{b} - \left(\frac{x}{b} \right)^3 \right] \tag{6.C-13}$$

where

$$v_z(0,z) = \frac{3Q_0}{4Wb} - \frac{3v_0}{2\alpha b}(1 - e^{-\alpha z}) \qquad \text{(6.C-14)}$$

$$v_x(b,z) = v_0 e^{-\alpha z}. \qquad \text{(6.C-15)}$$

Sketch the limiting solution for $\alpha b \ll 1$.

6.D Amoeboid Movement [27]

The movement of the pseudopod of an amoeba can be modeled as shown in Figure 6.D(a), in which the fluid "recirculates" between the inner and the outer portions. Allen and Roslansky [28] have measured the velocity profiles and found that the outer portion is truly rigid and thus has a flat (uniform) velocity distribution as shown in Figure 6.D(c) and that the fluid in the inner portion behaves pretty much like a Bingham plastic (see Section 3.5). This means that in the inner tubular section there is a central core with a zero velocity gradient (because the shear stress does not exceed the yield stress, τ_y) and a shear flow region between the core and the outer portion, as shown in Figure 6.D(b) and (c), in which the dotted line indicates the locus of the point $r = R_y$ where the shear stress exceeds the yield stress. For simplicity, the fluid is assumed to be Newtonian once it reaches the hemisphere.

(a)   Assuming that there is no pressure force or body force and that the movement of the amoeba is caused only by the action of the front of the pseudopod, which pulls the rigid Bingham core through the tube, reduce the steady state equation of motion to

$$\frac{d}{dr}(r\tau_{rz}) = 0. \qquad \text{(6.D-1)}$$

(b)   Solve the above equation so that

$$r\tau_{rz} = R_y \tau_y.$$

(c)   Using the Bingham model (Figure 3.5-1) to obtain the velocity distribution in the shear flow region,

$$v_z(r) = \frac{\tau_y R}{\mu}\left\{ \frac{R_y}{R}\ln\frac{r}{kR} + k - \frac{r}{R} \right\} - V, \qquad (6.D\text{-}2)$$

where $kR = R_1$ is the inner radius and V is the backward velocity, in relation to an arbitrary point of the amoeba, such as $z = L$ in Figure 6.D(b), of the rigid outer portion.
(d)   What is the velocity V_c of the rigid Bingham core?
(e)   Since our coordinate system moves with the amoeba, the net volumetric flow rate across any of its cross section must be zero; that is,

$$Q = \int_0^R 2\pi r v_z \, dr = 0$$

$$= \int_0^{R_y} 2\pi r V_c \, dr + \int_{R_y}^{kR} 2\pi r v_z(r)\,dr + \int_{kR}^{R} 2\pi r(-V)\,dr. \qquad (6.D\text{-}3)$$

By substitution of the appropriate previously obtained quantities into Eq. 6.D-3, show that

$$V = \frac{\tau_y R}{6\mu}\left\{ \xi_y^3 - 3k^2\xi_y + 2k^3 \right\}, \qquad (6.D\text{-}4)$$

where $\xi_y = R_y/R$.
(f)   Does Figure 6.D(c) seem to be a reasonably accurate representation of the above results?
(g)   The velocity distribution in the hemispheric portion can also be solved, as shown in Problem 2.C, where the boundary conditions are also listed. What is the physical interpretation of these boundary conditions?
(h)   The flow pattern in the hemisphere is depicted in Figure 6.D(d). Does it seem to be reasonable in view of the mathematical expressions obtained in Problem 2.C?

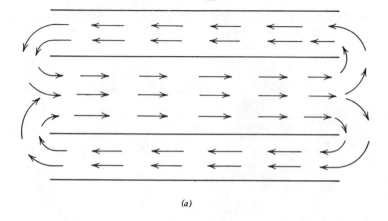

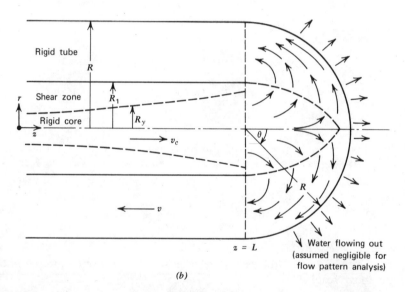

Figure 6.D. Hydrodynamic model of the motion of an amoeba. (a) Flow pattern in pseudopod. (b) Model of pseudopod and action at front of pseudopod. (c) Velocity profile in tubular section. (d) Velocities in hemispherical section. [27]

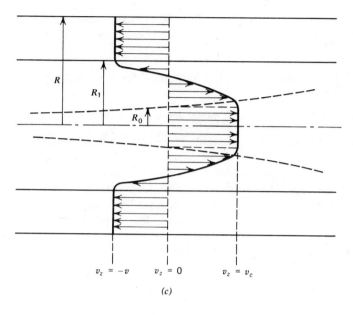

$$v_z = -v \qquad v_z = 0 \qquad v_z = v_c$$

(c)

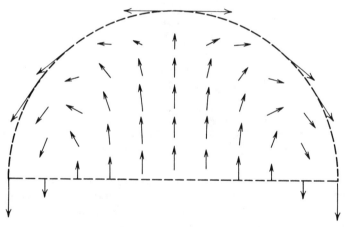

(d)

271

6.E Time-Averaged Volumetric Rate of Pulsatile Blood Flow

Interpret the result obtained in Problem 2.E on the time-average volumetric rate of pulsatile blood flow problem described in Example 6.1-3. What is the meaning of the extra Σ term? Does it seem to be reasonable?

6.F Vibration-Induced Hydrostatic Gradient in the Vascular System [29]

It has been suggested that longitudinal vibration of the human body can be used to create a hydrostatic gradient in the lower extremities in weightless environments. To show whether this would work or what form of motion it would require, let us use a horizontal open tube as our model. If vibration of the tube would generate a *net* flow, then there should be a hydrostatic gradient in the corresponding closed system.

(a)   Assuming there be no pressure or other body forces, show that the differential equation describing this situation is

$$\rho \frac{\partial v_z}{\partial t} = \frac{\mu}{r} \frac{\partial}{\partial r} \left(r \frac{\partial v_z}{\partial r} \right). \tag{6.F-1}$$

(b)   Write down the initial and boundary conditions for the three types of periodic motions shown in Table 6.F. What is the physical significance of these functions?

(c)   Solve Eq. 6.F-1 by Laplace transform. Following our "standard" notations, show that the intermediate and final results are as those shown in Table 6.F.

(d)   Explain the physical meanings of the above results, such as what wave form would be effective for our purpose and under what conditions.

(e)   What would the results be if there were a pressure gradient between two ends of the tube and hence a corresponding term in Eq. 6.F-1?

(f)   Devise a system in which this analysis can be experimentally tested.

(g)   What assumptions have been implicitly embodied in the solution of this problem? What correction must be considered when applied to the human body?

6.G Slippage Between a Fluid and Vibrating Tube

What would the result be for case (b) antisymmetric sawtooth motion in Problem 6.F if we consider that there exists a slippage (see Problem 6.A) between the fluid and the tube (blood is known to exhibit such a tendency over a biological surface)? The slippage velocity can be expressed as a function of the wall shear stress as in Problem 6.A.

Table 6.F*

	Sinusoidal	Antisymmetric Sawtooth	Asymmetric Sawtooth

(a)   (b)   (c)

Tube motion (within one cycle) $= v_z(R)$

- (a) $V\sin\omega t$
- (b) $2V(\omega t - 1/2)$
- (c) $V\omega t$

Average velocity $\langle v_z\rangle$

(a)
$$\frac{4\nu V}{R^2}\sum_{n=1}^{\infty}\left\{\frac{\sin(\omega t-\phi_n)}{\sqrt{\omega_n^2+\omega^2}}+\frac{\omega e^{-\omega_n t}}{\omega_n^2+\omega^2}\right\}$$

(b)
$$-\frac{8\nu V}{\pi R^2}\sum_{n=1}^{\infty}\sum_{k=1}^{\infty}\frac{1}{k}\left\{\frac{\sin(2k\pi\omega t-\phi_n)}{\sqrt{\omega_n^2+(2k\pi\omega)^2}}+\frac{2k\pi\omega e^{-\omega_n t}}{\omega_n^2+(2k\pi\omega)^2}\right\}$$

(c)
$$\frac{V}{2}\left(1-4\sum_{n=1}^{\infty}\frac{e^{-\omega_n t}}{\alpha_n^2}\right)-\frac{4\nu V}{\pi R^2}\sum_{n=1}^{\infty}\sum_{k=1}^{\infty}\frac{1}{k}\left\{\frac{\sin(2k\pi\omega t-\phi_n)}{\sqrt{\omega_n^2+(2k\pi\omega)^2}}+\frac{2k\pi\omega e^{-\omega_n t}}{\omega_n^2+(2k\pi\omega)^2}\right\}$$

Volumetric flowrate at long times $Q_\infty(t)=\pi R^2\langle v_z\rangle_{t\to\infty}$

(a)
$$4\pi\nu V\sum_{n=1}^{\infty}\frac{\sin(\omega t-\phi_n)}{\sqrt{\omega_n^2+\omega^2}}$$

(b)
$$-8\nu V\sum_{n=1}^{\infty}\sum_{k=1}^{\infty}\frac{1}{k}\cdot\frac{\sin(2k\pi\omega t-\phi_n)}{\sqrt{\omega_n^2+(2k\pi\omega)^2}}$$

(c)
$$\frac{V}{2}\pi R^2-4\nu V\sum_{n=1}^{\infty}\sum_{k=1}^{\infty}\frac{1}{k}\cdot\frac{\sin(2k\pi\omega t-\phi_n)}{\sqrt{\omega_n^2+(2k\pi\omega)^2}}$$

Flow per cycle $\int_t^{t+\frac{1}{\omega}}Q_\infty\,dt$

- (a) 0
- (b) 0
- (c) $\dfrac{V\pi R^2}{2\omega}$

*Where $\omega_n=\alpha_n^2\nu/R^2$; $\nu=\mu/\rho$, kinematic viscosity; $\alpha_n=$ roots of Bessel function $J_0(\alpha_n)=0$; $\phi_n=\tan^{-1}\omega/\omega_n$ for (a), and $\tan^{-1}2k\pi\omega/\omega_n$ for (b) and (c), the phase angle.

6.H Heat Conduction Through Cylindrical Body Layers

In Example 6.2-2 we have dealt with the heat conduction through flat layers of tissue, subcutaneous fat, skin, and clothing or fur. However, if we consider the upper extremities, a cylindrical-layer model seems to be more realistic, since the radius of curvature now is smaller than the torso or the lower extremities. Show that for this geometry the overall heat-transfer coefficient based on a constant effective thermal conductivity of the perfused-tissue layer is given by Eq. 6.2-22. Sketch temperature profiles through the layers. Are they still linear?

6.I Cylindrical Body Layers with Perfusion and Metabolic Heat

Using the same assumptions as in Example 6.2-3, obtain the expressions for temperature profiles and overall heat flux (corresponding to Eqs. 6.2-31 and 6.2-39) by assuming the upper extremities as being composed of infinite cylindrical layers of perfused tissue, fat, skin, and clothing or fur.

6.J Optimal Insulation Thickness

In Figure 6.J is depicted a layer of insulation (such as clothing) over a cylindrical body (such as an arm). The skin-surface temperature, T_{IS}, and the ambient temperature, T_a, are both known ($T_{IS} > T_a$). If we increase the insulation thickness, it will cut down the heat *flux*, but the enlarged surface area $2\pi R_2 L$ *may* cause a net *increase* in the *rate* of heat loss, up to a certain point.

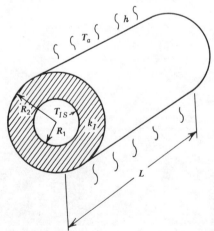

Figure 6.J. Insulation and heat loss around a cylindrical object.

(a)   Show whether there is an optimal insulation thickness $(R_2 - R_1)_{opt}$ (R_1 is constant) at which the rate of heat loss is a maximum. (Hint: You may wish to start with part of Eq. 6.2-22.) Is the result a conditional one?

(b)   For the case where such an optimum does exist, find the thickness $(R_2 - R_1)_0$ at which the rate of heat loss is the same as when there is no insulation at all. What does this mean physically? Is this likely to happen with the human body and the clothing over it?

(c)   Now turn around the problem by considering that there must be a constant amount of metabolic heat to be dissipated. Find the expression for the clothing thickness at which the skin-surface temperature, T_{IS}, will be a maximum or minimum (which is it?). What is the physical meaning of this result?

(d)   Does the optimal insulation thickness situation exist for rectangular and spherical geometries? Explain.

6.K   Unsteady-State Linear Two-Compartment Model of the Liver [30]

Frequently, compartments of the body or body parts are of such lengths that they cannot be regarded as a well-mixed lumped-parameter system such as those in Examples 8.8-1 and 8.8-2. In measuring the sinusoidal and extravascular volumes of the liver, Goresky [30, 31] considered the longitudinal variation of indicator concentration in both the sinusoids and the extravascular space. Here one could apply the *macroscopic* mass balance *to a differential length* in both compartments as in the case of placenta heat-exchanger analogy (Example 8.8-5). But in an unsteady-state situation, which we are dealing with in the present example, such an approach would be quite laborious, as shown in Goresky's original work, because time-differencing is also involved. The present example shows how the equation of continuity can simplify everything.

Figure 6.K is a schematic diagram of the model in which the sinusoidal and extravascular compartments are denoted by A and B, respectively. Following these notations the concentration of the injected indicator in these compartments are c_A and c_B, both being functions of both time, t,

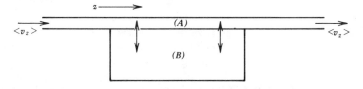

Figure 6.K.   Linear two-compartment model of the liver [30].

and the longitudinal coordinate, z.

(a) Show, by applying the appropriate equation of continuity from Table 5.5-1 we can write, for these compartments separately,

$$\frac{\partial c_A}{\partial t} + \langle v_z \rangle \frac{\partial c_A}{\partial z} = -\frac{kA_i(c_A - c_B)}{V_A}, \qquad (6.K\text{-}1)$$

$$\frac{\partial c_B}{\partial t} = \frac{kA_i(c_A - c_B)}{V_B}, \qquad (6.K\text{-}2)$$

where $\langle v_z \rangle$ is the average velocity of the blood flow in the sinusoid, k is the mass-transfer coefficient, A_i is the interfacial area between the two compartments, and V stands for compartmental volume. What assumptions have been made here? Where do the right-hand side terms come from? How can they be justified?

(b) Combine the two equations in the right proportion to yield the single partial differential equation

$$\frac{\partial c_A}{\partial t} + \gamma \frac{\partial c_B}{\partial t} + \langle v_z \rangle \frac{\partial c_A}{\partial z} = 0, \qquad (6.K\text{-}3)$$

where $\gamma = V_B / V_A$.

(c) Postulating that the intercompartmental transport is so rapid that the two concentrations c_A and c_B are equal at all points,* Goresky reduced Eq. 6.K-3 further to

$$\frac{\partial c_A}{\partial t} + \frac{\langle v_z \rangle}{1+\gamma} \frac{\partial c_A}{\partial z} = 0, \qquad (6.K\text{-}4)$$

which can most conveniently be solved by the Laplace transform method. Show that, if $f(z) = c_A(z,0)$ is the initial longitudinal distribution of the indicator at time $t = 0$ throughout the sinusoidal compartment, its distribution after time $t = 0$ is

$$c_A(z,t) = f\left(z - \frac{\langle v_z \rangle t}{1+\gamma}\right). \qquad (6.K\text{-}5)$$

(d) What does Eq. 6.K-5 mean? Attempt a few initial distributions and examine how the output changes. What happens when $\gamma = 0$?

By a similar approach, Goresky [31] modified this model to include a third (cellular) compartment, which the reader will find it even more interesting and challenging to tackle.

---

*Actually we need only to stipulate that equilibrium is quickly reached so that $c_B = Kc_A$ at all points, whence the equilibrium constant can be "absorbed" into the volume ratio γ.

6.L Perfusion–Diffusion of Inert Fat-Soluble Gases in a Slab of Adipose Tissue [32]

To assess the importance of perfusion versus diffusion in the adipose tissue, inert gases are administered to human subjects and the difference in the rates of uptake of different gases is explained using a suitable compartment model.

Consider a slab of adipose tissue of thickness b and with boundary surface in contact with lean tissue, which could be either poorly or highly perfused depending on where the adipose tissue is stored (such as in the subcutaneous or perirenal regions). The diffusional flux in the longitudinal (flow) direction is negligible compared with the perfusion; thus, the only variation of gaseous concentration that needs to be considered is in the lateral (x-) direction.

(a) Show that the equation of continuity for the adipose-tissue compartment yields

$$\frac{\partial c_A}{\partial t} = \mathfrak{D}_{AB}\frac{\partial^2 c_A}{\partial x^2} + Q\left(c_0 - \frac{\lambda_0}{\lambda}c_A\right), \tag{6.L-1}$$

where c_A is the local test-gas concentration in adipose tissue (moles per unit volume); c_0 is the systemic arterial test-gas concentration (moles per unit volume); $\mathfrak{D}_{AB}$ is the diffusivity of test gas in adipose tissue (area per time); Q is the rate of perfusion (volumetric capillary blood flow per unit tissue volume; time $^{-1}$); λ and λ_0 are solubility constants of the test gas in adipose tissue and in blood, respectively; x is the spatial (lateral) coordinate; and t is the time variable. Justify the inclusion of the last term in this equation.

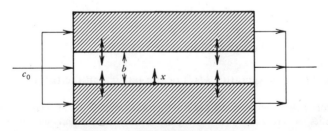

Figure 6.L. Compartment model of adipose and lean tissues [32].

(b) Defining

$$\phi = \frac{p}{p_0} = \frac{\lambda_0 c_A}{\lambda c_0}, \qquad \text{(where the arterial gas tension, } p_0, \text{ is constant)}$$

$$\xi = \frac{x}{b},$$

$$\tau = \frac{\mathcal{D}_{AB} t}{b^2},$$

$$\psi = \frac{\lambda_0 Q b^2}{\lambda \mathcal{D}_{AB}},$$

rewrite Eq. 6.L-1 as

$$\frac{\partial \phi}{\partial \tau} = \frac{\partial^2 \phi}{\partial \xi^2} + \psi (1 - \phi). \qquad (6.\text{L-2})$$

(c) The initial and boundary conditions have now become

$$\text{I.C.:} \quad \text{at } \tau = 0, \phi = 0$$

$$\text{B.C. 1: at } \xi = 0, \phi = 1$$

$$\text{B.C. 2: at } \xi = 1, \phi = 1.$$

What have we assumed about the solubility of the test gas in the *lean* tissue?

(d) Show that the solution to Eq. 6.L-2 for short times is*

$$\phi(\xi, \tau) = 1 - e^{-\psi\tau} + e^{-\psi\tau} \sum_{n=1}^{\infty} (-1)^n \left[\text{erfc} \left(\frac{n + \xi}{2\sqrt{\tau}} \right) + \text{erfc} \left(\frac{n + 1 - \xi}{2\sqrt{\tau}} \right) \right],$$

$$(6.\text{L-3})$$

where $\text{erfc}(x)$ is the complementary error function of x (see Example 2.13-1).

(e) Show that the quantity $M(t)$ of the test gas in the tissue slab at time t

---

*Perhaps the easiest way is by the Laplace transform method. Or alternately show that Eq. 6.L-3 satisfies the differential equation, Eq. 6.L-2, and the initial and boundary conditions.

is

$$M(t) = \int_V c_A \, dV \tag{6.L-4}$$

or

$$\Gamma(\tau) = \int_0^1 \phi \, d\xi$$

$$= 1 - e^{-\psi\tau} + 4\sqrt{\frac{\tau}{\pi}} \, e^{-\psi\tau} + 4e^{-\psi\tau} \sum_{n=1}^{\infty} (-1)^n \left[2\sqrt{\frac{\tau}{\pi}} \, e^{-n^2/4\tau} - n \operatorname{erfc} \frac{n}{2\sqrt{\tau}} \right],$$

$$\tag{6.L-5}$$

where

$$\Gamma(\tau) = \frac{\lambda_0 M(t)}{\lambda V c_0}.$$

(f) Show that for long times (i.e., t very large), solutions equivalent to Eqs. 6.L-3 and 6.L-5 are

$$\phi(\xi, \tau) = 1 - \frac{4}{\pi} e^{-\psi\tau} \sum_{n=0}^{\infty} \frac{\sin(2n+1)\pi\xi}{2n+1} e^{-(2n+1)^2 \pi^2 \tau} \tag{6.L-6}$$

$$\Gamma(\tau) = 1 - \frac{8}{\pi^2} e^{-\psi\tau} \sum_{n=0}^{\infty} \frac{1}{(2n+1)^2} e^{-(2n+1)^2 \pi^2 \tau} \tag{6.L-7}$$

Explain the results.

6.M The Effect of Longitudinal Diffusion in the Nephron on the Accuracy of Stop-Flow Measurements [33]

To examine whether the postoccludally collected urine samples are smeared by the longitudinal diffusion of the indicator, such as inulin, the nephron is modeled as a semi-infinitely long (i.e., $0 \leqslant z \leqslant \infty$) tube in which radial diffusion can be neglected. The stop-flow measuring procedure includes the intravenous injection of the indicator one-half minute prior to the release of the ureter clamp. By computing the indicator concentration distribution due to pure diffusion, we can obtain some idea as to its importance compared to that carried by the flow of the diuretically-formed fluid.

(a) By properly defining the various pertinent quantities such as concentration and diffusivity, show that the unsteady-state diffusion of the inulin can be described by a dimensionless partial differential equation such as Eq. 2.13-1 with the initial and boundary conditions.

(b) In Example 2.13-1 we obtained the solution in the form of Eq. 2.13-14 (or 2.13-19). Using a $\mathcal{D}_{inulin}$ of 1.4×10^{-4} cm^2/sec, calculate the distance downstream from the beginning of the proximal tubule at which the inulin concentration would be 10% of that filtered through the glomerular membrane 10 min after the release.

(c) Assuming the proximal tubule to be 1.28 cm in length, what is your conclusion regarding the reliability of the stop-flow data?

6.N Solute Transport by the Gall Bladder [34]

In determining the mechanism of solute transport by the gall bladder, it is modeled as an infinite plane sheet, since the thickness is very small compared with the radius. A salt (such as NaBr) available at surface concentration c_{A_1} is allowed to diffuse in the serosa with a diffusivity $\mathcal{D}_{AB}$. It is assumed that it cannot diffuse through the epithelial cells.

(a) Show that the salt buildup in the serosa can be described by

$$\frac{\partial c_A}{\partial t} = \mathcal{D}_{AB}\frac{\partial^2 c_A}{\partial x^2}, \tag{6.N-1}$$

with

$$\text{I.C.:}\quad \text{at } t=0, c_A = c_{A_0}$$

$$\text{B.C. 1: at } x=0, \frac{\partial c_A}{\partial x}=0$$

$$\text{B.C. 2: at } x = \delta, c_A = c_{A_1},$$

where c_{A_0} is the initial salt concentration in the serosa, x is the depth in the serosa measured from the interface with epithelial cells, and δ is the thickness of the serosa.

(b)   In actual experimentation, the volume-averaged salt concentration in the serosa is measured. Show that from Eq. 6.N-1, this quantity can be solved as

$$\frac{c_{A_1} - \langle c_A \rangle}{c_{A_1} - c_{A_0}} = \frac{2}{\pi^2} \sum_{n=0}^{\infty} \frac{(-1)^n}{(n+\frac{1}{2})^2} e^{-(n+\frac{1}{2})^2 \pi^2 \tau}, \qquad (6.\text{N-2})$$

where $\tau = \mathcal{D}_{AB} t / \delta^2$. This can be done either by separation of variables or by transform. Using the latter, however, one can avoid the task of obtaining the concentration profiles but directly perform the volume-averaging before inverting the transform.

6.O   Absorption–Diffusion of γ-Globulin by Lung Tissues

(a)   Obtain Eq. 6.3-13 by

$$m_i(t) = \int_0^t \mathcal{D}_{AB} \frac{\partial c_A}{\partial r} \bigg|_{r=R_i} 2\pi R_i L_i dt. \qquad (6.\text{O-1})$$

Why can one do this?

(b)   Rework Example 6.3-2 considering the case of transient diffusion with first-order chemical consumption of γ-globulin. Discuss the result along the lines of that example.

6.P   Axial-Velocity Maximum in Seepage Flow

(a)   In Example 6.1-4 show mathematically whether the maximum axial velocity occurs at the center of the tube or somewhere between the center and the wall. Explain physically.

(b)   Repeat (a) for exponentially decreasing wall seepage rate for both renal tubule and hemodialyzer slit. Again explain physically. (See Problems 6.B and 6.C.)

REFERENCES

1.   Poiseuille, J. L. *Compt. Rend.* **11**, 961, 1041 (1840), and **12**, 112 (1841); Hagen, G. *Ann. Phys, Chem.*, **46**, 423–442 (1839).
2.   Szymanski, P. *J. Math. Pures Appl.*, Ser. 9, **11**, 67–107 (1932).

3. Hershey, D., and Song, G. *A.I. Ch. E. J.*, **13** (3), 491–496 (1967).

4. McDonald, D. A. *J. Physiol.*, **127**, 533–552 (1955).

5. Womersley, J. R. *J. Physiol*, **127**, 553–563 (1955).

6. Womersley, J. R. *Wright Air Development Center Technical Report (TR56-614)*, Wright-Patterson AFB: Ohio, 1957.

7. Atabek, H. B., *Biophys. J.*, **8** (5), 626–649 (1968).

8. Ling, S. C. and Atabek, H. B., *ASME Paper 71-WA/BHF-3(1971)*, to be published in *J. Biomech.*

9. Macey, R. I. *Bull. Math. Biophys.*, **25**, 1–9 (1963).

10. Gottschalk, C. W., and Mylle, M. *Am. J. Physiol.*, **185**, 430 (1956).

11. Seagrave, R. C. *Biomedical Applications of Heat and Mass Transfer*, Iowa State Univ. Press: Ames, 1971, p. 97.

12. Gagge, A. P. In J. D. Hardy, A. P. Gagge, and J. A. J. Stolwijk, Eds., *Physiological and Behavioral Temperature Regulation*. Thomas: Springfield, Ill., 1970, Chapter 4.

13. Nevins, R. G., and Darwish, M. A. In J. D. Hardy, A. P. Gagge, and J. A. J. Stolwijk, Eds., *Physiological and Behavioral Temperature Regulation*, Thomas: Springfield, Ill., 1970, Chapter 21.

14. Wissler, E. H. In J. D. Hardy, A. P. Gagge, and J. A. J. Stolwijk, Eds., *Physiological and Behavioral Temperature Regulation*. Thomas: Springfield, Ill., 1970, Chapter 27.

15. Priebe, L. In J. D. Hardy, A. P. Gagge, and J. A. J. Stolwijk, Eds., *Physiological and Behavioral Temperature Regulation*, Thomas: Springfield, Ill., 1970, Chapter 20.

16. Bird, R. B.; Stewart, W. E.; and Lightfoot, E. N. *Transport Phenomena*, Wiley: New York, 1960.

17. Kreith, F. *Principles of Heat Transfer*, 2nd ed. International: Scranton, Pa., 1965.

18. Siegel, R.; Sparrow, E. M.; and Hallman, T. M. *Appl. Sci. Res.*, **A7**, 386 (1958).

19. Krogh, A. *J. Physiol.*, **52**, 391 (1918–1919).

20. Reneau, D. D.; Bruley, D. F.; and Knisely, M. H. In *Chemical Engineering in Medicine and Biology*, D. Hershey, Ed., Plenum: New York, 1967, p. 135.

21. Colquhoun, D. *J. Physiol.*, **181**, 781 (1965).

22. Hershey, D.; Miller, C. J.; Menke, R. C.; and Hesselberth, J. F. In *Chemical Engineering in Medicine and Biology*, D. Hershey, Ed., Plenum: New York, 1967, p. 117.

23. Scriven, L.E., and Pigford, R.L. *A. I. Ch. E. J.*, **4** (4), 439 (1958); **5** (3), 397 (1959).

24. Buckles, R. G.; Merrill, E. W.; and Gilliland, E. G. *A. I. Ch. E. J.*, **14** (5), 703 (1968).

25. Macey, R. I. *Bull. Math. Biophys.*, **27**, 117 (1965).

26. Kozinski, A. A.; Schmidt, F. P.; and Lightfoot, E. N. *Ind. Eng. Chem. Fund.*, **9** (3), 502 (1970).

27. Staats, W. R., and Wasan, D. T. In *Chemical Engineering in Medicine*, Chem. Eng. Prog. Symp. Series, No. 66, Vol. LXII, E. F. Leonard, Ed., American Institute of Chemical Engineers, New York, 1966, p. 132.

28. Allen, R. D., and Roslansky, J. D. *J. Biophys. Biochem. Cytol.*, **6**, 437 (1959).

29. Lih, M. M., and Newsom, B. D. "Vibration-Induced Hydrostatic Gradient in the Vascular System: A Theoretical Analysis of Fluid Flow in an Oscillating Tube" (in press).

30. Goresky, C. A. *Am. J. Physiol.*, **204**, (4), 626 (1963).

31. Goresky, C. A. *Am. J. Physiol.*, **207** (1), 13 (1964).

32. Perl, W.; Rackow, H.; Salanitre, E.; Wolf, G. L.; and Epstein, R. M. *J. Appl. Physiol.*, **20** (4), 621 (1965).

33. Kelman, R. B. *Bull. Math. Biophys.*, **27**, 53 (1965).

34. Diamond, J. M. *J. Physiol.*, **161**, 474 (1962).

CHAPTER SEVEN

Interphase Transport

In this chapter we shall expand on the interphase heat-transfer coefficient concept introduced in Section 6.2 to include the interphase transport of both mass and momentum as well. After presenting the phenomenological analogies between heat and mass transfer we will present a definition of generalized transfer coefficient on which theoretical, as well as empirical, treatment can be developed. With further manipulation we will see how the classical concept of friction factor and drag coefficient—both interphase momentum-transfer coefficients—can be included. Then we will survey some examples from the research literature to show how this concept has been, and can be, used practically in analyzing physiological systems. A hemodialyzer design problem using this concept will be given in Example 8.8-7.

7.1 ANALOGIES IN HEAT- AND MASS-TRANSFER COEFFICIENTS

In Section 6.2 we described the transfer of heat between a fluid phase and a solid phase via the use of a heat-transfer coefficient, h, in the definition

$$q_x = h(T_s - T_a), \tag{7.1.1}$$

where q_x is the heat flux normal to the solid–fluid interface or heat-transfer surface and T_s and T_a are the temperatures of the solid surface and of the bulk of the fluid phase, respectively (see Figure 6.2-1). For an analogous situation in mass transfer, such as that depicted in Figure 7.1-1, in which phase I may be a porous material soaked with species A at a surface concentration c_{A_s} and phase II a flowing fluid containing A but with a different bulk concentration c_{A_0}, a mass-transfer coefficient, k_c, can be defined in a similar manner:

$$J_{A_x}^* = k_c(c_{A_s} - c_{A_0}), \tag{7.1-2}$$

where $J_{A_x}^*$ is the molar mass flux with respect to the molar center of gravity as defined in Section 5.3. Subscript c for the coefficient denotes that it is based on the molar concentration difference, $(c_{A_s} - c_{A_0})$. (The implication here is that if we use molar fractions for the concentrations, we can define

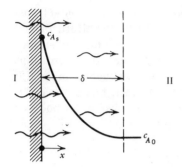

Figure 7.1-1.   Interphase mass transfer.

another mass-transfer coefficient, k_x, on that basis, and so on; see Ref. 1.) In fact, because of the variety of mass and molar concentrations and the fact that the coefficient is a function of mass-transfer rate (an added complication unique in mass transfer), its definition is more precise than that given here. However, to illustrate the analogies the present description is adequate.

From Eqs. 7.1-1 and 7.1-2 we can write the following generalized rule for these two types of transport

$$
\left\{ \begin{array}{c} \text{Flux of} \\ \text{transport} \end{array} \right\} = \left\{ \begin{array}{c} \text{Transfer} \\ \text{coefficient} \end{array} \right\} \times \left\{ \begin{array}{c} \text{Difference of} \\ \text{driving potentials} \\ \text{in the two phases} \end{array} \right\} \quad (7.1\text{-}3)
$$

or, if we prefer to use rate instead of flux,

$$
\left\{ \begin{array}{c} \text{Rate of} \\ \text{transport} \end{array} \right\} = \left\{ \begin{array}{c} \text{Transfer} \\ \text{coefficient} \end{array} \right\} \times \left\{ \begin{array}{c} \text{Transfer} \\ \text{area} \end{array} \right\} \times \left\{ \begin{array}{c} \text{Difference of} \\ \text{driving potentials} \\ \text{in the two phases} \end{array} \right\},
$$

$$(7.1\text{-}4)$$

wherein by *transfer area* we generally mean the area through which the mass or heat flux passes. Usually this is the actual contact area between the two phases. But, sometimes this can be an artificial quantity. For instance, in Example 6.2-4* we have a multilayer cylindrical tube, so that the interphase area increases as the radial distance increases. To simplify the overall computation we can use only one area. Thus, normally either the innermost or the outermost contact area is chosen and all of the interphase transfer coefficients are defined using this single area as the basis. Another case wherein this type of hypothetical surface area may be

*As well as Problems 6.H and 6.I, and Eq. 6.2-22.

necessitated is when a rough surface is involved. The actual contact area may be very difficult to determine or may not be worth the effort. In that case the macroscopic geometrical area is used, neglecting the surface roughness. This approach is fully justified when the coefficient thus evaluated is used on the same basis; that is, proper adjustment is made if it is used on a surface with a different roughness. In fact, this approach is implicit in all *in vivo* correlation in physiological systems, as no body or membrane surface is absolutely plain and smooth.

7.2 THE FRICTION FACTOR

In fluid flow where solid–fluid contact exists, a friction factor, f, can always be generally defined as follows

$$F_k = f \cdot A \cdot K, \qquad (7.2\text{-}1)$$

where F_k is the drag force associated with kinetic behavior of the fluid, such as friction on the solid–fluid interface due to the relative motion; A is the characteristic area; and K is the characteristic kinetic energy per unit volume. Definitions of the characteristic area and kinetic energy are deliberately left open, to be substituted by suitable quantities as the various specific cases call for. For example, the characteristic area is normally the solid–fluid interfacial contact area, much as in heat and mass transfer.

To illustrate how this general definition works, let us consider the common case of flow in a circular tube with radius R and length L. Here the most reasonable choice for the characteristic area is that of the cylindrical surface in contact with the fluid; that is,

$$A = 2\pi RL \qquad (7.2\text{-}2)$$

while the frictional drag force is exactly balanced by the pressure force (driving potential) under steady-state conditions. This gives us

$$F_k = \Delta p \cdot \pi R^2. \qquad (7.2\text{-}3)$$

Using the average velocity $\langle v_z \rangle$, as the characteristic velocity, the characteristic kinetic energy per unit volume should be

$$K = \tfrac{1}{2}\rho \langle v_z \rangle^2, \qquad (7.2\text{-}4)$$

since ρ (density) is the mass per unit volume. Substituting Eqs. 7.2-2, through 7.2-4 into 7.2-1, we have

$$f = \frac{\Delta p \cdot \pi R^2}{2\pi R L \cdot \frac{1}{2}\rho\langle v_z \rangle^2}$$

or

$$f = \frac{1}{4}\left(\frac{D}{L}\right)\frac{\Delta p}{\frac{1}{2}\rho\langle v_z \rangle^2}. \qquad (7.2\text{-}5)$$

Here $\Delta P/(\frac{1}{2}\rho\langle v_z \rangle^2)$ can be viewed as a "dimensionless pressure drop" and from Eq. 7.2-1 we can see that the friction factor is really the frictional drag force per unit contact surface area per unit kinetic energy. Note that Eq. 7.2-5 applies to flow in a circular tube, whether laminar or turbulent, Newtonian or non-Newtonian.

If we now further limit our case to the H–P flow (laminar, Newtonian) for which the average velocity over the cross-sectional area has been found to be

$$\langle v_z \rangle = \frac{\Delta p R^2}{8\mu L}, \qquad (6.1\text{-}8)$$

we can solve this equation for Δp and substitute it into Eq. 7.2-5 to get

$$f = \frac{16\mu}{D\langle v_z \rangle \rho}, \qquad (7.2\text{-}6)$$

where $D\langle v_z \rangle \rho/\mu$ is the well-known Reynolds number (Re) which is a measure of "fluidity".* Thus, we can simply write

$$f = \frac{16}{\text{Re}}, \qquad (7.2\text{-}7)$$

which appears as a straight line on a log–log plot of f versus Re with a negative slope, as appearing on Figure 7.2-1, for the region below Re $= 2100$.

Above this Reynolds number we enter the turbulent-flow region (but first a narrow, unstable transition region), where much of the work is empirical and experimental, or semitheoretical at best[3]. For example, for

---

*The reader can appreciate this by attempting to handle a trayful of water. He is likely to find it rather "unstable" or perhaps even spill it unless he reduces its size or compartmentalizes the tray, as in an ice tray. In doing so, he effectively reduces the D (and thus the Reynolds number if it were a flow system).

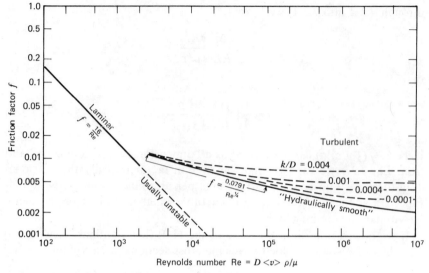

Figure 7.2-1. Friction factor chart for tube flow [2].

the turbulent velocity distribution[4]

$$v_z = v_{z_{max}} \left(1 - \frac{r}{R} \right)^{1/7}$$ (7.2-8)

the friction factor has been found to be

$$f = \frac{0.0791}{Re^{1/4}}$$ (7.2-9)

which is shown as the "hydraulically smooth" curve in Figure 7.2-1. In the nonideal case with varying degrees of surface roughness, curves with different "relative-roughness" (k/D) values are used. Here the roughness, k, whose values for various engineering materials can be obtained from standard references[5], is measured only relative to the tube size D, because the same degree of surface coarseness will be of lesser importance in a large tube than in a smaller one.

In a different situation where the fluid surrounds and flows past an object, such as a cell or sphere, or where an object moves within a large body of fluid, we can also evaluate the friction factor according to Eq. 7.2-1. In the case of a sphere, for instance, the characteristic kinetic energy is $\frac{1}{2} v_\infty^2$ where v_∞ is the approach velocity, or the fluid velocity far enough upstream for it to be unaffected by the sphere. For the characteristic area, the spherical surface area would seem to be the most logical choice, but it is customary to use the area of the great circle, πR^2. This is acceptable because the two differ only by a constant factor of 4. Thus, as long as we are consistent, it creates no substantial difference.*

In the creeping flow region ($Re = Dv_\infty \rho/\mu < 1$), the total drag force is given by the Stokes's law[6]:

$$F_k = 6\pi\mu R v_\infty. \tag{7.2-10}$$

Substituting these quantities into Eq. 7.2-1 gives

$$f = \frac{24}{Re}, \tag{7.2-11}$$

where $Re = Dv_\infty\rho/\mu$. This friction factor is more commonly known as the *drag coefficient*, sometimes denoted by C_D, and is shown as the straight line portion in Figure 7.2-2. For higher Reynolds numbers, curves are obtained whose expressions or values are given in the same chart.

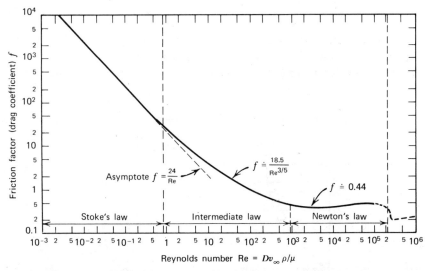

Figure 7.2-2. Drag coefficient for flow around a sphere [5].

*Similarly, there is another friction factor expression for H–P flow in which a slightly different definition based on $(\Delta p/\rho)$ is used, resulting in the correlation $f = 64/Re$ instead of Eq. 7.2-7. Thus, caution must be exercised to ascertain the precise definition before the equation is used.

7.3 FRICTION FACTOR AS MOMENTUM-TRANSFER COEFFICIENT

With a little insight and manipulation, we can see that the friction factor actually fits into the analogies discussed in Section 7.1 and that it can be considered as the *momentum-transfer coefficient*. Referring to Figure 7.3-1, which is a depiction of fluid flow adjacent to a solid surface, under the normal no-slip situation, the velocity, v_{z_0}, at the surface $(x=0)$ is zero. Within the boundary layer v_z varies with x, of course. At a sufficiently large distance δ from the surface the velocity, v_{z_δ}, assumes the mainstream velocity V. Then, analogous to saying that $\Delta T = T_s - T_a$ is the thermal energy driving potential in Eq. 7.1-1 and that $\Delta c_A = c_{A_s} - c_{A_0}$ is the mass driving potential in Eq. 7.1-2, we can say that V^2, or more precisely $\frac{1}{2}\rho V^2$, is the kinetic or momentum driving potential in Eq. 7.2-1 since

$$\Delta K = \Delta\left(\tfrac{1}{2}\rho v_z^2\right)$$

$$= \tfrac{1}{2}\rho v_{z_\delta}^2 - \tfrac{1}{2}\rho v_{z_0}^2$$

$$= \tfrac{1}{2}\rho V^2 - \tfrac{1}{2}\rho \cdot 0^2$$

$$= \tfrac{1}{2}\rho V^2. \tag{7.3-1}$$

Furthermore, since the drag force, F_k, can be regarded as the rate of momentum transport,* comparison of Eqs. 7.1-4 and 7.2-1 shows that the friction factor is the momentum-transport counterpart of the heat- and

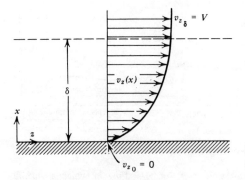

Figure 7.3-1.   Interphase momentum transport.

---

*This can be verified by their units: Rate of momentum $[=]\dfrac{g\text{-}cm/sec}{sec} = \dfrac{g\text{-}cm}{sec^2} = \text{dyne}[=]$ force.

mass-transfer coefficients. Hence, the friction factor can be regarded as a momentum-transfer coefficient.

To summarize, we present the defining equations of the three interphase transfer coefficients in Table 7.3-1.

Table 7.3-1   Analogies in the Defining Equations for Mass-, Momentum-, and Energy-Transfer Coefficients

			General Equation:		
Type of transfer	Rate of transfer	=	$\left\{\begin{matrix}\text{Transfer}\\\text{coefficient}\end{matrix}\right\}$ $\times$	$\left\{\begin{matrix}\text{Transfer}\\\text{area}\end{matrix}\right\}$ $\times$	$\left\{\begin{matrix}\text{Difference}\\\text{in}\\\text{driving}\\\text{potentials}\end{matrix}\right\}$
Mass	$W_A{}^*$	=	k_c	A	Δc_A
Momentum	F_k	=	f	A	ΔK
Energy	$Q^\dagger$	=	h	A	ΔT

* W_A = rate of mass transfer.

$\dagger Q$ = rate of heat transfer.

7.4 FUNCTIONAL DEPENDENCIES IN FORCED CONVECTION

If we consider the heat transfer between the walls of a tube and the fluid flowing through it, as shown in Figure 7.4-1, from the conduction point of view we can write for the rate of heat transfer at the wall

$$Q_R = \int_0^L \int_0^{2\pi} \left(-k \frac{\partial T}{\partial r} \right) \Bigg|_{r=R} R \, d\theta \, dz. \qquad (7.4\text{-}1)$$

On the other hand, if the temperature of the walls is T_s and the bulk of the fluid is at T_b, then according to the definition of the heat-transfer coefficient, we can write*

$$Q_R = h \cdot 2\pi RL (T_b - T_s). \qquad (7.4\text{-}2)$$

*Note from Eq. 7.4-1 that Q_R is written as in the $+r$-direction, so the temperature difference is $T_b - T_s$, also in the $+r$-direction. But this applies to Q_R opposing the r-direction as well in which case $T_b < T_s$ and $\partial T / \partial r$ will be positive. Also the bulk temperature actually varies longitudinally as heat is being lost, but we neglect this fact here. For a more precise treatment, see Section 13.1, Ref. 1.

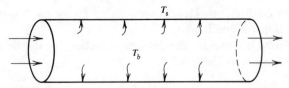

Figure 7.4-1. Forced convection.

Equating these two expressions and dividing by $2\pi RL(T_b - T_s)$, we have

$$h = \frac{1}{2\pi RL(T_b - T_s)} \int_0^L \int_0^{2\pi} \left(-k\frac{\partial T}{\partial r}\right)\bigg|_{r=R} R\, d\theta\, dz, \quad (7.4\text{-}3)$$

which can be further reduced to dimensionless form by multiplying both sides by D/k:

$$\text{Nu} = \frac{hD}{k} = \frac{1}{2\pi(L/D)} \int_0^{L/D} \int_0^{2\pi} \left(-\frac{\partial T^*}{\partial r^*}\right)\bigg|_{r^*=\frac{1}{2}} d\theta\, dz^*, \quad (7.4\text{-}4)$$

where all quantities with the asterisk are dimensionless—that is,

$$r^* = \frac{r}{D}, z^* = \frac{z}{L}, T^* = \frac{T - T_s}{T_b - T_s}.$$

The quantity $\text{Nu} = hD/k$ is called the Nusselt number, a dimensionless measure of the convective capacity of a fluid relative to its own conductive capacity.

For the functional dependency of Nu on the other parameters, we must find the general form of $T^*(r^*)$ in Eq. 7.4-4. For this, we go to the equation of energy first in the form of Eq. 4.9-7. To make our illustration simple without losing generality, let us assume an incompressible Newtonian fluid obeying the Fourier's conduction law. From the latter two conditions, we have Eq. 4.10-3. From the incompressibility assumption, we have $(\nabla \cdot \mathbf{v}) = 0$ (Eq. 4.3-9) and $\hat{C}_v = \hat{C}_p$, so that Eq. 4.10-3 becomes

$$\rho\hat{C}_p \frac{DT}{Dt} = k\nabla^2 T + \mu\Phi_v. \quad (7.4\text{-}5)$$

Dividing the entire equation by $\rho\hat{C}_p(T_b - T_s)V/D$, where V is a characteristic velocity (such as the average velocity of the flow) in terms of which the characteristic time is defined as D/V that is, $t^* = tV/D$, this equation becomes

$$\frac{DT^*}{Dt^*} = \frac{k}{\rho \hat{C}_p} \cdot \frac{1}{DV} \nabla^{*2} T^* + \frac{\mu D}{\rho \hat{C}_p (T_b - T_s) V} \Phi_v, \qquad (7.4\text{-}6)$$

where $\nabla^* = D\nabla$, the dimensionless differential operator of which $\partial / \partial r^*$ is one component. Noting from Eq. 4.10-2 that Φ_v has the same dimension as $(V/D)^2$ we can rewrite Eq. 7.4-6 into

$$\frac{DT^*}{Dt^*} = \frac{k}{\rho \hat{C}_p} \cdot \frac{1}{DV} \nabla^{*2} T^* + \frac{\mu V}{\rho \hat{C}_p (T_b - T_s) D} \Phi_v^*, \qquad (7.4\text{-}7)$$

where $\Phi_v^* = (D/V)^2 \Phi_v$, the dimensionless viscous dissipation function. The two clusters of parameters in Eq. 7.4-7 can be made more meaningful by multiplying some appropriate quantities to both the numerator and denominator, namely,

$$\frac{k}{\rho \hat{C}_p} \cdot \frac{1}{DV} = \frac{k}{\mu \hat{C}_p} \cdot \frac{\mu}{DV\rho} = \frac{1}{\Pr \mathrm{Re}}$$

and

$$\frac{\mu V}{\rho \hat{C}_p (T_b - T_s) D} = \frac{\mu V^2}{k(T_b - T_s)} \cdot \frac{k}{\mu \hat{C}_p} \cdot \frac{\mu}{DV\rho} = \frac{\mathrm{Br}}{\Pr \mathrm{Re}}$$

where the Prandtl (Pr) number is

$$\Pr = \frac{\mu \hat{C}_p}{k},$$

and the Brinkman (Br) number is

$$\mathrm{Br} = \frac{\mu V^2}{k(T_b - T_s)},$$

which is a measure of the relative importance of the viscous heating effect. The meaning of the Prandtl number will be explained in the following section. Now Eq. 7.4-7 can be more neatly written

$$\frac{DT^*}{Dt^*} = \frac{1}{\mathrm{Re}\,\Pr} \nabla^{*2} T^* + \frac{\mathrm{Br}}{\mathrm{Re}\,\Pr} \Phi_v^*. \qquad (7.4\text{-}8)$$

With adequate description of the system and initial and boundary conditions, we can in principle obtain the steady-state (time-independent) solution of this differential equation as a function of the position coordinates and the three dimensionless groups appearing here, or

$$T^* = T^*(r^*, \theta, z^*; \text{Re}, \text{Pr}, \text{Br}).$$ (7.4-9)

Substitution of this general relation into Eq. 7.4-4 eliminates r^*, θ, and z^* as variables but introduces L/D as an additional parameter, or

$$\text{Nu} = \text{Nu}\left(\text{Re}, \text{Pr}, \text{Br}, \frac{L}{D}\right).$$ (7.4-10)

Experiments have shown that except for highly viscous fluids at high shear rates, which rarely occur in biological systems, viscous dissipation is negligible, so that the Brinkman number is usually eliminated from consideration and thus

$$\text{Nu} = \text{Nu}\left(\text{Re}, \text{Pr}, \frac{L}{D}\right).$$ (7.4-11)

In many systems in which the length of the flow channel, L, is great, the correlations for various L/D ratios converge to a single relationship. On the other hand, if the bulk and surface temperatures differ greatly, the use of bulk temperature to evaluate the physical properties, such as μ and k, would cause considerable error unless a factor equal to $(\mu_b/\mu_s)^n$, where n is a constant evaluated from experimental correlation and subscripts b and s denote bulk and surface properties, respectively, is included.

In Section 7.6 we will present some specific correlations obtained from experimental data relating these parameters, Nu, Re, Pr, and L/D. It is obvious from here that there is a great advantage in working with these groups. For example, if we had to determine the effects of all seven parameters $(L, D, V, \rho, \mu, C_p, k)$ on h and the interactions thereof, the minimum number of experiments needed (assuming two levels for each) is $2^7 = 128$, while with these groups, we only have to change Re, Pr, and L/D as groups but not L, D, and the rest individually. In this example, we can run as few as $2^3 = 8$ experiments. With the great saving in effort we can afford to conduct more variations on experiments with the groups to yield better results, especially when the effects are nonlinear. We will see from Example 7.10-2 how unawareness of these dimensionless groupings has made researchers spend a great deal of experimental effort while studying only relative few parameters without generalizable results.

We should also mention here, at least in passing, that there is a simpler and more commonly used procedure to obtain the general relationship Eq. 7.4-11 and the analogous ones for mass and momentum transfer through the Buckingham's Π-theorem [7]. We will present this method in Appendix I. It must be emphasized that the success of this method depends on the selection of all of the correct parameters, which we have to do by intuition or other means. The theorem itself does not tell us how to make the selection. Therefore, the method described in this section is the more certain and fundamental one.

7.5 FUNCTIONAL DEPENDENCIES IN MASS AND MOMENTUM TRANSFER

Now that we have discussed so many things via analogies, it should not be surprising that we can find the general functional relation for mass transfer by making the multicomponent equation of continuity, Eq. 5.5-4, dimensionless. We leave this to the reader and will only point out, via a different analogy route, what the resultant general formula should be. Since convection means flow-aided transport, so the flow parameter, Re, should exist in the convective transport of both mass and energy (heat). We have already shown in the preceding section that in forced convection an important role is played by the Prandtl number, which can be segregated into two parts:

$$Pr = \frac{\hat{C}_p \mu}{k} = \frac{\mu}{\rho} \cdot \frac{\rho \hat{C}_p}{k} \qquad (7.5\text{-}1)$$

In this expression we recognize that μ/ρ is simply ν, the kinematic viscosity or momentum diffusivity, and $\rho \hat{C}_p / k$ is the reciprocal of thermal diffusivity α, or

$$Pr = \frac{\nu}{\alpha} = \frac{\text{momentum diffusivity}}{\text{thermal diffusivity}}. \qquad (7.5\text{-}2)$$

Thus, we see that if momentum transport is involved in both mass and energy convection, the quantity in mass transfer analogous to the Prandtl number, called the Schmidt number, Sc, should be

$$Sc = \frac{\text{momentum diffusivity}}{\text{mass diffusivity}} = \frac{\nu}{\mathcal{D}_{AB}}$$

$$= \frac{\mu}{\rho \mathcal{D}_{AB}}. \qquad (7.5\text{-}3)$$

Similarly, replacing the heat-transfer coefficient, h, by the mass-transfer coefficient, k_c, and thermal conductivity by the diffusional coefficient, the analogous quantity to Nu, called the Sherwood number, is

$$Sh = \frac{k_c D}{\mathfrak{D}_{AB}}. \qquad (7.5\text{-}4)$$

The reader can readily prove that both Sc and Sh are indeed dimensionless.

With these new quantities defined, we can confidently write, analogous to Eq. 7.4-11,

$$Sh = Sh\left(Re, Sc, \frac{L}{D}\right), \qquad (7.5\text{-}5)$$

which can also be obtained from Buckingham's Π-theorem (Problem 7.B).

We should keep in mind what we have said several times in this book: the analogies between the three types of transport are merely mathematical and in form. In substance these processes are quite different from one another. In fact, this is a good place to reinforce this statement, because while the heat transfer coefficient is generally considered to be independent of the heat-transfer rate, the mass-transfer coefficient does depend on the mass-transfer rate through the distortion of velocity and concentration profiles. This is why the rate-independent mass-transfer coefficient has a more precise definition (see Section 21.1, Ref. 1). However, we will not make such a fine distinction here, because the specific relationships will come from correlation of experimental data that masks this effect and thus takes care of this problem automatically.

By following the same reasoning or by making the equation of motion dimensionless, the friction factor can be shown to have the general dependency (see Section 6.2, Ref. 1 for details)*

$$f = f\left(Re, \frac{L}{D}\right), \qquad (7.5\text{-}5)$$

which, with the L/D effect negligible for long tubes, agrees with the results in Section 7.2.

Although the dimensional analysis here was performed using tube flow as an example, one can easily show that it and its results apply to other geometries as well. For example, for flows between parallel plates, the same functional relationship can be obtained with the thickness or half-

---

*In making the equation of motion dimensionless, we will obtain the Froude number, $Fr = V^2/gD$, where g is the gravitational constant. By considering a flowing fluid with no free surface (i.e., with the tube running full), there will be no influence of gravity and the body-force term can be dropped. Hence, the Froude number is also eliminated.

thickness in place of the diameter. For flows around spheres there is no L and the characteristic velocity is most likely the "approach velocity," v_∞, as discussed in Section 7.2.

7.6 HEAT-TRANSFER CORRELATIONS

Based on the general functional dependencies derived in the last section, there have been many experimental studies aimed at establishing the specific quantitative relationships between the various parameters for various geometries and flow conditions. We shall attempt to summarize here those that are potentially useful for biomedical work.

a. Flow Parallel to Tubes

Figure 7.6-1 is the correlation by Sieder and Tate [8] for flows parallel to, and either inside or outside* of, a circular tube or bank of tubes. In this correlation, which applies to both heating and cooling, the heat-transfer coefficient or Nu is combined with other previously discussed quantities to

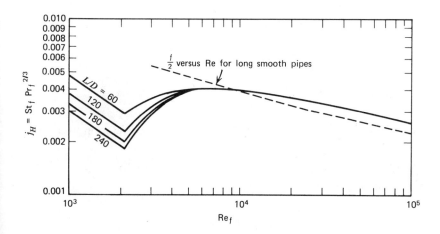

Figure 7.6-1.   Heat transfer coefficient in the form of the Chilton-Colburn j_H-factor; for flow parallel to circular tubes, both inside and outside, heating and cooling.

---

*For outside of tubes, such as that encountered in the dialysate side of the hollow-fiber type of hemodialyzer, the diameter, D, should be replaced by the equivalent diameter, defined as $D_e = 4 \times$ free cross-sectional-area perimeter.

form the j_H-factor:

$$j_H = \frac{h}{\rho \hat{C}_p V} \left(\frac{\hat{C}_p \mu}{k} \right)^{2/3}$$

$$= \frac{hD}{k} \cdot \frac{k}{\hat{C}_p \mu} \cdot \frac{\mu}{DV\rho} \left(\frac{\hat{C}_p \mu}{k} \right)^{2/3}$$

$$= \frac{\mathrm{Nu}}{\mathrm{Re\,Pr}} \mathrm{Pr}^{2/3}$$

$$= \mathrm{St\,Pr}^{2/3}, \tag{7.6-1}$$

where the Stanton (St) number is

$$\mathrm{St} = \frac{\mathrm{Nu}}{\mathrm{Re\,Pr}} = \frac{h}{\rho \hat{C}_p V}. \tag{7.6-2}$$

An alternate form is that in terms of the Graetz (Gr) number, defined as

$$\mathrm{Gz} = \mathrm{Re\,Pr} \frac{D}{L}, \tag{7.6-3}$$

and the interrelationship between Nu, St, and Gz is

$$\mathrm{Nu} = \mathrm{St\,Gz} \frac{L}{D}. \tag{7.6-4}$$

According to Figure 7.6-1 we can divide the correlation into three regions:

(1) *Laminar flow*: $\mathrm{Re} < 2100$
The mathematical representation is

$$j_H = 1.86\,\mathrm{Re}_f^{-2/3} \left(\frac{L}{D} \right)^{-1/3} \tag{7.6-5}$$

or

$$\mathrm{Nu} = 1.86\,\mathrm{Gz}_f^{1/3}, \tag{7.6-6}$$

where we have used physical properties evaluated at the film temperature, defined as $T_f = \frac{1}{2}(T_s + T_b)$, where T_s and T_b are themselves averages between the two longitudinal extremes. The use of these quantities, indicated by subscript f, allows us to eliminate the viscosity correction factor $(\mu_b/\mu_s)^n$. (In many correlations where properties at the bulk temperature T_b are used, the exponent for this correction factor is $n = 0.14$.)

(2) *Transition region*: $2100 < \text{Re} < 10{,}000$
There is considerable scatter of experimental data in this region, which is not well understood. Curves are usually hand-drawn, linking the laminar and turbulent flow regions. The correlation is thus not very reliable.

(3) *Turbulent region*: $\text{Re} > 10{,}000$
In this region, the L/D effect disappears and the mathematical representation is

$$j_H = 0.023 \text{Re}_f^{-0.2} \tag{7.6-7}$$

or

$$\text{Nu} = 0.023 \text{Re}_f^{0.8} \text{Pr}_f^{1/3}. \tag{7.6-8}$$

b. Flow Outside, and Normal to, Tube or Tube Bank (Transverse Flow)

The correlation is of the type

$$j_H = a \, \text{Re}^{-m} \tag{7.6-9}$$

or

$$\text{Nu} = a \, \text{Re}^{1-m} \text{Pr}^{1/3}, \tag{7.6-10}$$

where the Reynolds number is based on the approach velocity, v_∞.

(1) *A single tube*
The specific correlation is shown in Figure 7.6-2 [9]* from which we find $a \doteq 0.43$, $m \doteq 0.45$.

(2) *Banks of tubes more than 10 rows deep*
The values of a range from 0.15 to 0.33 for various types of tube arrangements, while $m = 0.4$ [10].†

c. Flow Around a Sphere

The correlation of Ranz and Marshall [11] is given in Figure 7.6-3, and the mathematical relationship is

$$\text{Nu} = 2 + 0.60 \text{Re}_f^{1/2} \text{Pr}_f^{1/3}. \tag{7.6-11}$$

*Also see Figure 111, p. 221, and Figure 113, p. 226, of Ref. 7 (2nd ed., 1942).
†Also see Ref. 5, p. 10-14; Ref. 7, pp. 229–230 (2nd ed., 1942).

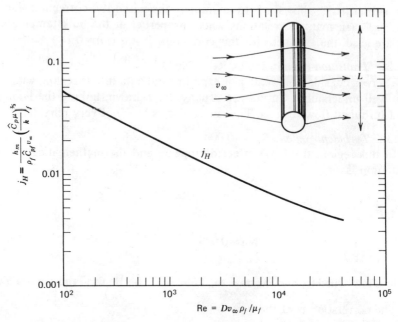

Figure 7.6-2. Forced convection outside of a single-tube, transverse flow [9] (from [1]). Subscript *m* stands for the *mean* value around the entire tube. Note that this chart can also be applied to mass transfer for j_D (see Section 7.7).

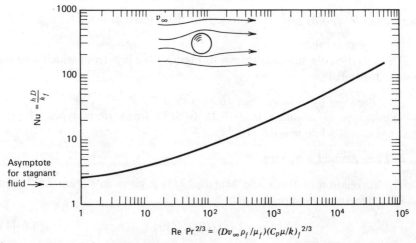

Figure 7.6-3. Forced convection in flow around a sphere [11] (from [1]). Note that it can also be applied to mass transfer (see Eq. 7.7-1).

300

It is of interest to note that the first term on the right-hand side is exact because it says that in the special case of velocity equal to zero, $Re=0$, and Nu must be theoretically equal to 2. This is indeed the case, as confirmed by considering the pure conduction around a sphere immersed in an infinite medium (see Problem 9.H_2, Ref. 1).

7.7 MASS-TRANSFER CORRELATIONS

As expected, mass-transfer-coefficient correlations are very much similar to those of the forced-convection heat-transfer coefficient. For example, for flow around a sphere we can use Figure 7.6-3 and write [11]

$$Sh = 2 + 0.60 Re_f^{1/2} Sc_f^{1/3}. \tag{7.7-1}$$

We can also define a j_D-factor, analogous to the heat-transfer j_H-factor, as

$$j_D = \frac{k_c}{V} \left(\frac{\mu}{\rho \mathcal{D}_{AB}} \right)^{2/3}$$

$$= \frac{Sh}{Re\,Sc} Sc^{2/3}$$

$$= St_{AB} Sc^{2/3}, \tag{7.7-2}$$

where St_{AB} is called the mass-transfer Stanton number. As a specific example, Figure 7.6-2 has actually been used for mass transfer in a transverse flow outside of a single tube. In this case, the j_D and j_H curves virtually coincide with each other [9]. Similarly, a mass-transfer Graetz number can also be defined

$$Gz_{AB} = Re\,Sc\,\frac{D}{L}, \tag{7.7-3}$$

and as a further example of the analogy to heat-transfer correlation of the type in Eq. 7.6-6, Wolf and Zaltzman [12] used the general relationship

$$Sh = a \cdot Gz_{AB}^m \tag{7.7-4}$$

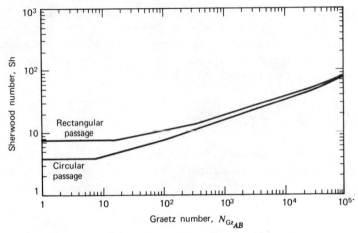

Figure 7.7-1. Convective mass-transfer correlation for rectangular and circular passages [12].

for the convective mass transfer in two types of hemodialyzer geometries. In the log–log correlation shown in Figure 7.7-1, the entire range is divided into three sections, each approximated by a straight line. For rectangular passage (flow between parallel plates):

$$a = 2.2 \quad m = 0.32 \quad \text{for } 330 < \text{Gz}_{AB}$$

$$a = 4.4 \quad m = 0.20 \quad \text{for } 15 < \text{Gz}_{AB} < 330$$

$$a = 7.6 \quad m = 0 \quad \text{for } \quad \text{Gz}_{AB} < 15.$$

For circular passage (flow inside tube):

$$a = 1.64 \quad m = 0.335 \quad \text{for } 100 < \text{Gz}_{AB}$$

$$a = 2.42 \quad m = 0.25 \quad \text{for } 7 < \text{Gz}_{AB} < 100$$

$$a = 3.9 \quad m = 0 \quad \text{for } \quad \text{Gz}_{AB} < 7.$$

These are but a few of the better-known relationships potentially useful for biomedical work. There are many other correlations for special purposes. However, some are for very high Reynolds numbers, which do not very likely occur in living bodies; thus, we will omit them here (but we refer the reader to Chapter 13 of Ref. 1).

The simultaneous interphase transport of mass and energy should also be mentioned. When evaporation (such as perspiration) or condensation is involved, heat is being lost or gained, respectively, in the form of latent heat in addition to the convectively transported sensible heat. This is why an exposed body or body part feels particularly cold when it is wet, as after a hot shower, and why a person catches cold most easily if he does not cover or dry his perspiring body after exercise. The use of alcohol to cool the body during a high fever is also based on this principle, because its high volatility facilitates evaporation and transport of latent heat from the body. In such cases the rate of heat transport is proportional to the partial pressure difference of the evaporating fluid in the two phases, in addition to being proportional to the temperature difference, as will be seen in Example 7.10-3.

7.8 FREE CONVECTION

Even when there is no externally driven flow, we may still have heat transfer by convection. This means that even when sitting still with no wind, a person can still lose heat at a rate higher than that predicted from merely considering conduction and thermal radiation, because of what we call free convection. Alternately called *natural convection*, this mode of heat transfer is aided by flow generated by the uneven distribution of density caused by the temperature differential. In this case the situation is quite complex, because although the heat-transfer rate depends on the flow rate, the latter in turn depends on the former. This interdependency makes rigorous theoretical analysis rather impractical. Therefore, we again resort to dimensionless-group correlations with the newly defined Grashof number

$$\text{Gr} = \frac{\rho^2 \beta g D^3 \Delta T}{\mu^2}, \tag{7.8-1}$$

where g is the gravitational constant, ΔT the difference between the surface and ambient temperatures, and β the thermal-expansion coefficient

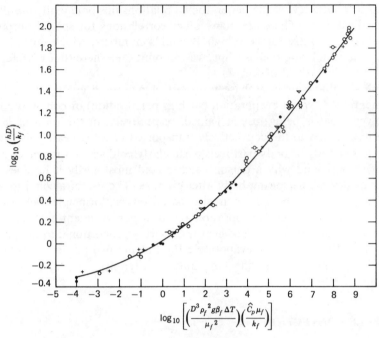

Figure 7.8-1. Free convection from long horizontal tubes to surrounding fluids [7] (from [1]). Note that all fluid properties are based on film temperature.

defined as the fractional volume change per unit temperature change, or

$$\beta = \frac{1}{V}\left(\frac{\partial V}{\partial T}\right)_p = -\frac{1}{\rho}\left(\frac{\partial \rho}{\partial T}\right)_p. \qquad (7.8\text{-}2)$$

As far as its specific correlation goes, some of the previous analogies still apply. For example, for free convection around a sphere, we have [11]

$$\mathrm{Nu} = 2 + 0.60 \ \mathrm{Gr}^{1/4}\mathrm{Pr}^{1/3}, \qquad (7.8\text{-}3)$$

while for outside of a long horizontal tube, many experimental data have

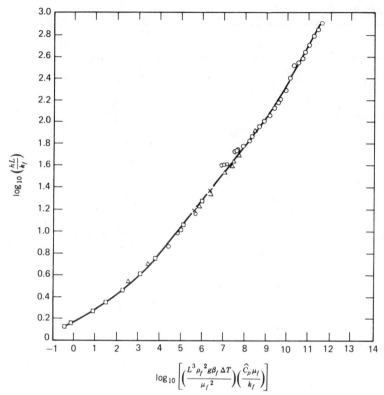

Figure 7.8-2. Free convection from a vertical plate [7] (from [1]). Note that all fluid properties are based on film temperature.

been found to lie on, or closely around, the curve shown in Figure 7.8-1, according to which we have [7, p. 176]

$$Nu = 0.525 (Gr\, Pr)^{1/4} \quad \text{for} \quad Gr\, Pr > 10^4. \tag{7.8-4}$$

For a vertical plate, we have Figure 7.8-2, which yields [7, p. 173]

$$Nu = 0.59 (Gr\, Pr)^{1/3} \quad \text{for} \quad 10^4 < Gr\, Pr < 10^9, \tag{7.8-5}$$

where L (height) replaces D as the characteristic length.

Table 7.9-1 Summary of Interphase Heat and Mass-Transfer Analogies [1]

	Heat-transfer quantities	Binary mass-transfer quantities
Profiles	T	x_A
Diffusivity	$\alpha = \dfrac{k}{\rho \hat{C}_p}$	$\mathcal{D}_{AB}$
Effect of profiles on density	$\beta = -\dfrac{1}{\rho}\left(\dfrac{\partial \rho}{\partial T}\right)_{p,x_A}$	$\zeta = -\dfrac{1}{\rho}\left(\dfrac{\partial \rho}{\partial x_A}\right)_{p,T}$
Flux	$q^{(c)}$	$J_A^* = N_A - x_A(N_A + N_B)$
Transfer rate	Q	$\mathcal{W}_A^{(m)} - x_{A_0}(\mathcal{W}_A^{(m)} + \mathcal{W}_B^{(m)})$
Transfer coefficient	$h = \dfrac{Q}{A\Delta T}$	$k_x = \dfrac{\mathcal{W}_A^{(m)} - x_{A_0}(\mathcal{W}_A^{(m)} + \mathcal{W}_B^{(m)})}{A\Delta x_A}$
Dimensionless groups which are the same in both correlations	$\mathrm{Re} = \dfrac{DV\rho}{\mu} = \dfrac{DG}{\mu}*$ $\mathrm{Fr} = \dfrac{V^2}{gD}$ $\dfrac{L}{D}$	$\mathrm{Re} = \dfrac{DV\rho}{\mu} = \dfrac{DG}{\mu}*$ $\mathrm{Fr} = \dfrac{V^2}{gD}$ $\dfrac{L}{D}$

7.9 CHILTON–COLBURN ANALOGY

Throughout the discussion in previous sections, we have seen how closely heat and mass transfers resemble each other in mathematical representations. We can thus summarize them in Table 7.9-1 and note that these relationships can be used interchangeably, especially when there are experimental data for only one type of transport available.

As expected, this type of analogy can be extended to cover momentum

Table 7.9-1 (*Continued*)

	Heat-transfer quantities	Binary mass-transfer quantities
Basic dimensionless groups which are different	$\mathrm{Nu} = \dfrac{hD}{k}$	$\mathrm{Sh} = \dfrac{k_x D}{c \mathcal{D}_{AB}}$
	$\mathrm{Pr} = \dfrac{\hat{C}_p \mu}{k} = \dfrac{\nu}{\alpha}$	$\mathrm{Sc} = \dfrac{\mu}{\rho \mathcal{D}_{AB}} = \dfrac{\nu}{\mathcal{D}_{AB}}$
	$\mathrm{Gr} = \dfrac{D^3 \rho^2 g \beta \Delta T}{\mu^2}$	$\mathrm{Gr}_{AB} = \dfrac{D^3 \rho^2 g \zeta \Delta x_A}{\mu^2}$
	$\mathrm{St} = \dfrac{\mathrm{Nu}}{\mathrm{Re}\,\mathrm{Pr}} = \dfrac{h}{\rho \hat{C}_p V}$	$\mathrm{St}_{AB} = \dfrac{\mathrm{Sh}}{\mathrm{Re}\,\mathrm{Sc}} = \dfrac{k_x}{cV}$
Special combinations of dimensionless groups	$\mathrm{Pé} = \mathrm{Re}\,\mathrm{Pr} = \dfrac{DV}{\alpha}$ †	$\mathrm{Pé}_{AB} = \mathrm{Re}\,\mathrm{Sc} = \dfrac{DV}{\mathcal{D}_{AB}}$ †
	$j_H = \mathrm{Nu}\,\mathrm{Re}^{-1}\,\mathrm{Pr}^{-1/3}$	$j_D = \mathrm{Sh}\,\mathrm{Re}^{-1}\,\mathrm{Sc}^{-1/3}$
	$= \dfrac{h}{\rho \hat{C}_p V} \left(\dfrac{\hat{C}_p \mu}{k} \right)^{2/3}$	$= \dfrac{k_x}{cV} \left(\dfrac{\mu}{\rho \mathcal{D}_{AB}} \right)^{2/3}$

*$G = V\rho$, the mass velocity, g/cm^2-sec.

†$\mathrm{Pé} = \mathrm{Re}\,\mathrm{Pr}$, the Péclet number, which is the ratio of bulk flow to molecular transport. When this number is large, it means that diffusive-type transport is negligible compared with the convective variety. Similar significance holds for $\mathrm{Pé}_{AB}$ in mass transfer.

transport, and there is a definite relation between the friction factor and the heat- and mass-transfer coefficients. This should not be too surprising because all of them are functions of the Reynolds number. In fact, Chilton and Colburn [13] found that in the turbulent flow region,

$$j_H = j_D = \tfrac{1}{2}f. \tag{7.9-1}$$

This offers a great advantage, because we need only one correlation chart, that of the friction factor, and need to compute the Reynolds number only.

7.10 BIOMEDICAL APPLICATIONS

Interphase transport represents an area in which the theories and correlations are potentially useful to, but remain relatively untapped by, the biomedical community because of lack of exposure to this subject on the part of researchers, which has caused a general lack of understanding of the subject matter. As a result, good examples in this area are difficult to find. In most of them, correlation of experimental data appears to be fragmented and to lack a theoretical, or even dimensional, basis such as that discussed in the preceding sections. For example, there have been many experimental studies of the effect of wind speed on the heat-transfer rate or coefficient, in the form of h versus v^a type of empiricism. Yet, practically no one has grouped the variables and parameters according to the Reynolds number and other dimensionless quantities. Thus, the knowledge gained from these studies and the general applicability of the results are not commensurate with funds and effort spent.

Thus, in the following examples, we will emphasize the success stories of some studies as well as indicate how and why others could be improved by the material discussed earlier in this chapter. It is the author's hope that this chapter will greatly aid research currently in progress as well as open up new frontiers for explorers in the biomedical sciences.

Example 7.10-1 Friction Factor for Pulsatile Blood Flow. The average velocity developed in Example 6.1-3 can be used in conjunction with Eq. 7.2-5 to yield the friction-factor expression for pulsatile blood flow. However, to make the Reynolds number a truly constant quantity, the time-averaged *average* velocity should be employed. This can be obtained by dividing Eq. (2.E-1) (see Problem 2.E), by πR^2. Thus, we have

$$\overline{\langle v_z \rangle} = \frac{\Delta p R^2}{8 \mu L} \left[1 + 16 \left(\frac{L}{\Delta p} \right) \frac{A \lambda^2}{\pi} \sum_{k=1}^{\infty} \frac{1}{\alpha_k^4} \cdot \frac{1 - \exp(-2\pi \alpha_k^2 / \lambda)}{\alpha_k^4 + \lambda^2} \right], \quad (7.10\text{-}1)$$

where

μ = viscosity of blood,
R = radius of tube,
$\Delta p / L$ = average pressure gradient between ends of tube,
A = amplitude of pressure-gradient fluctuation,
α_k = roots of $J_0(\alpha_k) = 0$,
$\lambda = R^2 \omega / \nu$,
ω = frequency of pressure-gradient fluctuation,
ν = kinematic viscosity (μ / ρ, ρ = density of blood).

Substituting Eq. 7.10-1 into Eq. 7.2-5, we have

$$f = \frac{1}{4}\left(\frac{D}{L}\right)\frac{\Delta p}{\frac{1}{2}\rho \langle v_z \rangle} \cdot \frac{8\mu L}{\Delta p R^2} \cdot \frac{1}{S},$$ (7.10-2)

where

$$S = 1 + 16\frac{A\lambda^2}{\pi(\Delta p/L)}\sum_{k=1}^{\infty}\frac{1}{\alpha_k^4}\cdot\frac{1-\exp(-2\pi\alpha_k^2/\lambda)}{\alpha_k^4+\lambda^2}.$$ (7.10-3)

Equation 7.10-2 can be quickly rearranged to yield

$$f = \frac{16}{\mathrm{Re}}\cdot\frac{1}{S},$$ (7.10-4)

where

$$\mathrm{Re} = \frac{D\langle v_z \rangle \rho}{\mu},$$

the Reynolds number based on the time-averaged flow.

We can thus see that the expression is a modified form of that for the H–P flow. In fact, if we let $\omega = 0$ or $A = 0$, it will be reduced to Eq. 7.2-7 for the H–P flow. This has actually been verified by Hershey and Song [14], who obtained excellent agreement between the theoretical expression and experimental data from blood at various pulsatory frequencies, as shown in Figure 7.10-1. Predicted pressure drops under various conditions were also found to be in close agreement with experimentally measured values.

Example 7.10-2 Effect of Air Speed on Body-Heat Gain and Loss. There have been a number of experimental studies relating the transfer coefficient of body-heat gain or loss to wind or ambient-air speed. A typical example is that of Colin and Houdas [15], who after making 69 measurements on 15 human subjects and deducting heat transfer attributed to evaporation and thermal radiation, reported the results as shown in Figure 7.10-2. From the slopes of the straight lines in Figure 7.10-2 they calculated the following values for heat-transfer coefficients at various ambient-air speeds:

v (m/sec)	h (kcal/m²-hr-°C)
0.2	4.8
0.5	7.0
1.2	10.8

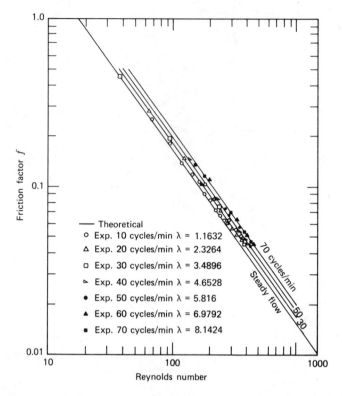

Figure 7.10-1. Friction factor versus Reynolds number at various pulsating frequencies; whole blood at 25°C [14].

from which they in turn obtained the correlation

$$h = 2.3 + 7.5v^{0.67}. \tag{7.10-5}$$

This and results from several other similar studies are summarized in Table 7.10-1.

To compare these results with the correlations in Section 7.6, we first note that none of the results in Table 7.10-1 is in the form of dimensionless groups. In fact, the film temperature, and hence the viscosity, density, heat capacity, and thermal conductivity, varied from run to run, even in the same series of experiment (see for example, Figure 7.10-2). In addition, D, the size or characteristic length of each human subject, was not reported. Therefore, there is no way to compute the Reynolds, Prandtl, and Nusselt numbers accurately. However, in order to examine the order-of-magnitude

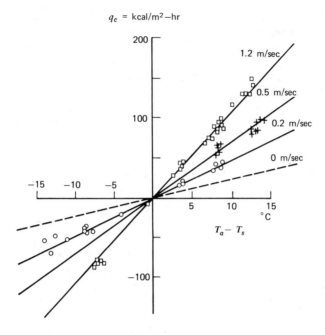

Figure 7.10-2. Convective heat gain or loss at various ambient-air speeds and temperature differences; heat gain at right and loss at left [15].

agreement between the experimental results and standard correlations, we shall use the following values throughout this example:

$$D = 10 \text{ cm.}^*$$

At 25°C,

$$\rho_{air} = 1.186 \times 10^{-3} \text{ g/cm}^3$$

$$\mu_{air} = 0.021 \text{ cp} = 2.1 \times 10^{-4} \text{ g/cm-sec}$$

$$\hat{C}_{p_{air}} = 0.25 \text{ cal/g-°C}$$

$$k_{air} = 6.25 \times 10^{-5} \text{ cal/cm-sec-°C.}$$

*While 10 cm hardly seems to be the size of a human subject, 7 cm and 15 cm have been suggested as the effective diameters of an adult human subject as a cylinder [20] and as a sphere [21], respectively. Since the human subject shapes somewhere in between, we use a round value 10 cm between the two. This range of values differs from the real situation and the suggestion in Section 7.6 (see footnote on p. 297) by a factor of approximately 4.

Table 7.10-1   Comparison of Experimental Correlations of Transfer Coefficient of Body-Heat Loss

Investigator(s)	Formula,[a] $h=$	Experimental conditions
Colin and Houdas [15]	$2.3 + 7.5v^{0.67}$	Subject reclining, air flow from feet to head
Buettner [16]	$2.12 + 6.3v^{0.5}$	Subject lying on table
Winslow and Herrington [17][b]	$3.9 + 17.9v$ $18v^{0.5}$	Subject reclining, air flow from feet to head
Nelson et al. [18]	$7.5v^{0.5}$	Subject standing, transverse air flow
Hall [19]	$10v^{0.8}$	Dummy subjects

[a]$h[=]$ kcal/m^2-hr-°C, while v $[=]$ m/sec.
[b]The authors termed the two expressions "low" and "high" air speed formulas. However, examination of Fig. 7.10-3 shows that they are from but the same set of data. See text discussions.

Hence,

$$\Pr_{air} = \left(\frac{\hat{C}_p \mu}{k} \right)_{air} = \frac{0.25 \times 2.1 \times 10^{-4}}{6.25 \times 10^{-5}} = 0.84.$$

First of all, the formulas by Colin and Houdas, Buettner, and Winslow and Herrington, except for the exponent 0.67 and coefficients 17.9 and 18, seem to fit the form of Eq. 7.6-11 meaning that in the experimental arrangement the human subject can probably be best approximated as a spherical object with the effective diameter D. Using the above values, Eq. 7.6-11 becomes

$$h = \frac{k}{D} \left[2 + 0.6 \left(\frac{Dv\rho}{\mu} \right)^{1/2} \Pr^{1/3} \right]$$

$$= \frac{6.25 \times 10^{-5}}{10} \left[2 + 0.6 \left(\frac{10 \times 1.186 \times 10^{-3}}{2.1 \times 10^{-4}} \right)^{1/2} v^{1/2} 0.84^{1/3} \right]$$

$$= (1.25 + 2.66v^{0.5})10^{-5} \text{cal/cm}^2 - \text{sec-°C} \qquad (v[=] \text{cm/sec})$$

$$= 0.45 + 9.58v^{0.5} \text{kcal/m}^2\text{-hr-°C} \qquad (v[=] \text{m/sec}). \tag{7.10-6}$$

We can thus see that except for the exponent in the Colin–Houdas formula, the coefficients on the velocity term from the three experimental results on human subjects are in good agreement* with that in the "standard" correlation, Eq. 7.10-6, which has come also from experimental studies but on nonliving objects such as metal pipes. The discrepancy between the exponents 0.67 and 0.5 on the velocity was partially explained [15] by the different ways the subject was placed between the Colin-Houdas work and the Buettner work. Although Winslow and Herrington's arrangement was the same as Colin and Houdas', the difference in their exponents does not seem to be a serious matter because the two formulas of Winslow and Herrington's had come from the same set of data, and the choice seems to be arbitrary because both appear to fit the data equally well within the range of experimentation, as shown in Figure 7.10-3. In fact, an intermediate value of exponent between 0.5 and 1.0 (such as 0.67) seems to fit even better.

However, the greatest discrepancy is in the constant term. The three experimental works on human subjects are in good agreement but considerably higher than that in Eq. 7.10-6. This is obviously because the

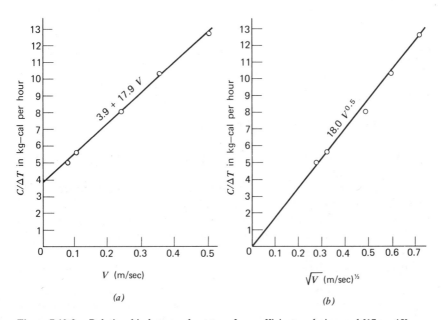

Figure 7.10-3.   Relationship between heat-transfer coefficient and air speed [17, p. 45].

*In this type of correlation, agreement is considered good within one order of magnitude.

human subjects were lying down or in a reclining position, against a chair or some form of support, effecting a large body-to-support contact area. This greatly increased the conductive heat loss,* and the author's assumption that conductive heat loss was negligible [15] seems to be unjustified in view of the present evidence.

The result of Nelson et al., should be compared with Eq. 7.6-9 or Eq. 7.6-10 with $a = 0.43$ and $m = 0.45$, because the experimental arrangement was clearly that of transverse flow. Equation 7.6-10 yields

$$h = \frac{k}{D} \left[0.43 \left(\frac{Dv\rho}{\mu} \right)^{0.55} \mathrm{Pr}^{1/3} \right]$$

$$= \frac{6.25 \times 10^{-5}}{10} \left[0.43 \left(\frac{10 \times 1.186 \times 10^{-3}}{2.1 \times 10^{-4}} \right)^{0.55} v^{0.55} 0.84^{1/3} \right]$$

$$= 2.34 \times 10^{-5} v^{0.55} \, \text{cal/cm}^2\text{-sec-}°\text{C} \, (v [=] \text{cm/sec})$$

$$= 10.56 v^{0.55} \, \text{kcal/m}^2\text{-hr-}°\text{C} \, (v [=] \text{m/sec}). \tag{7.10-7}$$

In addition to agreeing with Nelson et al., this result is consonant with that of Winslow and Herrington "at high speeds".

Finally, the result of Hall is in order-of-magnitude agreement with Eq. 7.6-8, which, after a computation similar to the above two, yields

$$h = 4.9 v^{0.8} \, \text{kcal/m}^2\text{-hr-}°\text{C} \, (v [=] \text{m/sec}). \tag{7.10-8}$$

However, we do not know whether the air flow was indeed parallel to the length of the dummy subject.

As pointed out earlier, none of the authors seemed to be aware of the effects of other parameters than that of the velocity, thereby leaving a number of the variables uncontrolled in their experiments. This may have been the cause of some of the discrepancies between their results and the "standard" correlations.

Throughout these comparisons we can appreciate one of the distinct

---

*The constant term was thought by the authors [15] to be the free-convection term, but from Eq. 7.6-11 and Problem 9.H$_2$, Ref. 1, it is clearly the conduction term. Moreover, from Eq. 7.8-3 the free-convection term is shown to consist of the Grashof number in conjunction with the Reynolds number. Thus, in a forced-convection situation, free convection occupies only a negligible place in the velocity term. Note in Eq. 7.8-3 that the conduction term, 2, also exists but is set apart from the free-convection (Gr) term.

advantages of the dimensionless correlations of the type shown in Sections 7.6–7.9, because they are independent of the system of units employed. Therefore, results obtained by different researchers could have been directly compared without having to compute for the particular set of parameters (such as subject size) under a certain set of conditions and to convert units back and forth. Such conversion could easily lead to errors that could be quite critical or costly especially when one applies the result.

What is worse in the reported correlations from human-subject data is that in most of them a mixture of units were employed, sometimes *even in the same formula*.* Such atrocity must end if scientists from various disciplines or engaged in different research programs are to communicate with each other efficiently and benefit from each others experience. We have seen that dimensionless correlations are indeed an unbiased solution regardless of which set of units one is familiar with. Its use in thermal regulation has actually long been advocated [22].

Example 7.10-3 The Wind-Chill Chart. Based on the discussions in Section 7.6 and Example 7.10-2, the speed of the ambient air has a great deal of effect on how cold a person feels in the winter through its effect on the heat-transfer coefficient and the rate of heat loss. An alternate explanation is that the wind reduces the thickness of the layer of warm air film around the body, so that the skin feels the ambient cold more directly. For example, if we have a person calibrated (trained) to judge various ambient temperatures in calm air, then with a 15-mph wind he would feel a 5°F ambient air as if it were at $-25°F$ in calm air. Such information, particularly useful in areas where the winter is bitter, is available in the form of the wind-chill chart, as shown in Figure 7.10-4 (based on data of Ref. 24), in which the equivalent calm-air temperature is plotted against the actual ambient temperature at various wind speeds. We can see that between 0 and 5 mph the "equivalent temperature depression" is very small, because at such low speed the air flow is in the laminar region and the heat-transfer coefficient is relatively insensitive to the wind speed ($Nu \propto Re^{1/3}$; see Eq. 7.6-6). Above 5 mph, turbulent flow takes over and the heat-transfer coefficient is almost proportional to the wind speed ($Nu \propto Re^{0.8}$, Eq. 7.6-8). This effect again diminishes beyond 25 mph.

Example 7.10-4 The effect of Wind Speed on Evaporation from Human Skin [25]. By defining a coefficient of evaporation heat transfer, h_e, with

$$q_e = h_e(p_s - p_a), \qquad (7.10-9)$$

where q_e is evaporative heat flux, p_s is the vapor pressure at the skin

*See for example, Ref. 23, where h is expressed in kcal/m²-hr-°C and v in ft/min.

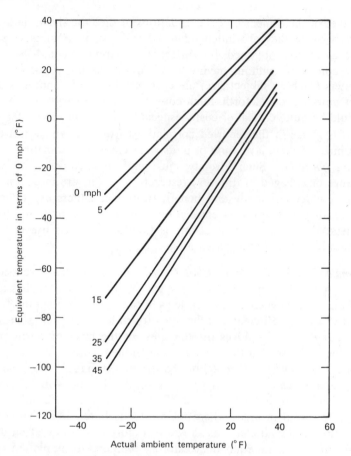

Figure 7.10-4. The wind-chill chart.

surface (saturated at skin temperature), and p_a is the vapor pressure in the ambient air, Clifford, et al. [25] obtained the following correlation:

$$h_e = 0.446v^{0.634 \pm 0.024}, \qquad (7.10\text{-}10)$$

where v is the wind speed in feet per minute, while h_e is in kcal/m²-hr-mmHg(!). Experimental data had been obtained with human subjects in varying sizes in standing position and whose skins were completely wet with sweat.

It is obvious that Eq. 7.10-9 is a hybrid form of the definitions of the heat- and mass-transfer coefficients, Eqs. 7.1-1 and 7.1-2. Such a definition is valid, since the heat dissipated is directly proportional to the amount of sweat evaporated. The h_e and v correspond to those in the Nusselt and Reynolds numbers and their relationship generally follows those presented in Sections 7.6 and 7.7 under the various conditions. However, further comparison with those correlations are not possible, as both the sensible and latent heats are lumped together and cannot be separated. In the present case, the ambient temperature is higher than the skin temperature; thus, the skin actually gains sensible heat only to be overcompensated by the loss of latent heat.

PROBLEMS

7.A Units of Mass-Transfer Coefficient

From the definition of the mass-transfer coefficient, k_c, and its analogy to the heat-transfer coefficient, show that the units for k_c are cm/sec and that Sh, the Sherwood number, is indeed dimensionless. What would be the units for, and definition of Sh in terms of, k_x?

7.B Functional Relationship for the Mass-Transfer Coefficient

Assuming that the mass-transfer coefficient, k_c, is a function of $\rho, \mu, \mathcal{D}_{AB}, D, L$, and V (which are, respectively, the density, viscosity, and diffusivity of the fluid, the diameter and length of the tube, and the velocity of the flow), find the dimensionless groupings relating these quantities. How many fundamental units are there? Is Buckingham's Π-theorem obeyed? (See Appendix I.)

7.C Correlation of Body-Heat Loss Data [28]

Show that the data on body-heat loss to air and water as reported by Colin and Houdas [15], Winslow and Gagge [26], and Witherspoon et al. [27] can be represented by the single equation

$$\mathrm{Nu} = 0.60 \cdot \mathrm{Re}^{0.52} \cdot \mathrm{Pr}^{0.23}. \qquad (7.B\text{-}1)$$

The data have been taken for unclothed subjects in reclining position in the range $10^4 < \mathrm{Re} < 10^5$.

7.D Body Temperature of Fish

(a)   By estimating the metabolic heat generation of a typical-size fish and its loss to surrounding water show that the heat loss through external body-surface area is *not* sufficient to account for its being cold-blooded.
(b)   From the oxygen requirement of fish and the content of oxygen physically dissolved in water, show that in order for the fish to extract enough oxygen through its gills to sustain life, its body temperature (through convection in the blood stream) must vary with the surrounding-water temperature.

Note. Not all fishes are cold-blooded. Certain species of tuna and mackerel shark manage to stay warm-blooded through the heat-shunt mechanism [29] (see also Example 8.8-4).

7.E Comparison of Body Heat Loss Correlations in Dimensionless Form

Using the air properties given in Example 7.10-2 to convert the data into dimensionless groups, compare the various correlations cited.

REFERENCES

1.   Bird, R. B.; Stewart, W. E.; and Lightfoot, E. N. *Transport Phenomena.* Wiley: New York, 1960.
2.   Moody, L. F. *Trans. ASME*, **66**, 671 (1944).
3.   Schlichting, H. *Boundary Layer Theory*, 6th ed. McGraw-Hill: New York, 1968.
4.   Hinze, J. O. *Turbulence.* McGraw-Hill: New York, 1959.
5.   Perry, J. H.; Perry, R. H.; Chilton, C. H.; and Kirkpatrick, S. D. *Chemical Engineers' Handbook,* 5th ed. McGraw-Hill: New York, 1973.
6.   Stokes, C. G., *Trans. Cambridge Phil. Soc.*, **9**, 8 (1850).
7.   McAdams, W. J. *Heat Transmission*, 3rd ed. McGraw Hill: New York, 1954, Chapter 5.
8.   Sieder, E. N., and Tate, G. E. *Ind. Eng. Chem.*, **28**, 1429 (1936).
9.   Sherwood, T. K., and Pigford, R. L. *Absorption and Extraction.* McGraw-Hill: New York, 1952.
10.   Colburn, A. P. *Trans. A.I.Ch.E.*, **29**, 174 (1933).
11.   Ranz, W. E., and Marshall, W. R., Jr. *Chem. Eng. Prog.*, **48**, 141, 173 (1952).
12.   Wolf, L., Jr., and Zaltzman S. In R. L. Dedrick, K. B. Bischoff, and E. F. Leonard, Eds., *The Artificial Kidney*, Chem. Eng. Prog. Symp. Series No. 84, Vol. LXIV American Institute of Chemical Engineers: New York, 1968, p. 104.

13. Chilton, T. H., and Colburn, A. P. *Ind. Eng. Chem.*, **26**, 1183 (1934).

14. Hershey, D.; and Song, G. *A. I. Ch. E. J.*, **13** (3), 491–496 (1967).

15. Colin, J. and Houdas, Y. *J. Appl. Physiol.*, **22**, 31 (1967).

16. Buettner, K. *Veröf. Preus. Meteorol. Inst. Abw.* (Berlin), **10** (5), (1934); cited in Ref. 15.

17. Winslow, C.-E. A., and Herrington, L. P. *Temperature and Human Life*, Princeton Univ. Press: Princeton, N.J., 1949.

18. Nelson, N.; Eichna, L. W.; Horvath, S. M.; Shelley, W. B.; and Hatch, T. F. *Am. J. Physiol.*, **151**, 626 (1947).

19. Hall, J. Memo Rept. MCREXD-696-1058, Wright-Patterson AFB, Ohio (1950); cited in Ref. 15.

20. Plummer, J. H. *Publ. Climat. Envir. Branch*, Off. Q. M. Gen., U. S. Army, August 25, 1944; cited in Ref. 22.

21. Soderstrom, G. F., and DuBois, E. F. *Arch. Int. Med.*, **19**, 931 (1917).

22. Newburgh, L. H. Physiology of Heat Regulation and the Science of Clothing, repr. of 1949 ed. Hafner: New York, 1968.

23. Winslow, C.-E. A.; Herrington, L. P.; and Gagge, A. P. *Am. J. Physiol.*, **120**, 288 (1937).

24. Folk, G. E., Jr. *Introduction to Environmental Physiology*. Lea and Febiger; Philadelphia, 1966, p. 96.

25. Clifford, J.; Kerslake, D. McK.; and Waddell, J. L. *J. Physiol.*, **147**, 253 (1959).

26. Winsolw, E. A., and Gagge, L. P. *Am. J. Physiol.*, **129**, 79 (1940).

27. Witherspoon, J. M.; Goldman, R. F.; and Breckenridge, J. R. *J. de Physiol.*, **63**, 459, (1971).

28. Pan, C. H. M.Ch.E. Thesis, Catholic Univ., Washington, D. C. (1973).

29. Carey, F. G. *Sci. Am.*, **228**(2), 36–44 (1973).

CHAPTER EIGHT

Macroscopic Transport

Heretofore we have discussed methods of studying transport processes at each individual point, as characterized by such differential equations as the equations of continuity, motion, and energy and the laws of Newton (viscosity), Fourier (heat conduction), and Fick (diffusion).

In this chapter we shall describe methods for studying how the human body, or a biological system, functions as a whole. Theoretically, we should sum up the behaviors of the individual points through the process of integration and arrive at the *macroscopic* mass, momentum, and energy balance equations. However, because of the mathematical complications that would be involved, we shall use a simpler, less rigorous approach.

In Section 8.1 we shall review the foundation of the macroscopic balance equations—the conservation laws—and their relationship with the differential and macroscopic balance equations. In the next four sections, we shall present the simplified approach in arriving at the macroscopic mass, momentum, energy, and mechanical energy equations, for *single component systems*. However, some of these equations by themselves are not very useful until we extend them to multicomponent systems, because biological systems are by nature *multicomponent*. We shall do this in Section 8.7 without going into great details, for the sake of avoiding mathematical complexity. In Section 8.6 we shall discuss the friction-loss term in the energy equation.

After these equations have been described, examples from such diverse areas as hemodialysis, body-heat conservation, placenta mass transport, and gastric-acid analysis will be presented in Section 8.8. It is hoped that the reader will be motivated to find an even larger variety in the application of these equations.

8.1 FROM DIFFERENTIAL TO MACROSCOPIC BALANCE EQUATIONS

In Chapter 4 we discussed the derivation of the equations of continuity, motion, and energy from the principles of mass, momentum, and energy conservation, respectively, using the Eulerian (fixed-element) approach. In the Lagrangian (following-motion) approach, these principles take the form of the "law of mass isolation," Newton's second law of motion, and the first law of thermodynamics, respectively, as shown in Table 8.1-1. We have also described the equation of mechanical energy in Section 4.8, but since mechanical energy is not conserved, it is not based on any conservation principle. Rather, it is obtained by taking the dot product between the velocity vector v and the equation of motion, as also shown in Table 8.1-1.

If we "lump" all points of the system together, we should have a macroscopic description of the system as a whole. Mathematically, this is done by integrating the three equations of change over the entire volume of the system, as indicated by the small arrows with the $\int dV$ symbols in Table 8.1-1. The results are the three macroscopic balance equations.

However, such an integration procedure is easier said than done. It involves mathematically advanced theorems [1] that would be quite time-consuming to develop. Thus, in the present chapter we adopt a simpler, less rigorous procedure—namely, the application of the three conservation laws directly to a macroscopic system, as indicated by the long and heavy arrows leading from the conservation statements directly to the macroscopic equations in Table 8.1-1. This simplified procedure will be described in the following sections. The only exception in applying this procedure is in the development of the macroscopic mechanical energy equation. Since there is no conservation principle from which this equation can be obtained and since it cannot be obtained by dot multiplication of the macroscopic momentum balance equation by the velocity vector v (as indicated by the crossed-out arrow), there is no other choice but to integrate the (differential) equation of mechanical energy (Eq. 4.8-1) over the entire volume.

It must be emphasized, however, that if one has the mathematical capability to do so, the rigorous integration procedure is definitely the more foolproof one, because, as will be seen from Section 8.7, the existence of many of the quantities, particularly those for multicomponent systems such as those concerning the transport across phase boundaries, is not so apparent in the simplified procedure (i.e., by merely considering the "in" and "out" streams). But they will automatically appear from integrating the multicomponent equations of change [1].

In the simplified approach, we will keep track of the input and output streams of a system along with any transport across system and phase boundaries as well as work done on or by the system. For this purpose, we might consider the box model of a biological system, such as the human body, as shown in Fig. 8.1-1. However, this schematic is *not* limited to the entire body. Any part of the body, such as the stomach or the circulatory system, can be considered a macroscopic system by itself. We shall see examples of these in Section 8.8.

Table 8.1-1. Interrelationship between Differential and Macroscopic Balanced Equations.

Basic principle	Statement	Applied to — Fixed element	Applied to — Element following motion	Yields → Differential equation of change	∫ dV System →	Macroscopic equation
Conservation of mass	$\left\{\begin{array}{c}\text{Rate of}\\\text{mass in}\end{array}\right\} - \left\{\begin{array}{c}\text{Rate of}\\\text{mass out}\end{array}\right\} = \left\{\begin{array}{c}\text{Rate of mass}\\\text{accumulation}\end{array}\right\}$	✓		Equation of continuity	$\int dV$	Mass balance
"Isolation" of mass	$\dfrac{D}{Dt}(\text{mass}) = 0$		✓			
Conservation of momentum	$\left\{\begin{array}{c}\text{Rate of}\\\text{mom. acc.}\end{array}\right\} = \left\{\begin{array}{c}\text{Rate of}\\\text{mom. in}\end{array}\right\} - \left\{\begin{array}{c}\text{Rate of}\\\text{mom. out}\end{array}\right\} + \sum \begin{array}{c}\text{Forces}\\\text{on system}\end{array}$	✓	✓	Equation of motion	$\int dV$	Momentum balance
Newton's second law of motion	$F = ma$					
(Lack of conser— vation of mech— anical energy)				Equation of mechanical energy $\xrightarrow{\ \cdot\ v\ }$	$\int dV$	Mechanical energy $\xleftarrow{\ \cdot\ v\ }$
Conservation of energy	$\left\{\begin{array}{c}\text{Rate of}\\\text{energy}\\\text{acc.}\end{array}\right\} = \left\{\begin{array}{c}\text{Rate of}\\\text{energy}\\\text{in}\end{array}\right\} - \left\{\begin{array}{c}\text{Rate of}\\\text{energy}\\\text{out}\end{array}\right\} + \left\{\begin{array}{c}\text{(Net) rate of}\\\text{heat conduc-}\\\text{tion in}\end{array}\right\} - \left\{\begin{array}{c}\text{Rate of}\\\text{work}\\\text{out}\end{array}\right\}$	✓	✓	Equation of energy	$\oint dV$	Energy balance
First laws of thermodynamics	$\Delta E = Q - W$					

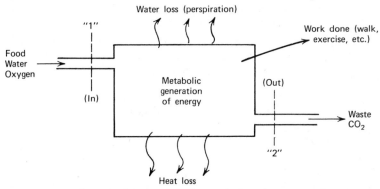

Figure 8.1-1.   Schematic diagram of the human body as a macroscopic system.

8.2   THE MACROSCOPIC MASS BALANCE: SINGLE-COMPONENT SYSTEMS

In Figure 8.1-1, if we denote the entrance plane of a system by "1" and the exit plane by "2" and restrict ourselves to the single-component system (so that there will be no consumption or transformation of material), the mass conservation statement that

$$\left\{ \begin{array}{l} \text{Rate of mass} \\ \text{accumulation} \end{array} \right\} = \left\{ \begin{array}{l} \text{Rate of} \\ \text{mass in} \\ \text{by flow} \end{array} \right\} - \left\{ \begin{array}{l} \text{Rate of} \\ \text{mass out} \\ \text{by flow} \end{array} \right\} \qquad (8.2\text{-}1)$$

can be symbolically written

$$\frac{dm_{\text{tot}}}{dt} = w_1 - w_2, \qquad (8.2\text{-}2)$$

where

$m_{\text{tot}} = $ total mass in the system between "1" and "2",

$w_k = \rho_k \langle v_k \rangle S_k, (k = 1 \text{ or } 2),$ the mass flow rate (mass/time) across plane k,

$\rho_k = $ density of fluid at plane k,

$\langle v_k \rangle = $ the average velocity (over the cross section) across plane k,

$S_k = $ the cross-sectional area of passage at plane k or simply

$$\frac{dm_{\text{tot}}}{dt} = -\Delta w, \qquad (8.2\text{-}3)$$

where Δ usually means the exit value minus the entrance value ("2" – "1").

At *steady state* (i.e., when the behavior of the system is time-independent), the macroscopic balance equation becomes simply

$$\Delta w = 0. \tag{8.2-4}$$

Here we have considered transport in a single phase ("by flow") only so that interphase mass transport, such as diffusion or perspiration, has not been taken into account. This is why the ultimate objective of presenting the multicomponent equations (Section 8.7) is so important.

8.3 THE MACROSCOPIC MOMENTUM BALANCE: SINGLE-COMPONENT SYSTEMS

If we apply the momentum conservation principle

$$\left\{\begin{array}{c} \text{Rate of} \\ \text{momentum} \\ \text{accumulation} \end{array}\right\} = \left\{\begin{array}{c} \text{Rate of} \\ \text{momentum} \\ \text{in by flow} \end{array}\right\} - \left\{\begin{array}{c} \text{Rate of} \\ \text{momentum} \\ \text{out by flow} \end{array}\right\} + \left\{\begin{array}{c} \text{Sum of forces} \\ \text{acting on the} \\ \text{system} \end{array}\right\}$$

$$\tag{8.3-1}$$

to the system in Figure 8.1-1, between planes "1" and "2", we have

$$\frac{d\mathbf{P}_{tot}}{dt} = \rho_1 \langle v_1^2 \rangle \mathbf{S}_1 - \rho_2 \langle v_2^2 \rangle \mathbf{S}_2 + (p_1 \mathbf{S}_1 - p_2 \mathbf{S}_2) - \mathbf{F} + m_{tot}\mathbf{g}, \tag{8.3-2}$$

where $\mathbf{P}_{tot}$ is the *total* momentum in the system; p_k, the pressure at plane k; $-\mathbf{F}$, the total surface forces* acting on the system; and $m_{tot}\mathbf{g}$, the total body force† on the system. The boldfaced quantities are vectors. For example, both $\mathbf{P}$, being mass (scalar) times velocity (vector), and $\mathbf{F}$ are vectors. The cross-sectional area would normally be a scalar, but in this case the orientation does give it a vectorial significance. The direction of a positive cross-sectional area is that of the average velocity at the cross section, the opposite direction being negative and those between these extremes being accounted for by a factor equal to the cosine of the angle deviating from the positive direction.

---

*Such as viscous and pressure forces acting on system boundaries. Also, since forces are summed vectorially, the total is also the *net* force.
†Such as gravitational, magnetic and centrifugal forces; the **g** here is the same as the one in Chapter 4.

As before "$\langle \ \rangle$" means "averaged over the cross section" and in this case the quantity to be so averaged is the square of the velocity, v_k^2 ($k = 1$ or 2). Note that this average of the square is *different* from the square of the average $\langle v_k \rangle^2$. For laminar flow, the velocity profile can usually be expressed analytically. Thus, both the average of its square and the square of its average can be obtained straightforwardly. For turbulent flow, however, this cannot be done easily, since many expressions are empirical. Fortunately, the use of the approximation

$$\langle v_k^2 \rangle \approx \langle v_k \rangle^2$$

is acceptable with only small error for this type of flow.

Alternately, Eq. 8.3-2 can be rewritten as, (remembering that $\Delta = $ "2" − "1")

$$\frac{d\mathbf{P}_{\text{tot}}}{dt} = -\Delta\left(\frac{\langle v^2 \rangle}{\langle v \rangle} \mathbf{w} + p\mathbf{S} \right) - \mathbf{F} + m_{\text{tot}}\mathbf{g} \tag{8.3-3}$$

for steady state, it becomes

$$\mathbf{F} = -\Delta\left(\frac{\langle v^2 \rangle}{\langle v \rangle} \mathbf{w} + p\mathbf{S} \right) + m_{\text{tot}}\mathbf{g}. \tag{8.3-4}$$

In the last two equations, $\mathbf{w}_k$ is again equal to $\rho_k \langle v_k \rangle \mathbf{S}_k$ as in Section 8.2. However, since $\mathbf{S}_k$ has vectorial significance here, so does $\mathbf{w}_k$.

8.4 THE MACROSCOPIC ENERGY BALANCE: SINGLE-COMPONENT SYSTEMS

In applying the principle of energy conservation to a macroscopic system, the potential energy has to be taken into account, since the entrance and the exit of the system may not be at the same level, so that the potential energy per unit mass at one may be different from that at the other. Thus, the principle states

$$\begin{Bmatrix} \text{Rate of} \\ \text{accumulation} \\ \text{of IE, KE, and} \\ \text{PE} \end{Bmatrix} = \begin{Bmatrix} \text{Rate of IE,} \\ \text{KE, and PE} \\ \text{in by flow} \end{Bmatrix} - \begin{Bmatrix} \text{Rate of IE,} \\ \text{KE, and PE} \\ \text{out by flow} \end{Bmatrix}$$

$$+ \begin{Bmatrix} \text{Net rate of heat} \\ \text{added to system} \\ \text{from surroundings} \end{Bmatrix} - \begin{Bmatrix} \text{Net rate of work} \\ \text{done by system} \\ \text{on surroundings} \end{Bmatrix}. \tag{8.4-1}$$

When this is applied to a macroscopic system, the following results:

$$\frac{dE_{\text{tot}}}{dt} = -\Delta\left[\left(\hat{U} + p\hat{V} + \frac{1}{2}\frac{\langle v^3\rangle}{\langle v\rangle} + \hat{\Phi}\right)w\right] + Q - W, \qquad (8.4\text{-}2)$$

where $E_{\text{tot}} = U_{\text{tot}} + K_{\text{tot}} + \Phi_{\text{tot}}$, the total energy (U, K, and Φ being the internal, kinetic, and potential energies, respectively); Q is the net rate of heat input; W is the net rate of work output and the caret denotes "per unit mass," as usual. Since w has the dimensions of mass per unit time, any of the energy terms with a caret multiplied by w has the meaning of the time rate of energy.

For steady state, Eq. 8.4-2 becomes

$$\Delta\left(\hat{U} + p\hat{V} + \frac{1}{2}\frac{\langle v^3\rangle}{\langle v\rangle} + \hat{\Phi}\right) = \hat{Q} - \hat{W}, \qquad (8.4\text{-}3)$$

in which all terms are in the "per unit mass" form. For example, the term $\frac{1}{2}\langle v^3\rangle/\langle v\rangle$, which for turbulent flow can be approximated by $\frac{1}{2}\langle v\rangle^2$, is clearly the kinetic energy per unit mass.

8.5 THE MACROSCOPIC MECHANICAL ENERGY EQUATION: SINGLE-COMPONENT SYSTEMS

Although all previous macroscopic balance equations can be derived either by integration of the equation of change over the entire volume of the system or by application of the appropriate conservation principles to the macroscopic system, the macroscopic equation for mechanical energy, which is *not* conserved, can be derived only by the former method—namely, integration of the equation of mechanical energy, of which we wrote the Lagrangian version in Section 4.8. However, the detailed procedure is rather complicated mathematically [1], and so, we shall only write down the result as follows:

$$\frac{d}{dt}(K_{\text{tot}} + \Phi_{\text{tot}} + A_{\text{tot}}) = -\Delta\left[\left(\frac{1}{2}\frac{\langle v^3\rangle}{\langle v\rangle} + \hat{\Phi} + \hat{G}\right)w\right] - W - E_v, \qquad (8.5\text{-}1)$$

where A_{tot} is the total Helmholtz free energy of the system; $\hat{G}$, the Gibbs free energy (per unit mass) of the system; and E_v, the rate at which mechanical energy is irreversibly converted into thermal energy, or the "friction loss."

For steady state, Eq. 8.5-1 becomes

$$\Delta\left(\frac{1}{2}\frac{\langle v^3\rangle}{\langle v\rangle}+\hat{\Phi}+\hat{G}\right)+\hat{W}+\hat{E}_v=0. \tag{8.5-2}$$

For isothermal systems, $dG = V\,dp$,*

$$\Delta\hat{G}=\int_{p_1}^{p_2}\hat{V}\,dp$$

$$=\int_{p_1}^{p_2}\frac{dp}{\rho}. \tag{8.5-3}$$

The latter is true, since $\hat{V}$ is the volume per unit mass and ρ is the mass per unit volume; thus, one is the reciprocal of the other. Furthermore, if in this case the fluid is incompressible,

$$\Delta\hat{G}=\frac{p_2-p_1}{\rho}, \tag{8.5-4}$$

since ρ is constant. On the other hand, if the fluid is an *ideal gas*, which means $\rho = Mp/RT$, where M is the molecular weight; R, the gas constant; and T, the absolute temperature, Eq. 8.5-3 becomes

$$\Delta\hat{G}=\frac{RT}{M}\ln\frac{p_2}{p_1}. \tag{8.5-5}$$

Equation 8.5-2, with the substitution of Eq. 8.5-3, gives

$$\Delta\frac{1}{2}\frac{\langle v^3\rangle}{\langle v\rangle}+\Delta\hat{\Phi}+\int_{p_1}^{p_2}\frac{dp}{\rho}+\hat{W}+\hat{E}_v=0. \tag{8.5-6}$$

This is the celebrated *Bernoulli equation*, which is the starting equation for most problems of fluid flow in tubes and pumping systems such as the circulatory system. However, extreme care must be exercised in using it, in regard to its limitations—steady state and isothermal conditions.

In these equations, $\Delta\frac{1}{2}\langle v^3\rangle/\langle v\rangle$ is the kinetic energy difference between the entrance and the outlet. $\Delta\hat{\Phi}$ is their potential energy difference and is

*Since $dG = V\,dp - S\,dT$ and, for the isothermal case, $dT = 0$.

directly proportional to the difference in elevation between these two points. In the circulatory system, W would be the work input required of the heart and have a negative value. If it had a positive value, that would mean work is being done by the system.

8.6 FRICTION LOSS

For the steady flow of an incompressible fluid in a straight tube of constant cross section without work input or output, Eqs. 8.3-4, 8.5-2, and 8.5-4 yield

$$F = -S\,\Delta p + m_{tot}g \qquad \text{(in scalar form)} \qquad (8.6\text{-}1)$$

$$g\,\Delta z + \frac{\Delta p}{\rho} + \hat{E}_v = 0, \qquad (8.6\text{-}2)$$

since $v_1 = v_2$, $\Delta\hat{\Phi} = g\,\Delta z$. Dividing Eq. 8.6-1 by ρS and noting that $m_{tot} = \rho S\,\Delta z$, Eq. 8.6-2 can be converted to

$$\hat{E}_v = \frac{F}{\rho S}. \qquad (8.6\text{-}3)$$

Further noting that $F = A \cdot K \cdot f$ (Eq. 7.2-1), $A = 2\pi RL$, (Eq. 7.2-2), $K = \frac{1}{2}\rho\langle v\rangle^2$ (Eq. 7.2-4), and $S = \pi R^2$, Eq. 8.6-3 can be rewritten

$$\hat{E}_v = f \cdot \tfrac{1}{2}\langle v\rangle^2 \cdot 4\left(\frac{L}{D}\right), \qquad (8.6\text{-}4)$$

which applies to both laminar and turbulent flows. We can thus see that the friction factor, f, is quite useful in estimating the energy loss by friction of the system.

For tubes with bends, valves, fittings, bifurcations, sudden expansions, and constrictions, such as those sometimes encountered in both *in vitro* and *in vivo* flow systems, $\hat{E}_v$ has to include the additional loss due to these irregularities. One common method is that of the equivalent length, L_e, defined as the hypothetical straight-tube length that would produce the same friction loss as the actual bends, valves, bifurcations, and the like. The sum of the equivalent lengths of these extra frictional losses is then added to the straight-tube length, L. Equivalent lengths of many of these for engineering systems can be found in standard handbooks [2]. Another method is expressing

$$\hat{E}_v = \sum_{i=1}^{m}\left(f \cdot \tfrac{1}{2}\langle v\rangle^2 \cdot 4\frac{L}{D}\right)_i + \sum_{j=1}^{n}\left(\tfrac{1}{2}\langle v\rangle^2 e_v\right)_j, \qquad (8.6\text{-}5)$$

where e_{v_j} is the *friction loss factor* of the jth bend, valve, fitting, or bifurcation. In this equation there are m straight-tube sections and n bends and valves, and so on. Values of e_v for some of these obstacles can be found in standard handbooks and texts [3, p. 217; 4, pp. 53–54].

More generally, it can be shown from the straight integration procedure [1] that E_v is related to the viscous dissipation function Φ_v defined in Section 4.10 by

$$E_v = -\int_V (\tau : \nabla \mathbf{v})\, dV$$

$$= \int_V \mu \Phi_v\, dV$$

which can be analytically evaluated for systems with regular geometries (such as flows in circular tube or between two disks).

8.7 MULTICOMPONENT SYSTEMS

As biological systems are by nature multicomponent, macroscopic equations in multicomponent form are extremely useful. While it would seem that they could be developed by applying the conservation principles to each component, taking into consideration the production or consumption due to chemical reaction, the more rigorous integration method is usually required so that some of the less obvious terms will not be omitted. For example, from integrating Eq. 5.5-4 over the entire volume of the system, the macroscopic mass balance equation for the ith species in a mixture of n components is

$$\frac{dm_{i_{\text{tot}}}}{dt} = -\Delta w_i + w_i^{(m)} + r_{i_{\text{tot}}}, \qquad i = 1, 2, \ldots, n \qquad (8.7\text{-}1)$$

where $r_{i_{\text{tot}}}$ is the rate of production of the ith component in the entire system while the superscript (m) denotes a quantity concerning the transport across bounding surfaces or phase boundaries. The term $w_i^{(m)}$ here is obviously the direct result of the integration of the diffusional term in Eq. 5.5-4, while $-\Delta w_i$ is that of the convective term, $(\nabla \cdot \rho_i \mathbf{v})$, in the same equation. This (m)-superscripted term would most likely be neglected if one were merely to consider the "in" and "out" streams. However, to avoid mathematical complexities, we omit the detailed integration procedure here, as will be the case for the rest of this section.

The macroscopic mass balance equation for the mixture as a whole can be obtained by summing the n equations represented in Eq. 8.7-1 for the n

components. The result is, noting that $\sum_{i=1}^{n} r_{i_{\text{tot}}} = 0$,

$$\frac{dm_{\text{tot}}}{dt} = -\Delta w + w^{(m)}. \tag{8.7-2}$$

Comparison of this equation with Eq. 8.2-3 shows that there is a difference between the mass balance of a pure substance and that of the mixture, even when we treat the latter as a whole.

The equations corresponding to Eqs. 8.7-1 and 8.7-2 in mole units are

$$\frac{d\,\mathfrak{M}_{i_{\text{tot}}}}{dt} = -\Delta\,\mathfrak{W}_i + \mathfrak{W}_i^{(m)} + R_{i_{\text{tot}}}, \qquad i = 1, 2, \ldots, n \tag{8.7-3}$$

for the ith species and

$$\frac{d\,\mathfrak{M}_{\text{tot}}}{dt} = -\Delta\,\mathfrak{W} + \mathfrak{W}^{(m)} + \sum_{i=1}^{n} R_{i_{\text{tot}}} \tag{8.7-4}$$

for the mixture; here the sum of all $R_{i_{\text{tot}}}$ does not vanish, since moles are *not* conserved in a chemical reaction.

Similarly, the other macroscopic equations are as follows:
Momentum:

$$\frac{d\mathbf{P}_{\text{tot}}}{dt} = -\Delta\left(\frac{\langle v^2 \rangle}{\langle v \rangle} \mathbf{w} + p\mathbf{S}\right) + \mathbf{F}^{(m)} - \mathbf{F} + m_{\text{tot}}\mathbf{g} \tag{8.7-5}$$

Energy:

$$\frac{dE_{\text{tot}}}{dt} = -\Delta\left[\left(\hat{U} + p\hat{V} + \frac{1}{2}\frac{\langle v^3 \rangle}{\langle v \rangle} + \hat{\Phi}\right)w\right] + Q^{(m)} + Q - W \tag{8.7-6}$$

Mechanical energy:

$$\frac{d}{dt}(K_{\text{tot}} + \Phi_{\text{tot}}) = -\Delta\left[\left(\frac{1}{2}\frac{\langle v^3 \rangle}{\langle v \rangle} + \hat{\Phi} + \frac{p}{\rho}\right)w\right] + B^{(m)} - W - E_v. \tag{8.7-7}^*$$

Here the extra terms with superscript m are again those associated with the transport of the respective superscripted entities across the bounding surfaces. For example, the term $B^{(m)}$ represents the potential and kinetic energies added to the system plus the work done against the system

*This equation is good for constant density ρ only.

pressure in forcing a fluid through the walls of the system. It should also be noted that it is meaningless to use the macroscopic momentum and energy equations for the individual species.

8.8 BIOMEDICAL APPLICATIONS

Because of the necessary limitation on the length of this book and each of its chapters, we can only present a few examples in this chapter, such as those dealing with macroscopic mass balance on a tracer in a network of blood vessels in the tissue and on hydrogen ions in the stomach, the occurrence of cavitation due to stenosis as explained by the macroscopic mechanical energy equation, and the design of hemodialyzers using a combination of macroscopic and interphase transport equations presented earlier.

There are a vast number of equally interesting examples that we do not have the space to cover. They include such problems as the estimation of energy losses and pressure drop in human bronchial passageways [5], the absorption of sodium chloride solutions from the gall bladder [6], and the parallel and series compartment models for the K^+ flux in the perfused rabbit heart [7], just to name a few.

Example 8.8-1 The Single-Compartment Model and the Tracer Method for Measuring Local Blood Flow. Although *in vitro* blood flow and that in individual vessels of reasonable sizes can be measured directly using hydrodynamical methods or others such as lasers, its flow rate in a network of small vessels, such as those in the tissue, can best be determined indirectly by measuring the rate at which a biologically and chemically inert, but easily detectable, substance is gained or lost from the blood stream. For this purpose we represent the locality through which the flow is to be measured by a single-compartment model, as shown in Figure 8.8-1. In the uptake method the substance is injected into the (arterial) inflow to the compartment initially devoid of the same substance. Its rate of appearance in the (venous) outflow is measured. On the other hand, the *clearance* method works on a reverse principle (i.e., an initial dose of the substance is present in the compartment, not in the inflow). Its gradual dilution in the outflow is measured. Substances frequently used as so-called "tracers" include dyes, radioactive materials, and such dissolved gases as nitrous oxide, krypton, and hydrogen. One of its important characteristics is that it should be highly diffusible so that perfect mixing is achieved in the compartment. This makes the concentration of this material in the compartment the same as that in the outflow, $c(t) = c_2(t)$, a desirable simplification for theoretical analysis.

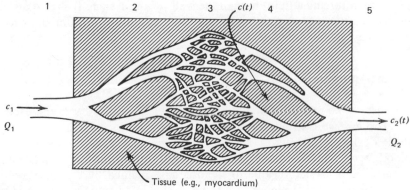

Figure 8.8-1. Single-compartment model of blood flow through capillary network in the tissue: 1, artery; 2, arterioles; 3, capillaries; 4, venules; 5, vein.

This method was first developed by Kety and Schmidt [8] and our present example deals with the determination of local blood flow through the capillary network in the myocardium from the clearance of dissolved hydrogen [9]. Let c_0 be the initial concentration of dissolved hydrogen and V the volume of blood in the myocardium. Since in the clearance method $c_1 = 0$ and the hydrogen does not go elsewhere via any mode of transport (such as interphase transfer or reaction) other than through the "in" and "out" streams, the macroscopic balance equation yields

$$\frac{d}{dt}(Vc) = Q_1 \cdot 0 - Q_2 c_2, \qquad (8.8\text{-}1)$$

where Q_1 and Q_2 are the volumetric rates of inflow to, and outflow from, the myocardium compartment. Since the blood also flows in and out through the "1" and "2" planes only, we can write, $Q_1 = Q_2 = Q$, and the volume of blood in the myocardium is constant. Thus, Eq. 8.8-1 becomes

$$\frac{dc}{dt} = -\frac{Q}{V}c_2. \qquad (8.8\text{-}2)$$

In the general case, similar to the present one, c_2 (hydrogen concentration in venous blood) is theoretically not the same as c (hydrogen concentration in tissue). Instead, they differ by a factor—the hydrogen solubility coefficient in equilibrium between blood and tissue, α, or

$$c_2 = \alpha c. \qquad (8.8\text{-}3)$$

But where rapid diffusion of the tracer occurs in the tissue, α has been found to be near unity, as in the present case [9]. Substituting Eq. 8.8-3 in Eq. 8.8-2 gives

$$\frac{dc_2}{dt} = -\frac{Q}{\alpha V} c_2, \tag{8.8-4}$$

which can be readily integrated with the initial condition to yield

$$c_2 = c_0 e^{-(Q/\alpha V)t}, \tag{8.8-5}$$

which is known as the exponential-decay function (Figure 8.8-2) and characteristic of many processes whose instantaneous rate is proportional to the concentration of the remaining material. A more useful plot is in semilog form, since

$$\ln \frac{c_2}{c_0} = -\frac{Q}{\alpha V} t, \tag{8.8-6}$$

from which the rate of blood flow can be calculated from the slope. Figure 8.8-3 is a typical example converted from Figure 8.8-2 in which the calculated Q-value differs with the actually measured venous flow by only 3%. This single-compartment type of analysis applies to many other physiological systems, of which the following example is one.

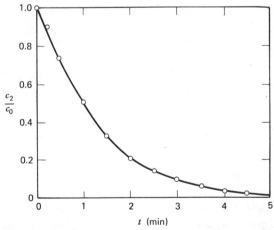

Figure 8.8-2. Exponential-decay clearance of dissolved hydrogen from the myocardium of the left ventricle [9] (taken from [17]). Note that this dimensionless concentration will drop to zero only at infinite time.

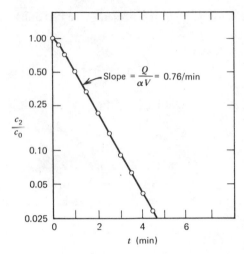

Figure 8.8-3. Semilog plot of Eq. 8.8-5 and Figure 8.8-2. Note that the distance from $c_2/c_0 = 1.00$ to .25 is the same as that from .10 to .025 and that this scale will never reach zero (from [17]).

Example 8.8-2 Transmucosal Migration of Water and Hydrogen Ions in the Stomach [10]. In order to determine the relative importance of diffusion and secretion in governing the acidity in the animal stomach, experiments were conducted in which an isolated flap of the ventral wall of a cat's stomach with intact blood supply was mounted between two chambers. A mixture of HCl and Na_2SO_4 was instilled in the "stomach" chamber (i.e., the one on the mucosal side) and the rates of change of the hydrogen and sulfate ions were measured. Since both water and H^+ can enter and leave only by diffusion and secretion and there is no food in the chamber, this is a model for the starved stomach with no input or output by flow. Consequently, both $\mathcal{W}_{i_1}$ and $\mathcal{W}_{i_2}$ (and hence $\Delta\mathcal{W}_i$, where i now represents the H^+ species) are zero and there is no consumption or production of H^+ by chemical reaction ($R_{i_{tot}} = 0$) in Eq. 8.7-3.

Let V and c be the instantaneous volume and H^+ concentration, respectively, of the H^+-containing juice at time t. Then the total number of H^{+*} at time t is $\mathfrak{M}_{i_{tot}}$. The interphase transport represented by $\mathcal{W}_i^{(m)}$ consists of two terms: (1) secretion ($= sc_a$, where s is the constant rate of secretion, volume per unit time and c_a is the H^+ concentration in the secreted fluid); and (2) diffusion ($= kc$, where k is the permeability of H^+

─────────

*Or the total mass or number of moles for Eq. 8.7-3; it does not matter in this case, since $R_{i_{tot}} = 0$ anyway.

ions, volume per unit time). The former represents an H^+ gain and thus carries a plus sign, while the latter represents an H^+ loss and thus a minus sign. Substituting these quantities in Eq. 8.7-3, we have*

$$\frac{d}{dt}(Vc) = sc_a - kc. \qquad (8.8\text{-}7)$$

Expanding the time derivative on the left-hand side into $V(dc/dt) + c(dV/dt)$ and noting that

$$V = V_0 + st \qquad (8.8\text{-}8)$$

where V_0 is the volume of instillate initially present (at $t=0$) in the stomach (chamber) so that

$$\frac{dV}{dt} = s \qquad (8.8\text{-}9)$$

Eq. 8.8-7 becomes**

$$\frac{dc}{dt} = \frac{sc_a - (k+s)c}{V_0 + st}. \qquad (8.8\text{-}10)$$

The integration of this expression looks much easier if we rearrange into

$$\frac{dc}{(k+s)c - sc_a} = -\frac{dt}{V_0 + st},$$

which, with the initial condition that $c = c_0$ (the H^+ concentration of the initial instillate) at $t=0$, gives

$$c(t) = \frac{c_a}{1+k/s}\left[1 - \frac{1-(c_0/c_a)(1+k/s)}{(1+st/V_0)^{(1+k/s)}}\right] \qquad (8.8\text{-}11)$$

---

*Although similar in form, this equation has a different basis from Eq. 8.8-1 or 8.8-4, because in the last example, there is no interphase transport but only bulk flow. Here the opposite is true. If we viewed the secretion and diffusion as flows in and out, then the two cases would seem to be similar.

**In the original paper [10], the macroscopic balance was on the concentration, not the total number, of H^+ ions, so that the term $sc/(V_0+st)$ was included as the "dilution" term. However, this may cause some confusion, since in the dilution process, no mass or number of ions is actually lost. Also, the balance equation should not be applied to concentration since it is not conserved. Since dilution comes about as a consequence of volume change, it is automatically reflected in the dV/dt term, which leads to the $sc/(V_0+st)$ term. Thus, the presentation here should be preferred over the original one.

which can be much more easily represented graphically if rearranged into

$$\Gamma(\tau) = \frac{c(t) - c_a/K}{c_0 - c_a/K}$$

$$= (1+\tau)^{-K} \qquad (8.8\text{-}12)$$

(where $\tau = st/V_0$ and $K = 1 + k/s$), which in addition to being in dimensionless form yields a family of straight lines on a log–log plot, as shown in Figure 8.8-4. Again, this is a typical illustration of the advantage of using dimensionless quantities because the seemingly very complex relationship can actually be represented on one single and simple graph.

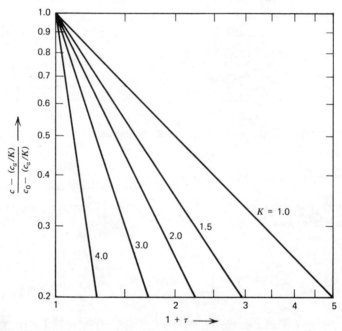

Figure 8.8-4. Dimensionless unified plot of the H^+ concentration change with time and diffusion parameters.

Further light may be shed on this problem if we consider the several possible limiting cases:

1.   If $V_0 = 0$,

$$c = \frac{c_a}{1 + k/s},$$

which means that if there is no initial hold-up volume, the H^+ concentration leaving the stomach is independent of time and initial concentration (since there is nothing present initially) and differs with incoming concentration by a factor related to the ratio of the permeability to the secretion rate, for these are the only two modes of mass transfer.

2. If $c_0 = 0$ (no initial H^+ concentration),

$$c = \frac{c_a}{1+k/s}\left[1 - \frac{1}{(1+st/V_0)^{(1+k/s)}}\right] \qquad (8.8\text{-}13)$$

3. If $t = \infty$ (steady-state):

$$c = \frac{c_a}{1+k/s}.$$

Note that the influence of initial concentration again disappears which means that regardless of the initial concentration in the instillate, the ultimate concentration depends only on the diffusion : secretion ratio and H^+ concentration in the secreted juice. This is because, as the volume increases infinitely with time, what is present initially becomes insignificant by comparison.

4. If $s = 0$ (no secretion), using L'Hopital's rule, we have

$$c = c_0 e^{-kt/V_0}. \qquad (8.8\text{-}14)$$

This is the same expression obtained by Teorell [11] earlier based on the diffusion theory. Data obtained by Öbrink and Waller in the present study [10] tended to confirm that diffusion is the predominant factor. Because of this, the result is also the same as the exponential decay in the single-compartment model discussed in the previous example.

Example 8.8-3 Gas Cavitation in Circulation [12]. The steady-state macroscopic mass balance and mechanical energy (Bernoulli) equations can be used to explain the phenomenon of gas cavitation in relation to poststenotic dilation. Referring to Figure 8.8-5, since there is no work done on or by the system and since the variation in elevation is usually small, the Bernoulli equation applied between prestenotic section (plane 1) and the stenosis (plane 2) is

$$\tfrac{1}{2}(v_2^2 - v_1^2) + \frac{p_2 - p_1}{\rho} + \hat{E}_{v_{1-2}} = 0. \qquad (8.8\text{-}15)$$

Similarly, that between the stenosis and the poststenotic section (plane 3) is

$$\tfrac{1}{2}(v_3^2 - v_2^2) + \frac{p_3 - p_2}{\rho} + \hat{E}_{v_{2-3}} = 0. \qquad (8.8\text{-}16)$$

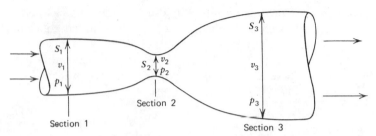

Figure 8.8-5. Reference diagram for areas, velocities, and pressures at a stenosis (from [12]).

Note that we have replaced $\langle v^3 \rangle / \langle v \rangle$ by $\langle v \rangle^2$ and written it as v^2. Actually, this is not exactly correct, for the average of the cubic is *not* the same as the cubic of the average (see Section 8.4). To account for the error, a factor N is used as a correction factor for the $\frac{1}{2}v^2$ term. The value for N is 1 for potential flow, 2 for laminar flow, and 1.06 for turbulent flow.

The friction loss $E_{v_{1-2}}$ is caused by contraction and can be found in standard engineering handbooks [2] to be equal to $0.04v_2^2/2$, while $E_{v_{2-3}}$, the expansion loss, is expressed as $C(v_2 - v_3)^2/2$, with C being related to the angle of conicity in graphical form.

Qualitatively, we can see from the Bernoulli equation that except for the energy loss, the sum of kinetic (velocity) energy and pressure energy should be conserved. This means that at the stenosis, the velocity is highest (because of smallest cross-sectional area) and the pressure must fall to very low. Then, as the poststenotic section expands, some of the pressure energy is recovered because of the decrease in velocity. If the pressure at the stenosis is low enough for vapor or gas bubbles to be formed, they will be carried along downstream from the stenosis until they reach the zone when the recovered pressure is high enough to make them collapse violently. The damage that can be caused by such action cannot be overestimated. In engineering work even metal or concrete components such as propeller blades or pipe wall may be eroded away in merely a few months by this *vapor pressure cavitation*.

Quantitatively, we can determine the criteria for gas cavitation by noting that p_2 must be equal to, or lower than, the atmosphere pressure ($\leqslant 0$ psig*) for it to happen. Then, inserting this and the above-mentioned quantities into Eq. 8.8-15 and noting that

$$v_2 = \frac{S_1}{S_2}v_1,$$

---

*psig = pounds per square inch, gauge; i.e., measured from standard atmospheric pressure—14.7 psia (absolute).

where S denotes cross-sectional area, we have[†]

$$\tfrac{1}{2}v_1^2\left(\frac{S_1^2}{S_2^2}N_2 - N_1\right) - \frac{p_1}{\rho} + 0.04 \cdot \tfrac{1}{2}v_1^2\frac{S_1^2}{S_2^2} = 0. \qquad (8.8\text{-}17)$$

Rearranging terms, we obtain:

$$\frac{S_1}{S_2} = \sqrt{\frac{N_1 + 2p_1/\rho v_1^2}{N_2 + 0.04}}, \qquad (8.8\text{-}18)$$

which means that if

$$\frac{S_1}{S_2} > \sqrt{\frac{N_1 + 2p_1/\rho v_1^2}{N_2 + 0.04}},$$

cavitation will occur at the stenosis, since this is when p_2 will be negative or lower than atmospheric pressure. A more complete representation of this criterion is shown in Figure 8.8-6.

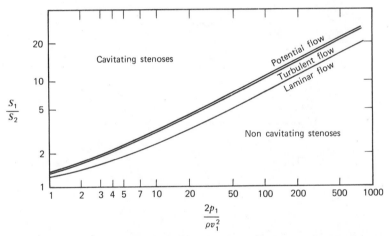

Figure 8.8-6. Values of S_1/S_2 necessary for the onset of vapor cavitation at different upstream conditions; computation is based on blood [12].

Applying the same type of analysis between planes 2 and 3, the reader can develop a set of conditions as to whether the bubble (vapor cavity) formed will collapse for both the cases of stenosis with $(S_1 < S_3)$ and without $(S_1 = S_3)$ dilation [12].

[†]Remember the correction factor N for replacing $\langle v^3\rangle/\langle v\rangle$ by $\langle v\rangle^2$. Note here also that pressure is in psig.

Example 8.8-4   Heat-Conserving Mechanism in Extremities. Although nature may have had other reasons for designing the arterial supply and venous return in the same extremity, this turns out to be an excellent heat-conserving mechanism, which can be demonstrated by using the same principles that engineers use to design countercurrent heat exchangers.

See Figure 8.8-7, where the arm and hand are schematically shown and where T_a is the temperature of the environment. Since both the kinetic and potential energy differences between the inlet and the outlet are negligible and no work is done in this section, a steady-state energy balance (Eq. 8.4-2) on the hand yields

$$\Delta(w\hat{H}) = Q, \tag{8.8-19}$$

where $\hat{H} = \hat{U} + p\hat{V}$, enthalpy per unit mass.

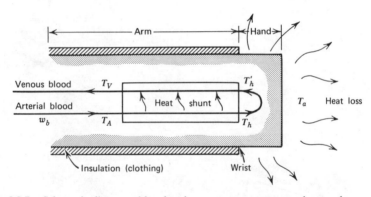

Figure 8.8-7. Schematic diagram of hand and arm as a countercurrent heat exchanger.

The enthalpy difference between the inlet and the outlet can be expressed in terms of the temperature difference between the arterial and venous blood at the wrist as*

$$-\Delta\hat{H} = \hat{C}_p(T_h - T_h'), \tag{8.8-20}$$

while $w_1 = w_2 = w_b$, the mass flow rate of the blood. With long-sleeve clothing and neglecting heat gain or loss by conduction at the wrist, the

*Here the negative sign is caused by the fact that $\Delta = $ "2" – "1" or out minus in.

quantity Q consists mainly of the surface heat loss from the hand, or*

$$-Q = hA(\overline{T}_h - T_a),\qquad(8.8\text{-}21)$$

where h and A are the heat transfer coefficient and area, respectively, of the exposed part of the hand, $\overline{T}_h$ is some appropriately defined average temperature of the hand between T_h and T_h'. We may take the arithmetic average as a simple illustration. From these two equations, we have

$$w_b \hat{C}_p (T_h - T_h') = hA(\overline{T}_h - T_a).\qquad(8.8\text{-}22)$$

It is quite obvious from Figure 8.8-7 that cooled blood in the venous return acts to precool the arterial blood to a lower temperature $T_h(<T_A)$ so as to lower the heat loss, $-Q$. At the same time, the cooled blood itself is rewarmed by the arterial supply in this internal countercurrent heat exchanger to a higher temperature $T_V(>T_h')$. A similar steady-state energy balance on the entire arm and hand system yields

$$w_b \hat{C}_p (T_A - T_V) = hA(\overline{T}_h - T_a).\qquad(8.8\text{-}23)$$

From Eqs. 8.8-22 and 8.8-23 or a steady-state energy balance on the internal countercurrent heat exchanger, we have

$$w_b \hat{C}_p (T_A - T_h) = w_b \hat{C}_p (T_h' - T_V).\qquad(8.8\text{-}24)$$

A numerical example will illustrate how this heat-exchanger principle works. Under normal conditions

$$T_A = 37°C$$

$$\hat{C}_p = 1 \text{ kcal}/l\text{-}°C \text{ (for blood)}^\dagger$$

$$A = 0.05 \text{ m}^2 \text{ (for the adult hand)}.$$

The flow rate of blood, w_b, is automatically regulated such that in a harsh environment (e.g., low T_a or high h), it will decrease so as to lower the surface temperature and prevent the heat loss from increasing excessively. Suppose for

$$T_a = 22°C$$

*Since Q is defined as heat gained, heat loss then carries a negative sign.
†See same footnote on p. 342.

and

$$h = 16 \text{ kcal/m}^2\text{-hr-}°C$$

we have

$$w_b = 0.04 \, l/\text{min}^\dagger$$

and the corresponding temperature attained at the end of the venous return is

$$T_V = 35°C.$$

Substitution of these values into Eqs. 8.8-23 and 8.8-24 yields

$$0.04 \times 60(37 - 35) = 16 \times 0.05 \left(\frac{T_h + T_h'}{2} - 22 \right)$$

$$0.04 \times 60(T_h - T_h') = 16 \times 0.05 \left(\frac{T_h + T_h'}{2} - 22 \right).$$

Solving these two equations simultaneously, we obtain*

$$T_h = 29°C$$

$$T_h' = 27°C.$$

The heat loss can be calculated to be

$$-Q = 16 \times 0.05(28 - 22) = 4.8 \text{ kcal/hr.}$$

On the other hand, if this countercurrent precooling of arterial supply and rewarming of venous return were *not* available and the entire amount of heat were forced to be dissipated via the hand, we should write

$$-Q = 0.04 \times 60(37 - T_h)$$

$$= 16 \times 0.05(T_h - 22),$$

so that

$$T_h = 33.3°C$$

†Strictly speaking $\hat{C}_p$ is on per unit mass basis and w_b is the mass flow rate. But since blood quantity is usually measured in volume, and as long as we change both, the units based on liter are justified.

*Of the three temperatures, T_V, T_h, and T_h', specifying one will fix the other two.

and the heat loss would be

$$-Q = 9\,\text{kcal/hr},$$

or almost double that with the countercurrent heat-conserving mechanism. In other words, with this mechanism 47% ([9.0 − 4.8]/9.0) of the heat that would otherwise be lost is conserved by being shunted to the venous return.

It has been suggested [13] that blood supplied to, and returned from, an organ also goes through a similar, but more complex, thermal regulation process through an internal and an external heat exchanger. Some fishes stay warm-blooded also through the heat-shunt mechanism (see [29] of Chap. 7).

Example 8.8-5   The Placenta-Heat Exchanger Analogy [*14, 15*].   The placenta is the first and most important lifeline of the human being and other mammals because it is the only pathway through which oxygen and carbon dioxide, nutrients and metabolic wastes can be transported between the fetus and the maternal body. Figure 8.8-8 is a schematic diagram

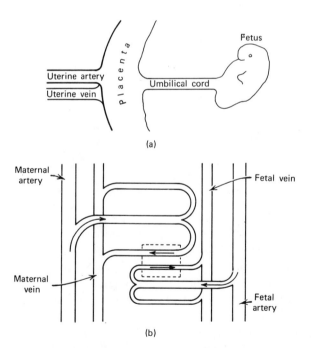

Figure 8.8-8.   The countercurrent model of animal placental circulation and mass exchange: (a) the fetal-uterine interface; (b) model: the countercurrent unit is shown in the dashed rectangle (from [17]).

of the placental circulation of such animals as rabbits and sheep. Here the maternal and fetal capillaries form countercurrent mass transfer units analogous to the countercurrent heat exchanger. Just as there are also co-current, tube-and-shell, and kettle-type heat exchangers, there are other animals with placentas of different anatomical structures as well. For example, it is generally known [16] that in the human placenta there are no capillaries on the maternal side. Instead the fetal capillaries are bathed in a "pool" of the maternal blood, as schematically shown in Figure 8.8-9. This is equivalent in operation to the tube-and-shell heat exchanger.

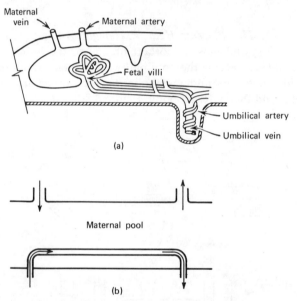

Figure 8.8-9. Schematic diagram of the human placental circulation: (a) capillary arrangement in the human placenta; (b) model of "pool" (from [17]).

In this example we shall show, with the details in Appendix F, how the macroscopic mass balance equation can be used to analyze and compare the transport efficiencies in these various anatomical arrangements, analogous to using the macroscopic energy balance equation for analyzing and designing industrial heat exchangers.

The schematic arrangements, concentration variations, and results in solving for the pertinent quantities for the various types of placental exchanges are summarized in Table 8.8-1 in which the symbolic stirrer (⊛) denotes that the material in the compartment is perfectly mixed, so that there is no spatial nonuniformity in its concentration. For example, in the maternal-pool model, the entering maternal blood has a concentration (say, of oxygen) of c_{MA}. Once it enters, it is instantaneously and thoroughly mixed with the content already in it which has been continuously exchanging material with the fetal stream. This means that the concentration abruptly rises or drops (depending on the material considered, whether it is oxygen, nutrients, or wastes) to its uniform level, which must also be the exit (venous) concentration, as indicated by the horizontal c_M line at a level of c_{MV} in the concentration diagram.

In the case of the double-pool model, both the maternal and fetal compartments are in well-mixed states, as indicated by the two horizontal lines in its concentration diagram.

One should note, however, that perfect mixing is only an idealization so that we can analyze the model mathematically. In many cases such an assumption turns out to be very close to reality. In cases where it does not, modifications can be made [3, p. 709].

Let us use the countercurrent placenta as an example and see how the various quantities are obtained. The other varieties follow a similar or even simpler procedure. First of all, the schematic countercurrent arrangement is enlarged and shown in Figure 8.8-10, where subscripts M and F on

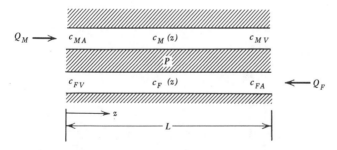

Figure 8.8-10. Detailed scheme of the countercurrent placenta (from [17]).

concentration c denote maternal and fetal sides, while A and V denote arterial and venous ends, respectively. Thus, c_{MA} means the concentration of the transported substance* in the blood stream entering the placenta on the maternal side, and so on. Applying the macroscopic mass balance

---

*In this example, we discuss only inert substances. For others that chemically react or associate with blood (such as O_2 and CO) modifications are needed [17, p. 277].

Table 8.8-1 Schematic Arrangements of Various Placental Circulation Models and Pertinent Mass-Exchange Quantities[a]

	Co-current	Countercurrent	Maternal pool	Double pool
$c_M =$	$\dfrac{R}{1+R}\left[\left(\dfrac{c_{MA}+c_{FA}}{R}\right) + (c_{MA}-c_{FA})e^{-Pz/Q_1L}\right]$	$\dfrac{R}{1-R}\left[\left(\dfrac{c_{MA}-c_{FV}}{R}\right) + (c_{FV}-c_{MA})e^{-Pz/Q_2L}\right]$	c_{MV}	c_{MV}
$c_F =$	$\dfrac{R}{1+R}\left[\left(\dfrac{c_{MA}+c_{FA}}{R}\right) - \dfrac{c_{MA}-c_{FA}}{R}e^{-Pz/Q_1L}\right]$	$\dfrac{R}{1-R}\left[\left(\dfrac{c_{MA}-c_{FV}}{R}\right) + \dfrac{c_{FV}-c_{MA}}{R}e^{-Pz/Q_2L}\right]$	$c_{MV} + (c_{FA}-c_{MV})e^{-Pz/Q_FL}$	c_{FV}

Ratio of end concentration differences	$\dfrac{c_{MV}-c_{FV}}{c_{MA}-c_{FA}} = e^{-P/Q_1}$	$\dfrac{c_{MV}-c_{FA}}{c_{MA}-c_{FV}} = e^{-P/Q_2}$	$\dfrac{c_{MV}-c_{FV}}{c_{MV}-c_{FA}} = e^{-P/Q_F}$	$\dfrac{c_{MV}-c_{FV}}{c_{MV}-c_{FV}} = 1$
$T_F = \dfrac{c_{FV}-c_{FA}}{c_{MA}-c_{FA}}$	$\dfrac{1-e^{-P/Q_1}}{R+1}$	$\dfrac{1-e^{-P/Q_2}}{R-e^{-P/Q_2}}$	$\dfrac{1-e^{-P/Q_F}}{1+R(1-e^{-P/Q_F})}$	$\dfrac{1}{Q_F}\cdot\left(\dfrac{1}{P}+\dfrac{1}{Q_1}\right)^{-1}$
$CI = \dfrac{wA}{c_{MA}-c_{FA}}$	$Q_1(1-e^{-P/Q_1})$	$Q_F\cdot\dfrac{1-e^{-P/Q_2}}{R-e^{-P/Q_2}}$	$Q_F\cdot\dfrac{1-e^{-P/Q_F}}{1+R(1-e^{-P/Q_F})}$	$\left(\dfrac{1}{P}+\dfrac{1}{Q_1}\right)^{-1}$
Special Cases — $Q_M \gg Q_F$ ($R\to 0$)	$Q_F(1-e^{-P/Q_F})$	$Q_F(1-e^{-P/Q_F})$	$Q_F(1-e^{-P/Q_F})$	$\left(\dfrac{1}{P}+\dfrac{1}{Q_F}\right)^{-1}$
$Q_M \ll Q_F$ ($R\to\infty$)	$Q_M(1-e^{-P/Q_M})$	$Q_M(1-e^{-P/Q_M})$	Q_M	$\left(\dfrac{1}{P}+\dfrac{1}{Q_M}\right)^{-1}$
$Q_M=Q_F=Q$ ($R=1$)	$\dfrac{Q}{2}(1-e^{-2P/Q})$	$\left(\dfrac{1}{P}+\dfrac{1}{Q}\right)^{-1}$	$Q\dfrac{1-e^{-P/Q}}{2-e^{-P/Q}}$	$\left(\dfrac{1}{P}+\dfrac{2}{Q}\right)^{-1}$
Membrane limited $Q_M \neq Q_F \gg P$	P	P	P	P
Flow Ltd. $P \gg Q_M, Q_F$ — $Q_M > Q_F$	Q_1	Q_F	Q_1	Q_1
$Q_M < Q_F$	Q_1	Q_M	Q_1	Q_1
$Q_M = Q_F = Q$	$Q/2$	Q	$Q/2$	$Q/2$
L_D — $Q_M > Q_F$	e^{-P/Q_1}	$\dfrac{R-1}{R-e^{-P/Q_2}}$	$\dfrac{1}{(R+1)e^{P/Q_F}-R}$	$\dfrac{Q_1}{P+Q_1}$
$Q_M < Q_F$	e^{-P/Q_1}	$\dfrac{R-1}{Re^{P/Q_2}-1}$	$\dfrac{1}{(R+1)e^{P/Q_F}-R}$	$\dfrac{Q_1}{P+Q_1}$
$Q_M = Q_F = Q$	$e^{-2P/Q}$	$\dfrac{Q}{P+Q}$	$\dfrac{1}{2e^{P/Q}-1}$	$\left(1+\dfrac{2P}{Q}\right)^{-1}$

^{a} Where $\dfrac{1}{Q_1}=\dfrac{1}{Q_M}+\dfrac{1}{Q_F}=\dfrac{1}{Q_M}\left(1+\dfrac{1}{R}\right)$, $\quad\dfrac{1}{Q_2}=\dfrac{1}{Q_M}-\dfrac{1}{Q_F}=\dfrac{1}{Q_M}\left(1-\dfrac{1}{R}\right)$, $\quad R=\dfrac{Q_F}{Q_M}=\dfrac{T_M}{T_F}$.

347

principle to a differential length dz of both capillaries, we have

$$Q_M \frac{dc_M}{dz} = \frac{P}{L}(c_F - c_M)$$

(8.8-25)

for the maternal stream and

$$Q_F \frac{dc_F}{dz} = \frac{P}{L}(c_F - c_M)$$

(8.8-26)

for the fetal stream, where Q is the volumetric flow rate of blood, P is the permeability (cm^3/sec) of the substance in the medium between the maternal and fetal capillaries, and L is the length of this countercurrent pair. Equations 8.8-25 and 8.8-26 are nothing but a statement that the rate of increase or decrease of the substance per unit length in either stream is equal to the rate of transport of this substance through the capillary walls and the medium. This pair of simultaneous differential equations can be most easily solved by using the Laplace transform method (see Appendix F) with the boundary conditions (B.C. 1) that at $z=0, c_M = c_{MA}$, and $c_F = c_{FV}$, to yield

$$c_M(z) = \frac{R}{1-R} \left\{ \left(\frac{c_{MA}}{R} - c_{FV} \right) + (c_{FV} - c_{MA}) \right.$$

$$\left. \times \exp \left[\frac{Pz}{Q_M L} \left(\frac{1}{R} - 1 \right) \right] \right\}$$

(8.8-27)

$$c_F(z) = \frac{R}{1-R} \left\{ \left(\frac{c_{MA}}{R} - c_{FV} \right) + \left(\frac{c_{FV} - c_{MA}}{R} \right) \right.$$

$$\left. \times \exp \left[\frac{Pz}{Q_M L} \left(\frac{1}{R} - 1 \right) \right] \right\}$$

(8.8-28)

where $R = Q_F/Q_M$ (see Eq. 8.8-34 later). From here we can obtain either the transport ratios as defined by the fetal transport fraction,

$$T_F = \frac{c_{FV} - c_{FA}}{c_{MA} - c_{FA}},$$

(8.8-29)

or by the maternal transport fraction,

$$T_M = \frac{c_{MA} - c_{MV}}{c_{MA} - c_{FA}}$$

(8.8-30)

or *diffusion clearance* (or, simply, clearance) as defined by

$$Cl = \frac{w}{c_{MA} - c_{FA}}, \tag{8.8-31}$$

where w is the rate of mass transfer (mass per unit time). The former two are measures of the efficiencies of transfer on the two sides, in terms of fractions of the initial difference between the concentrations in the two entering streams. Of these the fetal transport fraction is the more crucial quantity because it indicates how much the fetal blood has actually gained. Clearance, being the mass transfer rate per unit initial concentration difference between the two entering streams, can be shown to be related to the transport ratios as follows:

$$Cl = Q_M T_M = Q_F T_F, \tag{8.8-32}$$

since

$$w = Q_M(c_{MA} - c_{MV}) = Q_F(c_{FV} - c_{FA}), \tag{8.8-33}$$

which is actually a macroscopic balance on the entire length of the capillary pair. From the last two relationships, we can show that the fetal-to-maternal flow ratio is equal to

$$R = \frac{Q_F}{Q_M} = \frac{T_M}{T_F} = \frac{c_{MA} - c_{MV}}{c_{FV} - c_{FA}}. \tag{8.8-34}$$

Since T_F is an important quantity, let us obtain it by first noting (B.C.2) that at $z = L$, $c_M = c_{MV}$ and $c_F = c_{FA}$. Following that, we can obtain expressions for c_{MV} and c_{FA} from Eqs. 8.8-27 and 8.8-28 as (with $\alpha_1 = P/Q_M$)

$$c_{MV} = \frac{R}{1-R}\left[\left(\frac{c_{MA}}{R} - c_{FV}\right) + (c_{FV} - c_{MA})e^{\alpha_1(1/R-1)}\right] \tag{8.8-35}$$

$$c_{FA} = \frac{R}{1-R}\left[\left(\frac{c_{MA}}{R} - c_{FV}\right) + \left(\frac{c_{FV} - c_{MA}}{R}\right)e^{\alpha_1(1/R-1)}\right], \tag{8.8-36}$$

from which we can further obtain the ratio of the cross-placental fetal-to-maternal concentration differences between the two ends of the counter-current arrangement as

$$\frac{c_{MV} - c_{FA}}{c_{MA} - c_{FV}} = e^{-P/Q_2}, \tag{8.8-37}$$

where

$$\frac{1}{Q_2} = \frac{1}{Q_M} - \frac{1}{Q_F} = \frac{1}{Q_M}\left(1 - \frac{1}{R}\right), \qquad (8.8\text{-}38)$$

which, one should note, could be positive or negative, depending on whether Q_F is larger or smaller than Q_M, respectively.

Further algebraic manipulation yields (see Appendix F)

$$T_F = \frac{c_{FV} - c_{FA}}{c_{MA} - c_{FA}} = \frac{1 - e^{-P/Q_2}}{R - e^{-P/Q_2}} \qquad (8.8\text{-}39)$$

$$Cl = \frac{w}{c_{MA} - c_{FA}} = Q_F \frac{1 - e^{-P/Q_2}}{R - e^{-P/Q_2}} \qquad (8.8\text{-}40)$$

from which the various special limiting cases can be obtained, with occasional help from L'Hopital's rule when the situation of 0/0 occurs.

The derivational procedure for the co-current model is similar to that of the countercurrent one (see Problems 8.H) except that now the "reduced flow rate" of the pair of capillaries defined as

$$\frac{1}{Q_1} = \frac{1}{Q_M} + \frac{1}{Q_F} = \frac{1}{Q_M}\left(1 + \frac{1}{R}\right), \qquad (8.8\text{-}41)$$

appears. One can readily see that if we reversed the direction of flow of one stream, either Q_F or Q_M would take on a negative sign and Q_1 would become Q_2 (or $-Q_2$) and the subscripts A and V on the concentrations of that stream would be interchanged. Thus, with the same derivational procedure, we might say that each quantity in one case is an extension of the corresponding one in the other.

For the maternal-pool model, shown in Figure 8.8-11, the maternal blood is well mixed in the pool, so that its concentration abruptly drops from c_{MA} to c_{MV} as soon as it enters. After it enters, it stays at a constant level c_{MV} and exits at this same level. The differential-macroscopic balance now applies only to the fetal stream because the maternal side is no longer a distributed-parameter system. Mass balance on the fetal stream yields

$$Q_F \frac{dc_F}{dz} = \frac{P}{L}(c_{MV} - c_F), \qquad (8.8\text{-}42)$$

with the boundary condition that at $z = 0, c_F = c_{FA}$. This equation can easily be solved to yield (see Appendix F)

$$c_F = c_{MV} + (c_{FA} - c_{MV})e^{-Pz/Q_F L}, \qquad (8.8\text{-}43)$$

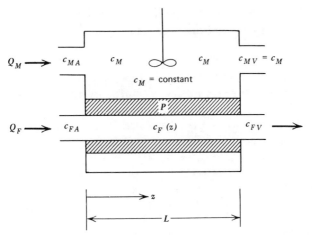

Figure 8.8-11. Detailed scheme of the maternal pool model of the human placenta.

and at $z = L$,

$$c_F = c_{FV} = c_{MV} + (c_{FA} - c_{MV})e^{-P/Q_F}$$

or

$$\frac{c_{MV} - c_{FV}}{c_{MV} - c_{FA}} = e^{-P/Q_F}. \qquad (8.8\text{-}44)$$

Along with the overall balance (Eq. 8.8-33), which still applies, the fetal transport fraction can be obtained as (also see Appendix F)

$$T_F = \frac{1 - e^{-P/Q_F}}{1 + R(1 - e^{-P/Q_F})} \qquad (8.8\text{-}45)$$

or

$$\frac{w}{c_{MA} - c_{FA}} = Q_F \frac{1 - e^{-P/Q_F}}{1 + R(1 - e^{-P/Q_F})} \qquad (8.8\text{-}46)$$

and the special cases follow.

From these special cases, we learn that, for example, when either flow rate is predominant over the other, it is the smaller flow rate that is the important factor in determining the transport efficiency. In fact, when that happens there is no difference between the expressions for clearances in the co-current and countercurrent systems, because the predominant-flow compartment in effect becomes a pool (because of the infinite relative flow rate). Consequently it does *not* have directional effect, which in turn makes the direction of the lesser flow immaterial. This is also why, when Q_M $\gg Q_F$, their expressions are also the same as that for the maternal-pool placenta.

The argument of predominance also holds between the membrane permeability and flow rates. When the flow rates are large compared with the permeability, the transport is diffusion-limited (or diffusion-controlled). When the reverse is true, the transport is flow-limited. This is because the fetal-side convective mass transfer, membrane diffusion and maternal-side convective mass transfer form a process with a series of steps. In a series operation, the slowest step is the bottleneck and the overall rate cannot be larger than that of this slowest step. These points will be verified later when the graphical solutions are plotted.

This concept of "limitation" is a useful one because it allows us to define such quantities as the "degree of diffusion limitation" (L_D). As the term implies, it is the fractional *decrease* of clearance when the diffusional resistance of the placental membrane relative to flow resistance increases from zero (P/Q_M and $P/Q_F = \infty$) to its actual value, or

$$L_D = \frac{Cl_{max} - Cl}{Cl_{max}}, \tag{8.8-47}$$

where, according to Table 8.8-1,

$$Cl_{max} = Q_1 = \frac{1}{1/Q_M + 1/Q_F} \tag{8.8-48}$$

for all cases* except countercurrent flow, in which case

$$Cl_{max} = Q_F \quad \text{if} \quad Q_M > Q_F$$

$$= Q_M \quad \text{if} \quad Q_M < Q_F.$$

It is of interest to note that the range of real values of L_D is between zero and unity. It could be as large as unity *only if* the membrane were infinitely impermeable ($P = 0$) and/or the flow were infinitely large and as small as zero *only if* the membrane were infinitely permeable ($P = \infty$) and/or there were no flow.

---

*This is true not only for the cases of $Q_M > Q_F$ and $Q_M < Q_F$ but also for $Q_M = Q_F = Q$, because in that case

$$Q_1 = \frac{1}{1/Q + 1/Q} = \frac{Q}{2};$$

thus, $Q/2$ is a special case of Q_1.

Another way to utilize these results is through the T_M-versus-T_F diagrams, with R and P/Q_M as parameters, as shown in Figure 8.8-12. It can be seen that for the same values of P/Q_M, and R, the countercurrent exchanger is the most efficient of all. This agrees with an analogous situation in heat-exchanger analysis and design because of the relatively uniform overall and large maternal-to-fetal concentration difference throughout the length of the capillary in the countercurrent situation.

Moreover, we can see that for the maternal-pool model, T_F is relatively insensitive to the P/Q_M value as long as the latter is larger than 10. For

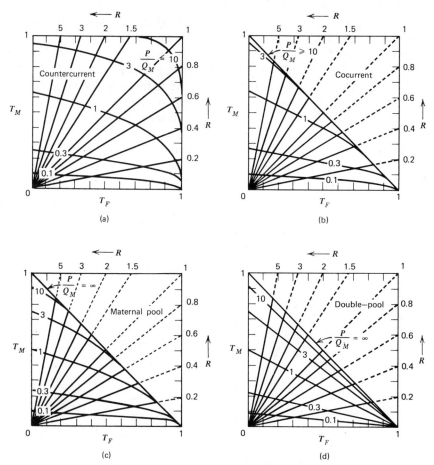

Figure 8.8-12. Efficiency of placental exchange [14]: (a) countercurrent model; (b) cocurrent model; (c) maternal-pool model; (d) double-pool model.

the co-current, countercurrent, and double-pool models, this value is about 3, 20, and 20, respectively. Above these values, T_F depends strongly and only on R. This is the flow-limited regime discussed earlier.

On the other hand, for small values of P/Q_M (where the R lines converge to the origin) the effect of R becomes relatively insignificant. Instead, the value of P assumes importance, for a certain value of Q_M. This is the diffusion-limited regime discussed above. In the diffusion-limited domain, especially when $P/Q_M < 0.1$, all of these models work almost identically. Between these two extremes, these charts are very useful in estimating the comparative effect on the transport efficiency from increasing the flow ratio and the permeability (see Problem 8.J).

The discussions above pertain only to ideal exchangers in that both permeability and flow ratios are uniform. Realistic analysis must include possible nonidealities, such as regional flow distribution and local variations of permeability. These were extensively discussed by Faber [14]. We shall mention only the extreme case of shunting in which parts of the fetal and/or maternal blood are shunted from their respective arterial to venous sides without passing through the placenta to give up or receive transported material. Similar derivation to the above results in diagrams with the curves being "squeezed" toward the origin, as shown in Figures 8.8-13 (a) and (b) for the countercurrent and maternal-pool models, respectively. It is quite obvious that in such cases, regardless of how large P/Q_M is or how we manipulate the flow ratio, neither T_M nor T_F can ever reach unity.

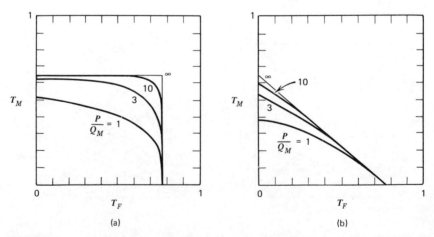

Figure 8.8-13. Efficiency of mass transport across placenta with maternal and fetal shunts of 36% and 23%, respectively [17] (a) countercurrent model; (b) maternal-pool model.

Example 8.8-6 The Log-Mean Difference. In analyzing the efficiency of steady-state macroscopic transport, such as that across the placenta in the preceding example, and in designing transfer apparatus such as the hemodialyzer to be discussed in the following example, the concentration or temperature difference between the two streams is of prime importance. In most cases such as the countercurrent and co-current exchangers, this difference varies from one end of the device to the other. The questions then arise: Which is the correct difference to use, the larger or the smaller one? Or, if it is an average or mean value, should we use arithmetic or geometric mean? The answer can be found in the following analysis.

In Figure 8.8-14 we schematically depict a countercurrent* mass exchanger in which a certain material is being transported from the A stream across a semipermeable membrane or tube wall to the B stream.† Let subscript 1 denote the starting point of the longitudinal coordinate that is also the point where the A stream enters and the B stream exits. The other end is denoted by 2. The rate of loss of this material from stream A is ‡

$$\Delta w^A = Q^A(c_2^A - c_1^A), \tag{8.8-49}$$

while that gained by stream B is§

$$\Delta w^B = Q^B(c_2^B - c_1^B), \tag{8.8-50}$$

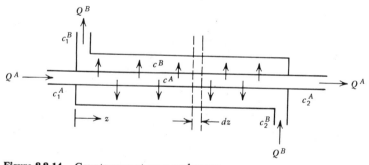

Figure 8.8-14. Countercurrent mass exchanger.

---

*The entire analysis also applies to the co-current and pool model, as will be seen later.
†Note that in this example we use A and B as *superscripts* to avoid the confusion that c_A might be construed as the "concentration of A," etc. Rather, the superscript denotes the stream in which this property or quantity exists.
‡Since in this case $c_2^A < c_1^A$, Δw^A will be negative, representing a loss.
§In this case $c_2^B < c_1^B$, but Q^B is negative, since the B stream is flowing in the negative z-direction. Thus, Δw^B is positive, representing a gain.

where w, c, and Q denote, respectively, the mass flow rate and concentration of the transported material and volumetric flow rate of either stream. Assuming that there is no loss of material from the B stream to the outside (in heat transfer this would mean perfect insulation from the outside), the relationship

$$\Delta w^A = -\Delta w^B \tag{8.8-51}$$

should always hold.

Now if we apply the mass balance to a differential section dz of the exchanger, we can write, for the two streams individually,

$$dw^A = Q^A dc^A \tag{8.8-52}$$

$$dw^B = Q^B dc^B, \tag{8.8-53}$$

while between them*

$$dw^A = U \cdot P \, dz (c^B - c^A) \tag{8.8-54}$$

or

$$dw^B = U \cdot P \, dz (c^A - c^B) \tag{8.8-55}$$

where U is the overall mass-transfer coefficient and P is the circumference of the tube separating the two streams. Substituting Eqs. 8.8-54 and 8.8-55 into Eqs. 8.8-52 and 8.8-53, respectively, we have

$$-\frac{dc^A}{c^A - c^B} = \frac{UP}{Q^A} dz \tag{8.8-56}$$

and

$$\frac{dc^B}{c^A - c^B} = \frac{UP}{Q^B} dz. \tag{8.8-57}$$

Combining the two, the following results:

$$-\frac{d(c^A - c^B)}{c^A - c^B} = UP \left(\frac{1}{Q^A} + \frac{1}{Q^B} \right) dz. \tag{8.8-58}$$

Strictly speaking, Q^A and Q^B may vary with temperature and hence with the longitudinal distance z. Thus, analytic integration appears impossible. However, with reasonable accuracy they may be assumed constant and Eq. 8.8-58 integrated to yield, after applying the boundary conditions

---

*Here we are writing from the viewpoint of material gained by the A stream. Since $c^A > c^B$ at any point, dw^A is negative, representing loss, and dw^B is positive, representing gain.

at $z = 0$ and $z = L$,

$$\ln \frac{c_1^B - c_1^A}{c_2^B - c_2^A} = U(PL)\left(\frac{1}{Q^A} + \frac{1}{Q^B}\right). \qquad (8.8\text{-}59)$$

Substituting Eqs. 8.8-49 and 8.8-50 into Eq. 8.8-59 while remembering Eq. 8.8-51 gives us

$$\ln \frac{c_1^A - c_1^B}{c_2^A - c_2^B} = \frac{UA}{\Delta w^A}\left[(c_2^A - c_1^A) - (c_2^B - c_1^B)\right],$$

where $A = PL$, the interphase transfer area. Rearranging we finally obtain

$$-\Delta w^A = UA \frac{(c_1^A - c_1^B) - (c_2^A - c_2^B)}{\ln\left[(c_1^A - c_1^B)/(c_2^A - c_2^B)\right]}$$

$$= \Delta w^B. \qquad (8.8\text{-}60)$$

Therefore, if we define a mean concentration difference $(c^A - c^B)_m$ according to the general formula, Eq. 7.1-4, so that

$$-\Delta w^A = \Delta w^B = UA\,(c^A - c^B)_m \qquad (8.8\text{-}61)$$

holds, comparing Eqs. 8.8-60 and 8.8-61, it must necessarily be

$$(c^A - c^B)_m = \frac{(c_1^A - c_1^B) - (c_2^A - c_2^B)}{\ln\left[(c_1^A - c_1^B)/(c_2^A - c_2^B)\right]}$$

$$= (c^A - c^B)_{\text{lm}} \qquad (8.8\text{-}62)$$

or

$$(\Delta c)_m = \frac{(\Delta c)_1 - (\Delta c)_2}{\ln\left[(\Delta c)_1/(\Delta c)_2\right]}$$

$$= (\Delta c)_{\text{lm}}, \qquad (8.8\text{-}63)$$

which we call the *log-mean concentration difference.*

The following points should be noted:

1.   The definition and use of the log-mean difference can be extended to heat transfer, in which case we simply replace the concentration by

temperature, or

$$(\Delta T)_{lm} = \frac{(\Delta T)_1 - (\Delta T)_2}{\ln[(\Delta T)_1/(\Delta T)_2]}, \qquad (8.8\text{-}64)$$

assuming constancy on the part of the heat-transfer coefficient U, densities ρ, and heat capacities $\hat{C}_p$ for both streams.

2. In cases where the interphase transfer coefficient varies greatly from one end to the other, so that it may not be assumed constant, Colburn [18] suggested the use of *log-mean flux*

$$(U\Delta c)_{lm} = \frac{(U\Delta c)_1 - (U\Delta c)_2}{\ln[(U\Delta c)_1/(U\Delta c)_2]} \qquad (8.8\text{-}65)$$

for mass transfer or

$$(U\Delta T)_{lm} = \frac{(U\Delta T)_1 - (U\Delta T)_2}{\ln[(U\Delta T)_1/(U\Delta T)_2]} \qquad (8.8\text{-}66)$$

for heat transfer.

3. The entire discussion above and the resulting Eqs. 8.8-63 and 8.8-64 apply to co-current (or parallel-flow) exchangers as well, because in this case, referring to Figure 8.8-15, $c_1^B < c_2^B$ but Q^B is now positive (flow in the

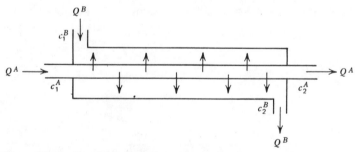

Figure 8.8-15. Parallel-flow (co-current) exchanger.

$+z$ direction). Thus, in Eq. 8.8-50, Δw^B still turns out to be positive. The rest, including the result, are still the same, because at no place does c^B exceed c^A.

4. Since countercurrent and co-current flows are two extreme cases, the ones in between—for example, single-pool (such as maternal-pool in Example 8.8-5) and double pool—can also be described by the above-derived

equations. In particular, if the B stream is a perfectly mixed pool, $c_1^B = c_2^B = c^B$ (constant) the log-mean becomes

$$(c^A - c^B)_{lm} = \frac{c_1^A - c_2^A}{\ln[(c_1^A - c^B)/(c_2^A - c^B)]} . \tag{8.8-67}$$

In double pool, since $c_1^B = c_2^B = c^B$ (constant) and $c_1^A = c_2^A = c^A$ (constant), it follows that $c^A - c^B = $ constant and there is no need to define any average value.

5.   As for which type of flow is more efficient for the heat or mass transfer, we can glean some qualitative arguments by examining the temperature or concentration progression for the two extreme types of flow depicted in Figures 8.8-16(a) and (b). In the parallel-flow (co-current) arrangement (a), the rich and lean (or hot and cold) streams are brought together head-on, so to speak, and the concentration (or temperature) difference is very large initially. But, after they pass through a relatively short distance, the difference quickly levels off to small values and further contact would not bring the two appreciably closer, because with a small potential difference the transport is rather inefficient. However, in the section where the large potential difference is available, the heat-transfer surface area requirement is very small. On the other hand, in the counter-flow arrangement (b), the lean (cold) stream can be brought to a maximum concentration (temperature) that is higher than the minimum of the rich (hot) stream. The potential difference is relatively uniform throughout. Therefore, it appears that if a large initial potential difference is available and the final goal is not too ambitious, using parallel flow saves a great

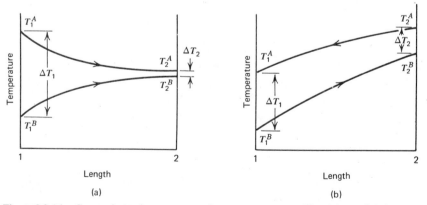

Figure 8.8-16.   Comparison of co-current and countercurrent profiles: (a) parallel (b) countercurrent.

deal of area requirement. On the other hand, if the available initial difference is not great and we wish to accomplish a lot, counterflow should be the choice.

6. If the potential difference at one end is twice that at the other, using the arithmetic mean would contribute an error of only approximately 4% compared with the use of the log-mean.

Example 8.8-7 Hemodialyzer Design [19]. In the last example in this general series before turning to such specialized topics as hemorheology and oxygen transport, a discussion of the design of an artificial kidney seems most appropriate, because in addition to involving all of mass, momentum, and energy transport, it also incorporates the interphase and macroscopic aspects of the transport processes. We shall see how previously discussed principles can be brought to bear on this design problem, which is the essence of engineering work.

Basically, hemodialysis involves the transport of metabolic wastes from the blood stream through a semipermeable membrane to the dialysate, as schematically shown in Figure 8.8-17. The unit or compartment wherein the transport actually occurs takes one of many forms, such as a rectangular passage with sandwiched blood and dialysate layers or a circular passage such as a bundle of hollow fiber tubes immersed in the dialysate bath. Even in one basic geometry, such as the rectangular type, we have variations such as Kolff's twin coils or Kiil's grooved plates. Here we shall

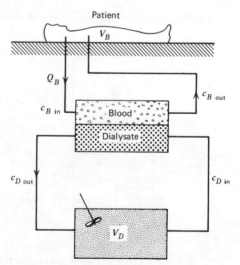

Figure 8.8-17. Hemodialysis (from [20]).

develop the design computation for the basic rectangular-passage type dialyzer and see what parameters to adjust in order to obtain the best performance.

In the rectangular-passage dialyzer, holes and notches are so cut in the frames supporting the membrane that the blood and the dialysate pass each other in alternate layers and countercurrently. Since the properties and flow rate of the dialysate can be varied to a large extent, depending on the particular need, parameters and variables for the blood stream are more critical, because of the various limiting factors. For instance, pressure drop, blood volume, and flow rate through the device are quite restricted by phsyiological capability of the human body. Thus, design calculations are based mainly on consideration of the blood compartment. The chief objective is to maximize the *dialysance*, a quantity to be defined later, subject to these constraints, many of which are closely interrelated. As an illustration, let us consider h, height of the blood flow channel. By making the dialyzer a multilayer one by reducing the height of each channel, as shown in Figure 8.8-18, we can greatly increase the total transfer area without increasing the blood priming (or hold-up) volume. However, narrow channels increase friction and pressure drop. There must be an optimum. We shall delve more deeply into this later, after developing some pertinent equations.

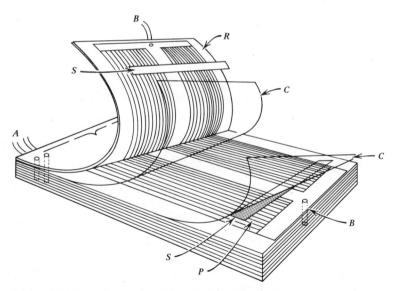

Figure 8.8-18. Multilayer hemodialyzer (from [21]): A, inlets and outlets for blood; B, inlets and outlets for dialysate; C, cellophane; R, frame; S, pad, P, grooves.

BLOOD FLOW RATE.   Figure 8.8-19 is a schematic diagram of the hemodialysis circuit, in which subscript 1 refers to the dialyzer unit in which the actual mass transport takes place and subscript 2 refers to the rest of the system, including the interconnecting tubing, cannulae, arteries, and veins, all of which combined, we assume, have a uniform equivalent diameter D and total equivalent length L_2. For a dialyzer unit with n layers that are each h in height, W_i in width, and L_1 in length, we can write, for the volumetric flow rate of blood in the dialyzer unit and in the rest of the system (see Table 6.1-1),

$$Q_1^B = n\left(\frac{\Delta p_1 h^3 W_i}{12\mu L_1}\right)$$ (8.8-68)

$$Q_2^B = \frac{\Delta p_2 \pi D^4}{128\mu L_2},$$ (8.8-69)

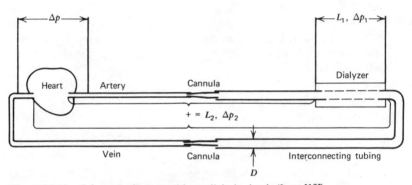

Figure 8.8-19.   Schematic diagram of hemodialysis circuit (from [19]).

where Δp is the pressure drop, μ is the viscosity of blood, and superscript B denotes the blood stream. If we consider

$$nW_i = W$$ (8.8-70)

to be the total width of all blood layers (or

$$h \cdot nW_i = hW = S$$ (8.8-71)

to be the total cross-sectional area of blood channels),* we can rewrite Eq.

*So that the result applies both to a dialyzer with n parallel layers each with a membrane area of hW_i and to one with a single layer but a large membrane area of hW.

8.8-68 as

$$Q_1^B = \frac{\Delta p_1 h^3 W}{12 \mu L_1}.$$ (8.8-72)

Since the same rate of blood has to flow through the entire system, $Q_1^B = Q_2^B = Q^B$ and we have, from Eqs. 8.8-69 and 8.8-72,

$$\Delta p = \Delta p_1 + \Delta p_2$$

$$= Q^B \left(\frac{12 \mu L_1}{h^3 W} + \frac{128 \mu L_2}{\pi D^4} \right)$$ (8.8-73)

or

$$Q^B = \Delta p \left(\frac{12 \mu L_1}{h^3 W} + \frac{128 \mu L_2}{\pi D^4} \right)^{-1}.$$ (8.8-74)

DIALYSANCE.   The rate of loss of a certain component (such as urea) of the metabolic wastes from the blood stream can be written, according to Eq. 8.8-49,*

$$-\Delta w^B = Q^B (c_i^B - c_0^B),$$ (8.8-75)

while that across the membrane is, according to Eq. 8.8-60,

$$-\Delta w^B = U_{\text{lm}} A (c^B - c^D)_{\text{lm}},$$ (8.8-76)

where the superscript D denotes the dialysate stream and U_{lm} the overall mass-transfer coefficient defined on the basis of the log-mean concentration difference. The mass-transfer area can be expressed as

$$A = 2nW_i L_1$$

$$= 2WL_1.$$ (8.8-77)

Following the definition for the log-mean concentration difference (Eq. 8.8-62), we can write

$$(c^B - c^D)_{\text{lm}} = \frac{(c_i^B - c_0^D) - (c_0^B - c_i^D)}{\ln[(c_i^B - c_0^D)/(c_0^B - c_i^D)]}.$$ (8.8-78)

*In the present case the superscript B takes the place of the A in Example 8.8-6 because it is the stream that loses the substance.

Normally, the dialysate can be circulated at a large flow rate, so that it is not the limiting factor. Under this circumstance, the change in the concentration of any metabolic waste component is very small, or*

$$c_i^D \doteqdot c_0^D = c^D, \tag{8.8-79}$$

so that Eq. 8.8-78 becomes

$$(c^B - c^D)_{\text{lm}} = \frac{c_i^B - c_0^B}{\ln\left[(c_i^B - c^D)/(c_0^B - c^D)\right]}. \tag{8.8-80}$$

combining Eqs. 8.8-75, 8.8-76, and 8.8-80 and canceling the common factor $c_i^B - c_0^B$, we have

$$\frac{c_0^B - c^D}{c_i^B - c^D} = e^{-U_{\text{lm}}A/Q^B} \tag{8.8-81}$$

or

$$c_i^B - c_0^B = (c_i^B - c^D)\left(1 - \frac{c_0^B - c^D}{c_i^B - c^D}\right)$$

$$= (c_i^B - c^D)(1 - e^{-U_{\text{lm}}A/Q^B}). \tag{8.8-82}$$

Dialysance is defined as the rate of transport of the metabolic waste per unit concentration difference at the blood stream entrance, or[†]

$$Di = \frac{(-\Delta w^B)}{c_i^B - c^D}. \tag{8.8-83}$$

Using Eqs. 8.8-75 and 8.8-82, Eq. 8.8-83 becomes

$$Di = Q^B(1 - e^{-U_{\text{lm}}A/Q^B}). \tag{8.8-84}$$

A dimensionless efficiency factor, the dialysance per unit rate of blood flow, can also be defined

$$E = \frac{Di}{Q^B} = 1 - e^{-U_{\text{lm}}A/Q^B}. \tag{8.8-85}$$

We can see that although Q^B is eliminated as a linear factor, it still affects

---

*This condition would be exactly met in a bath or pool type of device in which the dialysate is perfectly mixed.

[†]Sometimes the term "clearance" appears. It is the dialysance when the waste concentration in the dialysate is so low that $C^D = 0$.

the efficiency through the exponential factor, $U_{lm}A/Q^B$, which is a measure of the membrane permeability relative to the convective capacity of the blood. If the permeability is large and/or the convection small, then this factor is large, E approaches unity, and vice versa. Thus, the efficiency factor gives a combined measure of the membrane and flow efficiencies.

THE MASS-TRANSFER COEFFICIENT.   The overall mass-transfer coefficient consists of contributions from membrane, dialysate, and blood stream:

$$U_{lm} = \frac{1}{\dfrac{1}{k_{lm}^B} + R^M + R^D}, \qquad (8.8\text{-}86)$$

where k_{lm}^B is the mass-transfer coefficient, again defined according to a log-mean concentration difference, between the blood stream and the membrane; R^M and R^D are the mass-transfer resistances in the membrane and between the membrane and the dialysate, respectively. Of the three, R^D can usually be made negligible and k_{lm}^B obtained from the Graetz correlation (Eq. 7.7-4).

DESIGN OPTIMIZATION.   For the present example, we shall simply examine how dialysance can be optimized by the channel height, h. If we substitute Eq. 8.8-74 into Eq. 8.8-84, we have

$$Di = \frac{\Delta p}{(12\mu L_1/h^3 W) + (128\mu L_2/\pi D^4)}$$

$$\times \left\{ 1 - \exp\left[-\frac{U_{lm}A}{\Delta p} \left(\frac{12\mu L_1}{h^3 W} + \frac{128\mu L_2}{\pi D^4} \right) \right] \right\} \qquad (8.8\text{-}87)$$

At one extreme, if $h = 0$,

$$Di = \frac{\Delta p}{\infty} \{ 1 - e^{-\infty} \} = 0, \qquad (8.8\text{-}88)$$

while at the other, if $h = \infty$,

$$Di = \frac{\Delta p \pi D^4}{128\mu L_2} \left\{ 1 - \exp\left[-\frac{U_{lm}A}{\Delta p} \cdot \frac{128\mu L_2}{\pi D^4} \right] \right\}. \qquad (8.8\text{-}89)$$

Since the volume of the mass-transfer unit itself is limited by

$$V = A \cdot h,$$

which must remain below a certain value, then as $h \to \infty$, $A \to 0$ (i.e., we will have fewer and fewer layers) so that Eq. 8.8-89 becomes

$$Di = \frac{\Delta p \pi D^4}{128 \mu L_2} \{1 - e^0\} = 0. \qquad (8.8\text{-}90)$$

Therefore, at both extremes, dialysance drops to zero and there must be a maximum in between that could be obtained by setting $dD_i/dh = 0$ in Eq. 8.8-87. However, it is more instructive simply to present the numerical and graphical solution of Grimsrud and Babb [19], shown in Figure 8.8-20.

We can see that extreme values agree with the discussions above and that the maximum shifts with the dialyzer compartment length. Following hints provided by the earlier discussions, we can conclude that to the right side of the maximum, dialysance drops because of the decrease in area caused by the increase in channel height. This we call the *surface-area-*

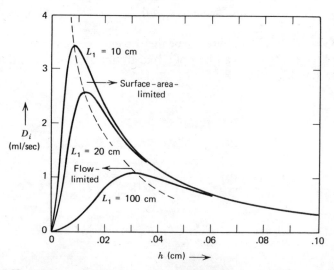

Figure 8.8-20. Dialysance as a function of channel height at varying dialyzer lengths [19]. Calculation is based on the following parameter values: $L_2 = 800$ cm, $D = 0.5$cm, $R^M = 2700$ sec/cm, $V = Ah = WL_1 h = 100$ ml (maximum allowable priming, or hold-up, volume), $\mu = 2$ cp, $\Delta p = 40$ mmHg, $k_c^B = 1/\gamma h = (1/30,000 \ h)$ cm/sec. (From Section 7.7 we have $k_c^B \propto v^{0.8}$ But to simplify the computation we can write $k_c^B \propto v$ with little loss of accuracy. Referring to a reference value k_0^B at v_0, we may write

$$k_c^B = k_0^B \frac{v}{v_0} = k_0^B \frac{Q^B}{Q_0^B} \cdot \frac{h_0}{h},$$

since $Q^B = vhW$ and $Q_0^B = v_0 h_0 W$. Thus, $\gamma = Q_0^B/Q^B k_0^B h_0$ and was found to be 30,000 sec./cm².)

limited regime. On the other hand, as the channel is narrowed below the maximum, the dialysance decreases because of the resultant decrease in blood flow. Hence, it is called the *flow-limited regime.*

Drawing on this work, Wolf and Zaltzman [22] generated sets of design charts for both the rectangular-passage and circular-passage types of dialyzer. It should also be emphasized that except for the fact that the mass-transfer coefficient and some related quantities are affected by the countercurrent or co-current arrangement, the concentrations themselves do not enter the expressions normally used (such as Eqs. 8.8-84 and 8.8-87). Therefore, these formulas should be applicable to any flow arrangement.* However, the most important lesson to be learned from this example is that dialysance is affected not by membrane permeability only but by flow characteristics as well.

PROBLEMS

8.A Renal Clearance and Blood Flow [23]

In the model of glomerulus and nephron depicted in Figure 8.A, let Q and c represent, respectively, volumetric flow rate and concentration of a certain substance (solute) in the flow, while the subscripts A, V, G, and U

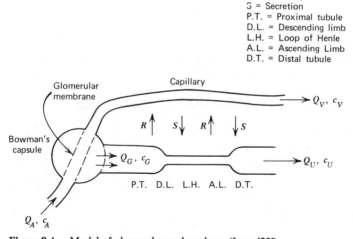

```
R   = Reabsorption
S   = Secretion
P.T. = Proximal tubule
D.L. = Descending limb
L.H. = Loop of Henle
A.L. = Ascending Limb
D.T. = Distal tubule
```

Figure 8.A. Model of glomerulus and nephron (from [23]).

*This is because the dialysate side is equivalent to a "pool"; see pp. 351–352.

denote the streams (renal artery, vein, glomerular filtrate, and urine, respectively) in which this substance appears. *Renal clearance* is defined as
the volume (per unit time) of plasma that would have to be *completely
cleared* of the substance to yield the amount excreted in the urine per unit
time.

(a)   From a material balance on the total fluid flow and one on the solute,
show that

$$Q_A = Q_U \frac{c_U - c_V}{c_A - c_V},$$ (8.A-1)

which is the renal plasma flow rate, since the solute is assumed to dissolve
in plasma only and the erythrocytes are not involved.

(b)   Show that from the definition above, renal clearance is

$$Q_{A_0} = \frac{Q_U c_U}{c_A}.$$ (8.A-2)

(c)   For a substance such as inulin, which can filter freely through the
glomerular membrane but is neither secreted from or reabsorbed by the
renal capillary, show that the glomerular filtration rate is

$$Q_G = \frac{Q_U c_U}{c_A},$$ (8.A-3)

which is the same as the renal clearance. This forms the basis of the inulin
technique for clearance measurement.

8.B Measurement of Overall Dialysis Coefficients [24]

To measure the overall dialysis coefficient, k_0, of a membrane, it is
stretched between two compartments, the fluid in each of which is well
mixed so that the use of surface concentrations can be avoided.

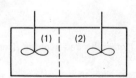

(a)   Show that a mass balance on the solute for each chamber yields

$$V_1 \frac{dc_1}{dt} = k_0 A (c_2 - c_1), \qquad (8.B-1)$$

$$V_2 \frac{dc_2}{dt} = k_0 A (c_1 - c_2), \qquad (8.B-2)$$

where A, c, and V denote, respectively, membrane surface area, solute concentration and compartmental volume. What assumptions are implicit? (b)   Solve the above two equations and show what experimental data must be taken in order to evaluate k_0, assuming A, V_1, V_2 and the initial concentrations are known? Must we measure the concentrations c_1 and c_2 as well as know the initial concentrations c_{1_0} and c_{2_0} individually?

8.C   Differential-Macroscopic Mass Balance in the Sequential Diaphron Hemodiafilter [25]

The diaphron hemodiafilter is an improved artificial kidney based on the principle of ultrafiltration. There are three types of operation: single-stage and multistage sequential, and simultaneous. Figure 8.C is a schematic diagram of the single-stage sequential diafiltration, in which the reconstituting fluid is premixed with the blood before entering the diafiltration unit. Inside the unit, it is assumed, the fluid is perfectly mixed laterally but is completely unmixed longitudinally. Q and c are, respectively, the flow rate and the solute concentration of the various streams as indicated by the subscripts. Also let J denote the ultrafiltration flux, A the membrane area, and L the length of the diafiltration unit.

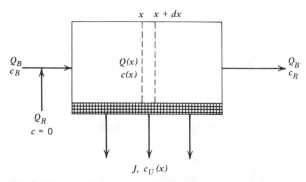

Figure 8.C.   One-stage sequential diafiltration (from [25]).

(a)   Show that for a section of the unit at a distance x from the entrance, the steady-state macroscopic mass balance on the blood is

$$Q = Q_B + Q_R - JA\frac{x}{L}. \tag{8.C-1}$$

(b)   Show that a differential-macroscopic solute balance at x yields

$$\frac{d}{dx}(Qc) + \frac{JA}{L}c_U = 0, \tag{8.C-2}$$

where c_U can be related to $c(x)$ through the partitioning constant of the solute across the membrane:

$$c_U = Kc. \tag{8.C-3}$$

(c)   Show that

$$\left(Q_B + Q_R - JA\frac{x}{L}\right)\frac{dc}{dx} + (K-1)\frac{JA}{L}c = 0 \tag{8.C-4}$$

(d)   If we define fractional solute rejection by the membrane as

$$R = \frac{c - c_U}{c} = 1 - K, \tag{8.C-5}$$

and if there is no rejection of solute show that

$$\frac{dc}{dx} = 0.$$

What does this mean? From a solute balance over the entire unit, show that

$$c_R = \frac{Q_B c_B}{Q_B + Q_R} \tag{8.C-6}$$

and

$$\epsilon = \frac{\delta}{1 - \delta}, \tag{8.C-7}$$

where $\epsilon = JA/Q_B$ and is the blood dilution ratio, while $\delta = 1 - c_R/c_B$.
(e)   Now if the solute is partially rejected by the membrane so that $c_U < c$, show that

$$\epsilon = \left(\frac{1}{1-\delta}\right)^{\frac{1}{K}} - 1. \tag{8.C-8}$$

(f)   Plot ϵ versus δ for a certain value of K (between 0 and 1). Explain the meaning of the graph.

8.D Multistage Sequential Diafiltration [25]

Following a procedure similar to that in the previous problem, show that for the stage-wise arrangement of diafiltration system depicted in Figure 8.D the overall blood dilution ratio after n stages is

$$\epsilon = n\left[\left(\frac{1}{1-\delta}\right)^{1/nK} - 1\right]. \qquad (8.D\text{-}1)$$

Plot this on the same graph with the result from Problem 8.C (with the same K value) for various numbers of stages (such as $2, 4, 6\ldots$). What is the meaning of the results? Would the result change (in comparison with single-stage operation) if we specified that the total membrane area or total volume of diafiltration units must be the same as that in the single-stage operation?

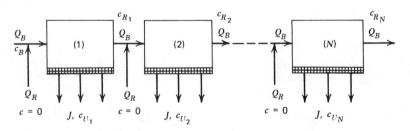

Figure 8.D. Multistage sequential diafiltration system (from [25]).

8.E Simultaneous Diafiltration [25]

(a)   For the diafiltration arrangement shown in Figure 8.E, where there is no longitudinal mixing, show that if there is complete lateral mixing, a differential macroscopic solute balance yields

$$Q_B\frac{dc}{dx} + \frac{JA}{L}c_U = 0, \qquad (8.E\text{-}1)$$

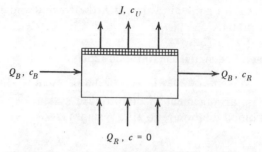

Figure 8.E. Simultaneous diafiltration unit (from [25]).

which can be solved to obtain

$$\epsilon = \frac{1}{K}\ln\left(\frac{1}{1-\delta}\right). \tag{8.E-2}$$

(b) Show that if there is no lateral mixing, the solute balance yields

$$Q_B\frac{d\langle c\rangle}{dx} + \frac{JA}{L}c_B = 0 \tag{8.E-3}$$

and subsequently

$$\epsilon = \frac{\delta}{K}, \tag{8.E-4}$$

where $\langle c\rangle$ is the average concentration that would exist across any cross section if the blood were well mixed at that point.

(c) Compare the effectiveness of the simultaneous diafiltration unit with the single-stage and multistage operations through graphing as suggested in the preceding problem.

8.F Cellular Transport of Calcium in Rat Liver [26]

The efflux and influx transfer coefficients between rat liver slices and a suitable buffered medium were studied by incubating the ^{47}Ca-labeled slice in the medium, initially free of the radiocalcium.

(a) Using the notation

$c_S, c_B = {}^{47}$Ca concentrations in liver slice and medium, respectively,

$V_S, V_B =$ volumes of liver slice and medium, respectively,

$k, k' =$ efflux and influx transfer coefficients respectively,

Show that the following differential equation governs the ^{47}Ca-concentration variation in the medium

$$\frac{dc_B}{dt} = \frac{V_S}{V_B}(kc_S - k'c_B),$$ (8.F-1)

in which both c_S and c_B are functions of time.
(b)   Explain why we can write

$$V_S c_{S_0} = V_S c_S + V_B c_B,$$ (8.F-2)

where subscript 0 denotes initial condition (i.e., at $t=0$).
(c)   Combining Eqs. 8.F-1 and 8.F-2, show that we have

$$\frac{dc_B}{dt} + K \cdot c_B = \frac{V_S}{V_B} kc_{S_0},$$ (8.F-3)

in which c_B is the only variable, since $K = k + k'(V_S/V_B)$.
(d)   Solve Eq. 8.F-3 to yield

$$c_B(t) = \frac{V_S}{V_B} \cdot \frac{kc_{S_0}}{K}(1 - e^{-Kt}).$$ (8.F-4)

What initial condition have we used?
(e)   Show why Eq. 8.F-4 can be rewritten as

$$\frac{c_B(t)}{c_{B_\infty}} = 1 - e^{Kt}.$$ (8.F-5)

What is the expression for and the meaning of c_{B_∞}?
(f)   Show why we can plot $\ln(1 - c_B/c_{B_\infty})$ versus t and obtain $[k + k'(V_S/V_B)]$ as the slope of the straight line if one results?
(g)   Show how one can calculate the efflux, and then influx, transfer coefficients from the value of this slope and the known and measureable quantities.

8.G Accumulation of Radioiodine in Thyroid Gland [27]

To determine the kinetics of the accumulation of protein-bound iodine (^{131}PBI) in the thyroid gland, the radioiodide substrate concentration in the thyroid and the rate of formation of ^{131}PBI are measured. In such experiments a three-compartment model is used in which the serum radioiodide, at concentration c_B, in the blood compartment exchanges with that in the thyroid compartment via a one-way "clearance constant" Q (milliliters of

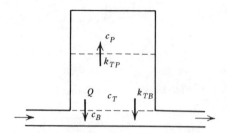

Figure 8.G. Compartemental model of thyroid gland (from [27]).

blood serum per minute). Thyroid radioiodide, at concentrations c_T, goes (1) to the PBI via a binding rate constant k_{TP} and (2) back to the blood stream via an exit rate constant k_{TB}. It is assumed that the time period of study is so brief that no ^{131}PBI is metabolized to release the labeled material. All compartments are assumed to have uniform concentrations.

(a)   Show that the following simultaneous ordinary differential equations govern:

$$\frac{dc_P}{dt} = k_{TP}c_T, \tag{8.G-1}$$

$$\frac{dc_T}{dt} = \frac{Q}{V}c_B - (k_{TP} + k_{TB})c_T, \tag{8.G-2}$$

so that

$$\frac{d}{dt}(c_P + c_T) = \frac{Q}{V}c_B - k_{TB}c_T = \frac{Q_{\text{eff}}}{V}c_B, \tag{8.G-3}$$

where V is the volume of the thyroid gland and Q_{eff} is an *effective clearance* that is, milliliters of serum per minute that *appears* to be rid of radioiodide to account for the observed increasing ^{131}I in the thyroid.

(b)   Show that if Q_{eff} is constant and if $c_T = c_P = 0$ initially, we have

$$c_T + c_P = \frac{Q_{\text{eff}}}{V} \int_0^t c_B \, dt, \tag{8.G-4}$$

$$c_T = \frac{Q}{V} e^{-k_{\text{eff}} t} \int_0^t c_B e^{k_{\text{eff}} t} \, dt, \tag{8.G-5}$$

where $k_{\text{eff}} = k_{TP} + k_{TB}$, the fraction of the radioiodide removed from the thyroid gland per minute *by all routes*.

(c)   At large t, the radioiodide content in the blood approaches a constant level. Show that in this case,

$$c_T = \frac{c_B Q}{k_{eff} V}(1 - e^{-k_{eff}t}).\qquad(8.G\text{-}6)$$

From this obtain the steady-state expression for c_T/c_B. What does this mean?
(d) At what times would the c_T/c_B values be 50% and 95% of the steady-state value?

8.H The Co-current Placental Model

Following a procedure similar to that of Example 8.8-5, develop the various expressions pertinent to the mass exchange between the maternal and fetal streams in a co-current placental capillary model as shown in Table 8.8-1. Compare the results, especially the transport efficiencies, both in general and for the special limiting cases, with those of the countercurrent and maternal-pool models.

8.I The Double-Pool Placental Model

Develop expressions pertinent to the mass exchange between the maternal and fetal pools in a double-pool placental model as shown in Table 8.8-1. Compare the results with those of the other types.

8.J The Use of Placental-Exchange Charts

Consider the human placenta as a maternal-pool exchanger. On Figure 8.8-12(c) locate point A at $P/Q_M = 1, R = 3$. Remembering that in Eqs. 8.8-30 and 8.8-33

$$w = Q_M(c_{MA} - c_{MV})$$

$$= Q_M \cdot T_M(c_{MA} - c_{FA})\qquad(8.J\text{-}1)$$

and assigning a value of k (in moles per minute for example) to $Q_M(c_{MA} - c_{FA})$, we have

$$w = T_M \cdot k.\qquad(8.J\text{-}2)$$

(a) Read from Figure 8.8-12(c) the values of T_M and T_F. Calculate the mass transfer rate w.
(b) On the same $R = 3$ line, locate point B at its intersection with the

$P/Q_M = \infty$ line. Read off T_F and T_M values. Calculate w again, as well as the percent increase of w as we go from A to B. What is the significance of this case?.

(c)   On the $P/Q_M = 1$ line, locate point C at its intersection with $R = \infty$ line (the vertical axis). Read off the T_M value and calculate w as well as the increase of w as we go from A to C. What is the significance of this case?

(d)   Now let us increase the maternal flow tenfold. This would mean an R value of 0.3 and P/Q_M of 0.1 as well as a change of Eq. 8.J-2 to $w = 10 T_M k$. Locate this point as D and calculate w as well as the percent increase of w between A and D.

(e)   Discuss the result by further increasing Q_M to very large values. What is the conclusion one can draw from this exercise? The reader may attempt his own exercise by using other the placental models and charts.

REFERENCES

1.   Bird, R. B. *Chem. Eng. Sci.*, **6**, 123 (1957).

2.   Perry, J. H.; Perry, R. H.; Chilton, C. H.; and Kirkpatrick, S. D., Eds. *Chemical Engineers' Handbook*, 5th ed., McGraw-Hill: New York, 1973.

3.   Bird, R. B.; Stewart, W. E.; and Lightfoot, E. N. *Transport Phenomena*. Wiley: New York, 1960.

4.   Kramers, H. *Physische Transportverschijnselen*. Technische Hogeschool: Delft, Holland, 1958.

5.   Pedley, T. J.; Schroter, R. C.; and Sudlow, M. F. *Resp. Physiol.*, **9**, 371, 387 (1970).

6.   Grim, E. *Am. J. Physiol.*, **205**, 247 (1963).

7.   Humphrey, E. W., and Johnson, J. A. *Am. J. Physiol.*, **198**, 1217 (1960).

8.   Kety, S. S., and Schmidt, C. F. *Am. J. Physiol.*, **143**, 53 (1945); *J. Clin. Invest.*, **27**, 476 (1948).

9.   Aukland, K.; Bower, B. F.; and Berliner, R. W. *Circ. Res.*, **14**, 164 (1964).

10.   Öbrink, K. J., and Waller, M. *Acta Physiol. Scand.*, **63**, 175–185 (1965).

11.   Teorell, T. *Skand. Arch. Physiol.*, **66**, 225 (1933).

12.   Fox, J. A., and Hugh, A. E. *Phys. Med. Biol.*, **9**, 359 (1964).

13.   Magilton, J. and Swift, C. Unpublished MS. Iowa State University; see Ref. 20, Example 6.9.

14.   Faber, J. J. *Circ. Res.*, **24**, 221 (1969).

15.   Meschia, G.; Battaglia, F. C.; and Bruns, P. D., *J. Appl. Physiol.*, **22** (6), 1171–1178 (1967).

16.   Guyton, A. C. *Textbook of Medical Physiology*, 3rd ed. Saunders: Philadelphia, 1966, p. 1154.

17. Middleman, S. *Transport Phenomena in the Cardiovascular System*. Wiley-Interscience: New York, 1972.

18. Colburn, A. P. *Ind. Eng. Chem.*, **25**, 873 (1933).

19. Grimsrud, L., and Babb, A. L. *Trans. Am. Soc. Artif. Int. Organs*, **10**, 101 (1964).

20. Seagrave, R.C. *Biomedical Applications of Heat and Mass Transfer*. Iowa State Univ. Press: Ames, Iowa, 1971.

21. Skeggs, L. T., Jr., and Leonards, J. R. *Science*, **108**, 212 (1948).

22. Wolf, L., Jr., and Zaltzman, S. In Ref. 28, p. 104.

23. Abbrecht, P. H., In Ref. 28, p. 1.

24. Smith, K. A.; Colton, C. K.; Merrill, E. W.; and Evans, L. B. In Ref. 28, p. 45.

25. Bixler, H. J.; Nelson, L. M.; and Besarab, A. In Ref. 28, p. 90.

26. Wallach, S.; Reizenstein, D. L.; and Bellavia, J. V. *J. Gen. Physiol.*, **49**, 743 (1966).

27. Wollman, S. H., and Reed, F. E. *Am. J. Physiol.*, **202** (1), 182 (1962).

28. Dedrick, R. L.; Bischoff, K. B.; Leonard, E. F., Eds., *The Artificial Kidney*, Chem. Eng. Prog. Symp. Series, No. 84, Vol. LXIV. American Institute of Chemical Engineers: New York, 1968.

CHAPTER NINE
Phenomenological Hemorheology

In the fluid-flow examples in Section 6.1 we have deliberately used fairly general problems, because we were planning to discuss all specific hemodynamic problems in the present chapter. Since blood possesses such a unique viscosity behavior, it deserves special treatment. In Section 3.5 we have already explained when and why blood can be considered a Newtonian or non-Newtonian fluid. Here we shall first show an example using a particular (and well-accepted) non-Newtonian model.

Then, we shall turn to a different aspect of hemodynamics—blood as a suspension of cells in plasma. In particular, we will discuss the apparent reduction of viscosity and the migratory tendency of blood cells in narrow flow channels. We will also see how this well-known Fåhraeus–Lindqvist effect is tied in with the hematocrit reduction phenomenon through the concept of capillary and organ hematocrits, respectively analogous to the concept of volume and flow-averaged temperatures discussed in Example 6.2-6.

It must be emphasized that in order to keep this book and chapter within reasonable length, the author has had to make some choices and sacrifices in the material covered, sometimes by its relative importance and at other times according to the author's own interest and knowledge. However, it is hoped that enough references will be cited to allow those interested in any area to explore further. Also, as mentioned in the Preface and practiced throughout the previous chapters, our main objective is to describe phenomena formally and interpret the result of the mathematical analysis with the application aspect in mind. Thus, instead of analyzing the precise mechanism, which in many cases has not been fully understood anyway, we shall simply present the experimental facts and accept the investigator's interpretation, except in cases where it is obviously erroneous. More fundamental analyses and theoretical discussions can be found in several recently published papers and texts [1–4].

9.1 THE CASSON MODEL

As explained in Section 3.5, the fibrinogen molecules on the surface of the red corpuscles cause them to stack together in the *rouleaux formation*. This phenomenon is most pronounced in pathological conditions as well as

378

during pregnancy, when a large number of fibrinogen molecules are present at the erythrocyte surfaces. Such aggregation of cells in certain blood vessels, such as arterioles and small arteries, are responsible for the blood's deviation from Newtonian behavior.

Although the precise mechanism for such particular behavior has not been fully established, the non-Newtonian behavior exhibited by whole blood is a combination of the Bingham and pseudoplastic characteristics, as shown in Figure 3.5-1. That is, it has a yield stress, but even above this yield stress, the stress (τ)–shear-rate ($\dot{\gamma}$) relationship* is nonlinear. Using Casson's general model† [5], which linearizes the rheogram by making square-root plots, Merrill and Pelletier [6] found experimentally that for $\dot{\gamma} < 20$ sec^{-1},

$$\sqrt{\tau} = \eta^{\frac{1}{2}}\sqrt{\dot{\gamma}} + \sqrt{\tau_y} \qquad (9.1\text{-}1)$$

and for $\dot{\gamma} > 100$ sec^{-1},

$$\tau = \mu\dot{\gamma}, \qquad \text{(Newtonian)} \qquad (9.1\text{-}2)$$

with the region 20 sec$^{-1} < \dot{\gamma} < 100$ sec^{-1} being the transition region. In their particular experiments with human blood at hematocrit of 40% and temperature of 37°C, they obtained excellent correlation at each of the fibrinogen concentration levels tested between 0.14 and 0.27 g/100 ml, regardless of the type of viscometer used. A typical plot is given in Figure 9.1-1. Although the variation of the non-Newtonian viscosity (η in Eq. 9.1-1 for low-shear-rate region) and the Newtonian viscosity (μ in Eq. 9.1-2 for high-shear-rate region) with fibrinogen concentration does not follow a definite trend, the yield stress τ_y does seem to diminish to zero at zero fibrinogen content, as shown in Figure 9.1-2. This gives credence to the belief that at least the yield-stress phenomenon is caused by the presence of fibrinogen molecules.

Example 9.1-1   Steady-State Velocity Profiles for Blood as a Casson Fluid.   Using Eq. 9.1-1, the velocity profiles for steady-state laminar flow of blood in small channels or blood vessels of such a size that the Casson model prevails can be derived. Since a yield stress, τ_y, is involved, the formal description of this model is

$$\dot{\gamma} = 0 \qquad (9.1\text{-}3)$$

*Where $\tau = \tau_{rz}, \dot{\gamma} = -dv_z/dr$, etc.

†Casson's original work was for pigment-oil suspensions such as printing inks.

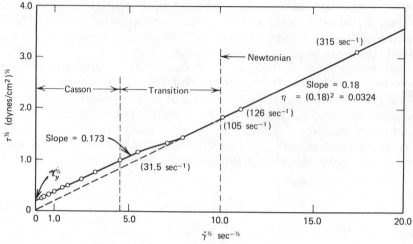

Figure 9.1-1. Typical hemorheological data [6], taken with coaxial-cylinder viscometer. (Hematocrit: 40; temperature: 37°C; fibrinogen concentration: 0.27 g/100 ml).

for the core region in which $\tau_{rz} < \tau_y$, and

$$\sqrt{\tau_{rz}} = \sqrt{\tau_y} + \eta^{\frac{1}{2}}\sqrt{\dot{\gamma}} \tag{9.1-1}$$

for the annular region in which $\tau_{rz} > \tau_y$, where

$$\dot{\gamma} = -\frac{dv_z}{dr}, \tag{9.1-4}$$

as footnoted earlier. Referring to Figure 9.1-3, in the central core, where the shear stress does not exceed the yield stress, the fluid itself does not flow but is merely carried along by the fluid in the annular region, in which the shear stress does exceed the yield stress. Thus, the central core has a flat velocity profile, designated as $v_z^<$, and this kind of flow is called the *plug flow*. In the annular region the velocity distribution is a function of r, designated by $v_z^>(r)$.

Since the fluid is non-Newtonian, the appropriate starting equation is Eq. C in Table 4.12-2(b). With the usual elimination of vanishing terms, the following results:

$$\frac{1}{r}\frac{d}{dr}(r\tau_{rz}) = \frac{\Delta p}{L}, \tag{9.1-5}$$

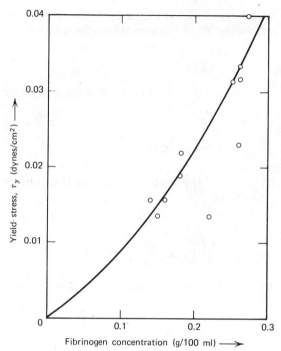

Figure 9.1-2. Dependence of yield stress of Casson fluid on fibrinogen content (Based on blood data by Merrill and Pelletier [6]; hematocrit: 40; temperature: 37°C).

with the boundary conditions

$$\text{B.C. 1: at } r = 0, \quad \tau_{rz} = 0 \quad \text{finite}$$

$$\text{B.C. 2: at } r = R_y, \quad \tau_{rz} = \tau_y$$

$$\text{B.C. 3: at } r = R, \quad v_z = 0,$$

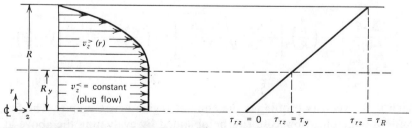

Figure 9.1-3. Schematic of Casson flow. Note the continuous momentum flux at the boundary between the two regions.

where R_y is the radial position at which the shear stress starts to exceed the yield stress. Integrating Eq. 9.1-5 with the application of B.C. 1 gives

$$\tau_{rz} = \frac{\Delta pr}{2L} .$$ (9.1-6)

Application of B.C.2 results in the following relationship between the yield stress and R_y:

$$\tau_y = \frac{\Delta pR_y}{2L} .$$ (9.1-7)

For the velocity profile of the annular region, we first substitute Eq. (9.1-6) into Eq. 9.1-1, to obtain

$$\sqrt{\frac{\Delta pr}{2L}} = \sqrt{\tau_y} + \eta^{1/2}\sqrt{\dot{\gamma}} .$$ (9.1-8)

To avoid incurring half-power terms of dv_z/dr, it is wise to rearrange Eq. 9.1-8 into

$$\sqrt{\dot{\gamma}} = \eta^{-1/2}\left(\sqrt{\frac{\Delta pr}{2L}} - \sqrt{\tau_y}\right)$$ (9.1-9)

before squaring both sides of Eq. 9.1-9. Also, remembering that $\dot{\gamma} = -dv_z^>/dr$, we have

$$-\frac{dv_z^>}{dr} = \frac{1}{\eta}\left[\frac{\Delta pr}{2L} - 2\sqrt{\frac{\Delta pr\tau_y}{2L}} + \tau_y\right].$$ (9.1-10)

Integrating now with the application of B.C. 3 yields

$$v_z^> = \frac{\Delta pR^2}{4\eta L}\left[1-\left(\frac{r}{R}\right)^2\right] - \frac{4R}{3\eta}\sqrt{\frac{\Delta p\tau_y R}{2L}}\left[1-\left(\frac{r}{R}\right)^{3/2}\right] + \frac{\tau_y R}{\eta}\left(1-\frac{r}{R}\right),$$

(9.1-11)

which is the velocity profile for the annular region where $\tau_{rz} > \tau_y$. The flat profile for the plug-flow core can be obtained by evaluating the above at $r = R_y$, since at this point $v_z^< = v_z^>$. Doing this and remembering Eq. 9.1-7, we have

$$v_z^< = \frac{\Delta p R^2}{4\eta L}\left[1-\left(\frac{R_y}{R}\right)^2\right] - \frac{4R}{3\eta}\sqrt{\frac{\Delta p \tau_y R}{2L}}\left[1-\left(\frac{R_y}{R}\right)^{3/2}\right] + \frac{\tau_y R}{\eta}\left(1-\frac{R_y}{R}\right)$$

$$= \frac{\Delta p R^2}{4\eta L}\left[1-\frac{8}{3}\sqrt{\frac{R_y}{R}} + 2\frac{R_y}{R} - \frac{1}{3}\left(\frac{R_y}{R}\right)^2\right]. \tag{9.1-12}$$

We can see from Eq. 9.1-1 that if the yield stress, τ_y, is zero, we have a Newtonian fluid. Checking this with Eqs. 9.1-10 and 9.1-11 for $\tau_y = 0$ and $R_y = 0$, we indeed obtain the parabolic velocity profile characteristic of the H–P flow. On the other hand, from Eq. 9.1-6, we can obtain the shear stress at the walls $(r = R)$ as

$$\tau_R = \frac{\Delta p R}{2L}. \tag{9.1-13}$$

Thus, if $\tau_R < \tau_y$, which means that nowhere in the flow channel, including the wall where stress is largest, does the shear stress exceed the yield point, then $R_y > R$, which means that the entire cross section is plugged up; hence, no flow occurs. This is the other extreme but does not often occur in blood, as the yield stress of blood is usually quite small.

The dimensionless velocity profile in terms of τ_y, τ_R, and the maximum-core velocity can be obtained from Eqs. 9.1-11 and 9.1-12, with the help of Eqs. 9.1-7 and 9.1-13:

$$\frac{v_z^>}{v_{z\,max}} = \frac{v_z^>}{v_z^<}$$

$$= \frac{(\tau_R/2)\left[1-(r/R)^2\right]-\frac{4}{3}\sqrt{\tau_R \tau_y}\left[1-(r/R)^{3/2}\right]+\tau_y[1-r/R]}{(\tau_R/2)\left[1-(\tau_y/\tau_R)^2\right]-\frac{4}{3}\sqrt{\tau_R \tau_y}\left[1-(\tau_y/\tau_R)^{3/2}\right]+\tau_y[1-\tau_y/\tau_R]},$$

$$\tag{9.1-14}$$

which is plotted in Figure 9.1-4 for various τ_R/τ_y values. We can see that the flow approaches Newtonian as the shear stress is increased to very large values.

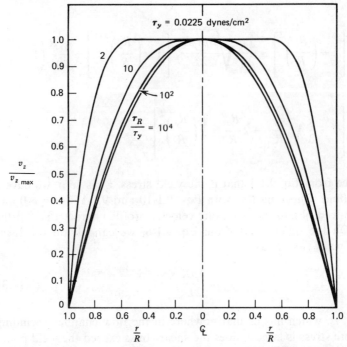

Figure 9.1-4. Velocity profiles of blood as a Casson fluid at various wall shear stresses [7].

The volumetric flow rate can be obtained by integrating the velocity profiles over the two cross-sectional regions:

$$Q = 2\pi \int_0^{R_y} v_z^< r \, dr + 2\pi \int_{R_y}^R v_z^> r \, dr. \qquad (9.1\text{-}15)$$

However, both the integration procedure and the aftermath (canceling terms) are usually very complicated. For Bingham and Casson types of non-Newtonian fluid, a special procedure is preferred. If we use the integration by parts formula, Eq. (c) of Table 2.4-1, we can write

$$Q = 2\pi \int_0^R v_z r \, dr$$

$$= 2\pi \left\{ \left[\tfrac{1}{2} v_z r^2 \right]_{r=0}^{r=R} - \int_{r=0}^{r=R} \tfrac{1}{2} r^2 \, dv_z \right\}. \qquad (9.1\text{-}16)$$

Since $v_z = 0$ at $r = R$, the first term of this expression vanishes at both limits, leaving only the second term, which becomes*

$$Q = -\pi \int_0^R r^2 \left(\frac{dv_z}{dr} \right) dr \qquad (9.1\text{-}17)$$

and can now be separated into the two regions:

$$Q = -\pi \left\{ \int_0^{R_y} r^2 \left(\frac{dv_z^<}{dr} \right) dr + \int_{R_y}^R r^2 \left(\frac{dv_z^>}{dr} \right) dr \right\}. \qquad (9.1\text{-}18)$$

We note from Figure 9.1-3 that in the core region, $dv_z^< / dr = 0$ (flat profile), so that the first term on the right-hand side of Eq. 9.1-18 again vanishes. Also noting from Figure 9.1-3 that

$$\frac{r}{R} = \frac{\tau_{rz}}{\tau_R}, \qquad (9.1\text{-}19)$$

the remainder of Eq. 9.1-18 can be rewritten

$$Q = -\pi \left(\frac{R}{\tau_R} \right)^3 \int_{\tau_y}^{\tau_R} \tau_{rz}^2 \left(\frac{dv_z^>}{dr} \right) d\tau_{rz}. \qquad (9.1\text{-}20)$$

From Eq. 9.1-1 we have

$$-\frac{dv_z^>}{dr} = \frac{\left(\sqrt{\tau_{rz}} - \sqrt{\tau_y} \right)^2}{\eta}, \qquad (9.1\text{-}21)$$

which is substituted into Eq. 9.1-20 with the resulting expression integrated to yield

$$Q = \frac{\Delta p \pi R^4}{8 \eta L} \left[1 - \frac{16}{7} \sqrt{\frac{2 L \tau_y}{\Delta p R}} + \frac{4}{3} \left(\frac{2 L \tau_y}{\Delta p R} \right) - \frac{1}{21} \left(\frac{2 L \tau_y}{\Delta p R} \right)^4 \right]. \qquad (9.1\text{-}22)$$

*This equation is very useful for computing the volumetric flow rate for cases in which the velocity cannot be analytically obtained from the velocity gradient. It applies to both Newtonian and non-Newtonian fluids.

or, noting that

$$\tau_R = \frac{\Delta p R}{2L},$$

we can express the *reduced* (but *not* dimensionless) average velocity as

$$\langle U \rangle = \frac{4\langle v_z \rangle}{R} = \frac{4Q}{\pi R^3}$$

$$= \frac{\tau_R}{\eta}\left[1 - \frac{16}{7}\sqrt{\frac{\tau_y}{\tau_R}} + \frac{4}{3}\left(\frac{\tau_y}{\tau_R}\right) - \frac{1}{21}\left(\frac{\tau_y}{\tau_R}\right)^4 \right]. \qquad (9.1\text{-}23)$$

Again we can see that these expressions have Newtonian fluids as their special case for $\tau_y = 0$. Even for a non-Newtonian fluid with a nonzero yield stress τ_y, the flow approaches Newtonian behavior at high wall stress —that is, $\tau_R \gg \tau_y$. They have the interesting properties that $\langle U \rangle$ and its first and second derivatives with respect to τ_R are all zero at $\tau_R = \tau_y$. They can also be shown to be consistent with the general expression (see Problem 9.C)

$$\dot{\gamma}_{r=R} = \frac{Q}{\pi R^3}\left[3 + \left(\frac{\partial \ln Q}{\partial \ln \Delta P}\right)_{R,L} \right]. \qquad (9.1\text{-}24)$$

These expressions are quite useful in obtaining rheological information from pressure flow data [7].

9.2 BLOOD AS A SUSPENSION: HEMATOCRIT-DEPENDENCY OF VIS-COSITY

It is a well-known fact that blood actually consists of various kinds of cells such as red blood cells (RBC or erythrocytes), white blood cells (WBC or leukocytes), and platelets (or thrombocytes) suspended in plasma. By far, RBC's outnumber the others and assume major hemorheological impor-tance in addition to playing the vital role of carrying oxygen to all parts of the body. They are in the shape of biconcave disks approximately 7.2 μ in diameter and 2.2 μ thick. Hemoglobin is contained in the cell by the red-cell membrane. It is said that such a shape is an optimum between diffusing efficiency and mechanical strength against hemolysis. They are flexible to a certain extent.

The percent volume concentration of RBC's in the whole blood is called the *hematocrit*. Hematocrit values for human normally range between 40 (for the female) and 45 (for the male). As expected for a suspension, the hematocrit value has a definite effect on the viscosity of blood. As early as 1906, Einstein [8] developed a simple linear relationship for rigid spherical particles which has later been used to relate the two:

$$\eta_r = \frac{\mu}{\mu_0}$$

$$= 1 + Bh, \qquad (9.2\text{-}1)$$

where μ and μ_0 are the viscosities of the whole blood and plasma, respectively; h is the hematocrit; and B is a constant found to be equal to 2.5. According to Einstein's theory, this expression holds only for a very dilute suspension of spherical particles (1% particle concentration), although Trevan [9] found experimentally that linearity appeared to hold up to 40% but with a different constant, 6.3. On the other hand, Hatschek [10] deduced mathematically that the linearity factor should be 4.5 for spheres. The discrepancies between these studies can only be partially accounted for by the shape of the particles used,* and the Einstein formula, Eq. 9.2-1, seems to have enjoyed the widest acceptance. Above 40% hematocrit, Trevan's experimental data [9] confirmed the following formula derived mathematically by Hatschek [10] for emulsions:

$$\eta_r = \frac{\mu}{\mu_0}$$

$$= \left(1 - \sqrt[3]{h}\right)^{-1}. \qquad (9.2\text{-}2)$$

Considering the particle–particle interactions in suspension at higher concentrations than that allowed by the Einstein model, Vand [11] derived theoretically

$$\ln \eta_r = \frac{Bh + A_1(A_2 - B)h^2 + \cdots}{1 - A_3 h}, \qquad (9.2\text{-}3)$$

where A_1, A_2, and A_3 are constants related to particle-particle collision and hydrodynamic interaction. Using values for rigid spheres, Eq. 9.2-3 can be shown to yield

$$\eta_r = 1 + 2.5h + 7.349h^2 + \cdots. \qquad (9.2\text{-}4)$$

*For example, Trevan used RBC's from cats and dogs suspended in plasma and saline.

Later, Oliver and Ward [12] proposed the following:

$$\eta_r = 1 + Bh + (Bh)^2 + (Bh)^3 + \cdots$$

$$= \sum_{i=1}^{\infty} (Bh)^i, \qquad (9.2\text{-}5)$$

where the second-, third-, and higher-order terms represent, respectively, the binary, ternary, and so on, interaction of particles or cells. Using data taken by a number of other investigators, as well as their own, they determined that the constant B was essentially the same as the Einstein constant—namely, 2.5—and that this model was good up to 35% hematocrit.

Under this condition, since the maximum value for Bh (2.5×0.35) is less than unity, Eq. 9.2-5, can be rewritten

$$\eta_r = \frac{1}{1 - Bh}, \qquad (9.2\text{-}6)$$

based on the series expansion Eq. (10) of Table 2.5-1. This formula is particularly useful for obtaining the analytical solution for the velocity profiles and subsequent quantities (see Problem 9.D).

Noting the effect of particles on the viscosity of a suspension, Casson originally proposed the following model, of which both Eqs. 9.1-1 and 9.2-6 are special cases:

$$\sqrt{\tau} = \sqrt{\frac{\mu_0 \dot\gamma}{(1-h)^B}} + \frac{C\sqrt{\dot\gamma}}{B}\left[\sqrt{\frac{1}{(1-h)^B}} - 1\right] \qquad (9.2\text{-}7)$$

or, alternately,

$$\eta^{1/2} = \sqrt{\mu_0}\,(1-h)^{-B/2} + \frac{C}{B}\left[(1-h)^{-B/2} - 1\right], \qquad (9.2\text{-}8)$$

where B has been found to be approximately 2.3 [13] (very close to the Einstein value, $B = 2.5$) and $C/B = 0.2 \text{ g}^{1/2}\text{cm}^{-1/2}\text{sec}^{-1}$.

9.3 THE FÅHRAEUS–LINDQVIST EFFECT AND ASSOCIATED PHENOMENA

In this section, we shall discuss two interrelated phenomena that occur when blood flows through narrow vessels such as arterioles.* One is the tendency of erythrocytes to migrate toward the center of the flow channel, leaving a relatively cell-free, slower-moving layer of plasma [14]. The other is the reduction of apparent viscosity as calculated from the slope of the Q-versus-Δp line based on the H–P equation (Eq. 6.1-7) and compared with the same quantity obtained in a larger vessel [15]. Both phenomena were observed and reported first by Fåhraeus and Lindqvist, who perceived that the former was the cause of the latter. This belief was later supported by the works of many, both experimental and theoretical, but most notably those of Maude and Whitmore [16, 17]. It has been subsequently developed into the so-called marginal-zone theory, which will be discussed in the following section. Fåhraeus and Lindqvist also suggested that the higher cell concentration in the central core, where the fluid flows faster, would make the concentration of cell *appear* to decrease while flowing through. This is the hematocrit reduction phenomenon to be discussed in Section 9.6. Thus, the entire set of phenomena is universally known as the Fåhraeus–Lindqvist effect.†

Dix and Scott Blair** [18] suggested that since in a narrow tube the blood cell diameter is sufficiently close in order of magnitude to the tube diameter, blood could no longer be considered a continuum. Thus, thickness of the fluid layers is finite, as depicted in Figure 9.3-1, which is the basic cause of the apparent viscosity decrease. Simply explained, we recall that if blood could be considered a continuum and Newtonian fluid, we could substitute

$$\frac{dv_z}{dr} = -\frac{\Delta p r}{2\mu_a L} \qquad (9.3\text{-}1)$$

into Eq. 9.1-17 to obtain

$$Q = \frac{\pi \Delta p}{2\mu_a L} \int_0^R r^3 \, dr, \qquad (9.3\text{-}2)$$

---

*But not the smallest capillaries, in which the blood cell, having a diameter approximately equal to that of the vessel, flows through in single file. The dynamics of this belong to an entirely separate category.

†For this and many other pioneering scientific contributions, some of which had been overlooked for a number of years, Fåhraeus was appropriately awarded the first Poiseuille Medal of the International Society of Hemorheology, just two years before his death in 1968.

** For his many important contributions in hemorheology, Scott Blair won the second Poiseuille Medal in 1969.

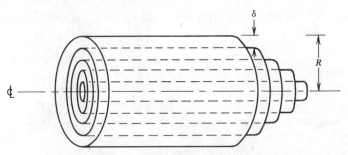

Figure 9.3-1. Blood flow model with layers of finite thickness.

which upon integration would yield the H–P equation:

$$Q = \frac{\pi \Delta p R^4}{8\mu_a L},$$
(9.3-3)

where the subscript a on μ denotes that this is the *apparent* viscosity based on the continuum assumption. However, if this assumption is no longer valid, as Dix and Scott Blair suggested, the integration in Eq. 9.3-2 cannot be performed. Instead a summation procedure should be used in which the flow channel is radially divided into n layers, each having a finite thickness δ. Thus, dr is replaced by δ and r by $k\delta$, which means that the arbitrary kth layer is at an arbitrary distance of r from the axis. Equation 9.3-2 is then replaced by

$$Q = \frac{\pi \Delta p}{2\mu_t L} \sum_{k=1}^{n} (k\delta)^3 \delta$$

$$= \frac{\pi \Delta p \delta^4}{2\mu_t L} \sum_{k=1}^{n} k^3,$$
(9.3-4)

where subscript t denotes the "true" viscosity after the noncontinuum is properly considered. The sum of the cubic of the first n integers is [19]*

$$\sum_{k=1}^{n} k^3 = \frac{n^2(n+1)^2}{4},$$

*A few illustrations should help convince the reader of its validity: If $n=3$, $1+2^3+3^3=36$ $=3^2(3+1)^2/4$; if $n=5$, $1+2^3+3^3+4^3+5^3=225=5^2(5+1)^2/4$; etc.

so that, remembering that $R = n\delta$,

$$Q = \frac{\pi \Delta p R^4}{8 \mu_t L} \left(1 + \frac{\delta}{R} \right)^2. \tag{9.3-5}$$

Comparing Eqs. 9.3-3 and 9.3-5, we find

$$\frac{1}{\mu_a} = \frac{1}{\mu_t} \left(1 + \frac{\delta}{R} \right)^2, \tag{9.3-6}$$

since both δ and R are scalars, $[1 + (\delta/R)]^2 > 1$ so that $\underline{\mu_a < \mu_t}$; that is, the apparent viscosity in a narrow vessel computed from the H–P equation is always lower than the "true" viscosity or that in a large vessel.† Since this effect arises from the summation procedure, it is generally referred to as the *sigma* (Σ) *phenomenon*.

9.4 THE MARGINAL-ZONE THEORY

In 1959 Haynes and Burton [20] reported experimental data* on the deviation from the linear relationship between the flow rate and pressure gradient characteristic of the H–P flow. As shown in Figure 9.4-1, where data at two different tube radii are arbitrarily selected, when the asymptotic (high-shear) lines at different hematocrit values are extrapolated to the low shear region, they meet at a single point on the Q-axis, and this intercept is a function of tube radius R. As illustrated in Figure 9.4-2(a) for a series of runs at a single radius and hematocrit, the true flow rate (solid curve) can be related to the apparent flow rate Q_a as computed from the apparent viscosity via the H–P equation by first subtracting from Q_a a correction term that is equal to I(R) at infinite $\Delta p/L$ but diminishes to zero as the origin is approached. Mathematically, we can write

$$Q = \frac{\pi \Delta p R^4}{8 \mu_a L} - f\left(R, h, \frac{\Delta p}{L} \right) \cdot I(R), \tag{9.4-1}$$

where

$$f = 1 - \exp\left[-k(R, h) \cdot \frac{\Delta p}{L} \right]. \tag{9.4-2}$$

---

†We see that in a large vessel, $\delta \ll R$ so that $\mu_a \doteq \mu_t$; i.e., this effect disappears.

*Taken with human erythrocytes resuspended in standard acid-citrate-dextrose (ACD) solution the rheological properties of which were believed to be the same as those of whole blood.

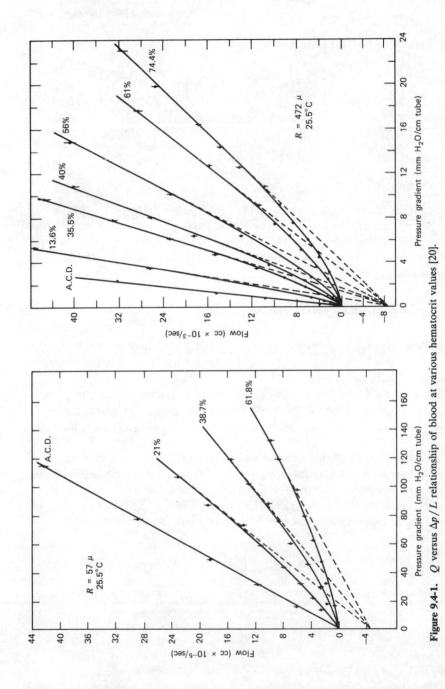

Figure 9.4-1. Q versus $\Delta p / L$ relationship of blood at various hematocrit values [20].

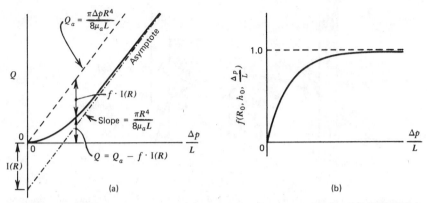

Figure 9.4-2. Relationship between apparent and true flow rates and viscosities: (a) definition of $f(R, h, \Delta p/L)$ and $I(R)$; (b) $f(R, h, \Delta p/L)$ for a certain set of $R = R_0$, $h = h_0$.

Furthermore, $k(R, h)$ should be such that it approaches infinity at infinite R or zero h.

The marginal-zone theory, as originally proposed by Haynes [21] and later modified by Gordon [22] considers an annular zone of thickness ϵR near the tube wall in which there is no blood cell. The viscosity of the plasma in this zone is μ_p, while the central core, of radius κR ($\kappa = 1 - \epsilon$) and uniform cell concentration, has a viscosity of μ_c. The plasma in the marginal zone (II) is Newtonian, so that we may write (see Eq. 6.1-3)*

$$-\mu_p \frac{dv_z^{\mathrm{II}}}{dr} = \frac{\Delta p r}{2L}. \tag{9.4-3}$$

With the boundary conditions that at $r = R$, $v_z = 0$, Eq. 9.4-3 can be integrated to yield

$$v_z^{\mathrm{II}} = \frac{\Delta p R^2}{4\mu_p L}(1 - \xi^2), \tag{9.4-4}$$

where $\xi = r/R$, the dimensionless radial coordinate. The volumetric flow

---

*This would involve evaluating the first integration constant $C_1 = 0$. Formally, this would entail solving the central-core and annular-zone velocity profiles simultaneously and "pasting" the two solutions at $r = \kappa R$, as shown in Example 9.6-1. However, we can take a shortcut by extrapolating the velocity profile back to $r = 0$ (see broken curve in Figure 9.4-3) and using the symmetry argument that τ_{rz} would be equal to zero at $r = 0$ even though $r = 0$ is outside of the annular plasmatic zone.

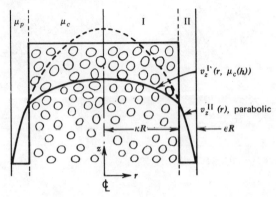

Figure 9.4-3. The central core of uniform cell distribution and the marginal plasmatic zone.

rate of plasma in the marginal zone is

$$Q_p^{II} = 2\pi \int_{\kappa R}^{R} v_z(r)r\,dr$$

$$= 2\pi R^2 \left(\frac{\Delta p R^2}{4\mu_p L}\right) \int_{\kappa}^{1} (1-\xi^2)\xi\,d\xi$$

$$= \frac{\pi \Delta p R^4}{8\mu_p L}(1-\kappa^2)^2. \tag{9.4-5}$$

For the central core (I), an additional force term representing the coherence resistance between fluid layers due to the interaction or collision of cells should be included. This force, denoted by F and for "per unit cylindrical surface area" (not cross-sectional area), is of a similar nature to the shear stress but in addition to that caused by the regular rate of shear, so that we may simply substitute

$$\tau_{rz} = F - \mu_c \frac{dv_z}{dr} \tag{9.4-6}$$

to the already simplified equation of motion for this case (see Eq. 9.1-6), thus yielding

$$\frac{dv_z^I}{dr} = \frac{1}{\mu_c}\left(F - \frac{\Delta p r}{2L}\right) \tag{9.4-7}$$

which upon integration with the boundary condition that at $r = \kappa R$, $v_z^I = v_z^{II}$

$= (\Delta p R^2 / 4\mu_p L)(1 - \kappa^2)$, gives

$$v_z^{\mathrm{I}} = \frac{\Delta p R^2}{4\mu_p L}\left[(1 - \kappa^2) + \frac{\mu_p}{\mu_c}(\kappa^2 - \xi^2)\right] - \frac{FR}{\mu_c}(\kappa - \xi). \qquad (9.4\text{-}8)$$

The combined volumetric flow rate of cells and plasma in the central core is given by

$$Q_{c,p}^{\mathrm{I}} = 2\pi \int_0^{\kappa R} v_z(r) r\, dr$$

$$= \frac{\Delta p \pi R^4}{8\mu_p L}\left[2\kappa^2(1 - \kappa^2) + 4\frac{\mu_p}{\mu_c}\int_0^{\kappa}(\kappa^2 - \xi^2)\xi\, d\xi\right]$$

$$- 2\pi \frac{FR^3}{\mu_c}\int_0^{\kappa}(\kappa - \xi)\xi\, d\xi$$

$$= \frac{\Delta p \pi (\kappa R)^4}{8\mu_p L}\left(2\frac{1 - \kappa^2}{\kappa^2} + \frac{\mu_p}{\mu_c}\right) - \frac{\pi F(\kappa R)^3}{3\mu_c}. \qquad (9.4\text{-}9)$$

Combining Eqs. 9.4-5 and 9.4-9, we have the total flow rate through the entire blood vessel being

$$Q_M = Q_{c,p}^{\mathrm{I}} + Q_p^{\mathrm{II}}$$

$$= \frac{\Delta p \pi R^4}{8\mu_c L}\left[\eta_{rc} + (1 - \eta_{rc})\kappa^4\right] - \frac{\pi F(\kappa R)^3}{3\mu_c}, \qquad (9.4\text{-}10)$$

where $\eta_{rc} = \mu_c / \mu_p$, the viscosity of the central core relative to that of the plasma, and Q_M denotes the volumetric flow rate as calculated by the marginal-zone theory, *not* the actually measured flow rate Q in Eq. 9.4-1. Since the plasmatic zone is usually quite thin, μ_c can be taken as the viscosity of the whole blood, and, comparing Eq. 9.4-10 with Figure 9.4-2(a), we conclude that

$$\mathrm{I}(R) = \frac{\pi F(\kappa R)^3}{3\mu_c} \qquad (9.4\text{-}11)$$

or

$$\log \mathrm{I} = \log k + 3\log R, \qquad (9.4\text{-}12)$$

where $k = \pi F \kappa^3 / 3 \mu_c$. Plotting the I-versus-$R$ data by Haynes and Burton [20] on a log–log scale (see Figure 9.4-4), Gordon [22] indeed found that the slope $= 2.92 \pm 0.06$, very close to the theoretically predicted value of 3. From the intercept of I-versus-R chart at $\log R = 0, k$ can be determined. Knowing the viscosity, the coherence resistance F for human red cells resuspended in acid citrate dextrose has been in turn estimated at $2{\sim}3$ dynes/cm^2 [22]. This is quite close to the value obtained by Cokelet, et al. [13] with a grooved viscometer.

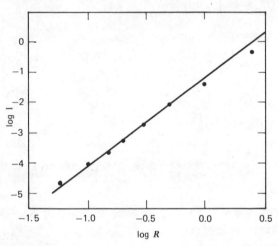

Figure 9.4-4. *I*-versus-*R* data by Haynes and Burton; log–log scale (from [22]).

9.5 AXIAL MIGRATION OF CELLS AND THE TUBULAR PINCH EFFECT

Ever since Fåhraeus first made the monumental observation of the centrally migratory tendency of erythrocytes in narrow flow channels, there have been many investigations on this subject, both theoretical and experimental, phenomenological and hydrodynamical. Most of the observations and analyses have shown a maximum cell concentration at the center of the tube with a reasonably flat profile through a large portion of the central corè and the local hematocrit drops to zero rapidly near the wall, such as shown in Figure 9.5-1 as curve *A*. A useful empirical formula describing this distribution is

$$h(r) = h_m \left[1 - \left(\frac{r}{R} \right)^n \right], \qquad (9.5\text{-}1)$$

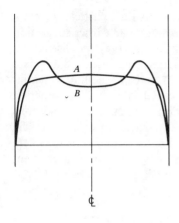

Figure 9.5-1. Concentration distribution of red blood cells in narrow flow channels; A, general observation; B, Segré and Silberberg's tubular pinch effect or "necklace formation."

in which $h(r)$ is the local hematocrit, h_m is the maximum hematocrit at the center of the tube, and n is a parameter determining the exact shape of the profile. As the flow rate changes, the particle distribution is expected to change, and thus, this index is expected to change with it. This has indeed been confirmed by Lih [23] in a correlation wherein n is shown to be a function of Reynolds number (see Example 9.6-2).

However, it has been observed by Segré and Silberberg [24,25] that the maximum concentration does not occur at the center but at approximately 0.6 fractional radial distance from the center, shown in Figure 9.5-1 as curve B. This so-called "tubular pinch" effect is thought to be caused by the inertia of the moving fluid. However, their detailed analysis is subject to some revision, as a parabolic velocity profile was used. As will be shown later in Section 9.6, because of its dependency on cell concentration, the viscosity is no longer uniform over the cross section, and consequently, the velocity profile is not exactly parabolic, as it would be in the case of H–P flow with no particles present.

As far as the underlying theory behind the axially migratory phenomenon is concerned, Goldsmith, Mason, and their co-workers, in a series of distinguished papers [26–30], have made detailed theoretical and cinematographic analyses of the behavior of particles of various shapes, both singly and in formation, and its relationship to the above-mentioned experimental observations. Their work reinforced Brenner's belief [4] that the inertial and wall effects, together with the deformation of cells, were mainly responsible for their migratory tendency. Furthermore, the translational and rotational slip near the wall has also been found to be of major importance [30].

Instead of making a detailed analysis of the individual particles or cells, Deakin [31, 32] used a different tactic by following up an earlier suggestion by Jeffery [33] that the cells were so arranged, or they so distributed themselves, that the mechanical energy dissipation (see Section 8.6) was a minimum. Using the Casson model (Eq. 9.2-8) and the method of variational calculus, he attempted to obtain a particle distribution function corresponding to a minimum energy dissipation subject to the conditions that a constant flow rate and an overall particle rate were to be maintained. With a series of approximations to render this problem mathematically tractable, the result showed only qualitative agreement with Segré and Silberberg's observation [24,25]. Further work by Tesfagaber and Lih [38] using the simpler Einstein model (Eq. 9.2-1) has been inconclusive.

We shall present examples of using some of the models described herein after we introduce the concept of hematocrit reduction in the following section.

9.6 LARGE-VESSEL AND CAPILLARY HEMATOCRITS

If we accept the nonuniform red-cell distribution as an experimental fact, whether or not we agree on or understand the theoretical basis for it, an immediate consequence is the existance of two types of overall hematocrits, analogous to the volume-averaged and flow-averaged temperatures discussed in the forced convection problem in Example 6.2-6.

In Figure 9.6-1 we model the flow of blood through a capillary (a) by a tube-and-tank arrangement (b). The model is made a closed system by the continuous recirculation of the suspension using a pump. As the entire system is at steady state, it would seem that the particulate concentration ought to be uniform throughout.

In the model (b), the reservoir feeding the tube and the tank receiving the flow from the tube are equivalent, respectively, to the feeding artery and receiving vein in the true picture (a). The particulate concentrations in the two tanks are indeed the same, which is equivalent to the large-vessel hematocrit, h_l, in the circulatory system. However, if we measured the particulate concentration in the tube by suddenly closing off the two valves at both ends of the tube and removing the content between them, we would find that its particulate concentration is in general lower than that in the tanks. This particulate concentration is equivalent to the capillary hematocrit, h_c, in the *in vivo* situation (a). The question then arises, since both the circulatory system (a) and its model (b) are closed, how can the hematocrit, or particulate concentration, suddenly drop in one section of

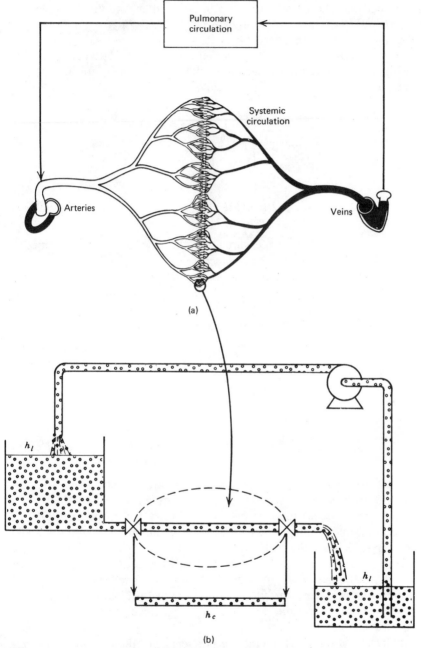

Figure 9.6-1. Blood flow through a single capillary: (a) reality; (b) model.

the system? Should it not be uniform throughout as "intuitively" suggested earlier?

The answer becomes very clear if we realize that we are actually measuring the hematocrit in two different ways. In the latter method (i.e., by isolating the content in a section of the tube), we measure the volume-averaged concentration, or, for a circular tube,

$$h_c = \frac{\int_0^L \int_0^{2\pi} \int_0^R h(r)\,r\,dr\,d\theta\,dz}{\int_0^L \int_0^{2\pi} \int_0^R r\,dr\,d\theta\,dz}$$

$$= 2\int_0^1 h(\xi)\xi\,d\xi, \qquad (9.6\text{-}1)$$

where $\xi = r/R$. Since this is measured with the flow having been stopped, it is also referred to as the *static hematocrit*.

In the former scheme, however, we are measuring the content of the tube as it continuously flows into the receiving tank, much the same way we measure the flow-averaged or cup-mixing temperature in the forced convection case in Example 6.2-6. Formally, we write

$$h_l = \frac{\int_0^{2\pi} \int_0^R h(r)v_z(r)\,r\,dr\,d\theta}{\int_0^{2\pi} \int_0^R v_z(r)\,r\,dr\,d\theta}$$

$$= \frac{2}{\langle v_z \rangle} \int_0^1 h(\xi)v_z(\xi)\xi\,d\xi, \qquad (9.6\text{-}2)$$

where $v_z(r)$ is the velocity distribution of the blood or suspension. Comparing Eqs. 9.6-1 and 9.6-2 just as we compared Eqs. 6.2-75 and 6.2-77, we can see that $v_z(r)$ assumes the role of a weighting factor in the averaging or integration process. Since it has a maximum at the tube axis and gradually decreases to zero at the wall (although not necessarily parabolically for non-Newtonian fluid) and since in most cases $h(r)$ assumes the form of curve A in Figure 9.5-1 (i.e., the velocity is higher where the local concentration is higher), the average thus obtained is biased in favor of the central portion of the tube. In plain language, since the local particulate concentration is higher where the local velocity is higher, the flow-averaged concentration is more representative of the higher than lower local concentration. Since this concentration is measured while the blood or suspension is continuously flowing out of the tube, it is also called the *dynamic hematocrit*, which has indeed been found to be generally larger than the static or capillary hematocrit.

The hematocrit reduction as blood flows through a small vessel can thus be defined

$$h_R = \frac{h_l - h_c}{h_l}.$$ (9.6-3)

Following are examples in which some of the concepts and models presented above have actually been applied to the analysis of hemorheological problems.

Example 9.6-1 The Stepwise Red-Cell Distribution. As an approximation to curve A of Figure 9.5-1 Thomas [34] proposed an erythrocyte distribution essentially the same as that in the marginal-zone theory. As shown again in Figure 9.6-2, the red cells are assumed to have a uniform concentration, and hence there is a uniform viscosity, μ_c, across the central core, while the cell-free plasmatic layer has a uniform viscosity, μ_p. In effect, we are considering the flow as consisting of two coaxial zones each of which is a Newtonian fluid with its own viscosity. To solve this formally, we start with Eq. F in Table 4.12-2(b) and write, for steady state,

$$\frac{\mu_c}{r} \frac{d}{dr} \left(r \frac{dv_z^{I}}{dr} \right) = -\frac{\Delta p}{L}$$ (9.6-4)

for the central core with cells and

$$\frac{\mu_p}{r} \frac{d}{dr} \left(r \frac{dv_z^{II}}{dr} \right) = -\frac{\Delta p}{L}$$ (9.6-5)

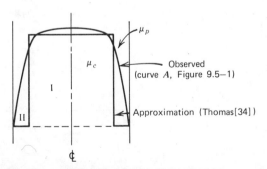

Figure 9.6-2. The stepwise approximation of red-cell distribution.

for the cell-free plasma layer, with the following boundary conditions:

$$\text{B.C.1: at } r=0, \quad v_z^I=\text{finite}, \quad \frac{dv_z^I}{dr}=0$$

$$\text{B.C.2a: at } r=\kappa R, \, v_z^I=v_z^{II}$$

$$\text{B.C.2b: at } r=\kappa R, \, \mu_c\frac{dv_z^I}{dr}=\mu_p\frac{dv_z^{II}}{dr}$$

$$\text{B.C.3: at } r=R, \quad v_z^{II}=0.$$

Equations 9.6-4 and 9.6-5 can be easily solved individually with B.C. 1 and B.C. 3, with the solutions being "pasted" together by B.C. 2a and B.C. 2b at the boundary. The results are (see also Section 9.4 but without the F term)

$$v_z^I=\frac{\Delta p R^2}{4\mu_p L}\left[\frac{1}{\eta_{rc}}(\kappa^2-\xi^2)+(1-\kappa^2)\right], \qquad (9.6\text{-}6)$$

$$v_z^{II}=\frac{\Delta p R^2}{4\mu_p L}(1-\xi^2), \qquad (9.6\text{-}7)$$

where $\eta_{rc}=\mu_c/\mu_p$. The velocity profiles are plotted in Figure 9.6-3 where the would-be parabolic profile (if there were no cells in the central core) is shown by the interrupted curve as an extension of the profile of the plasmatic layer.

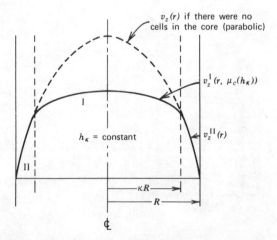

Figure 9.6-3. Velocity profile with uniform cell distribution in central core.

Now if we let h_κ be the uniform cell concentration in the central core, the capillary hematocrit can be easily calculated:

$$h_c = \frac{h_\kappa \pi (\kappa R)^2}{\pi R^2} = h_\kappa \kappa^2, \qquad (9.6\text{-}8)$$

while for the large-vessel hematocrit the integration has to be divided into the two zones:

$$h_l = \frac{\int_0^{2\pi} \int_0^R h_\kappa v_z(r) r \, dr \, d\theta}{\int_0^{2\pi} \int_0^R v_z(r) r \, dr \, d\theta}$$

$$= \frac{\int_0^{\kappa R} h_\kappa v_z^I(r) r \, dr + \int_{\kappa R}^R 0 \cdot v_z^{II}(r) r \, dr}{\int_0^{\kappa R} v_z^I(r) r \, dr + \int_{\kappa R}^R v_z^{II}(r) r \, dr}$$

$$= h_\kappa \frac{\int_0^\kappa v_z^I(\xi) \xi \, d\xi}{\int_0^\kappa v_z^I(\xi) \xi \, d\xi + \int_\kappa^1 v_z^{II}(\xi) \xi \, d\xi}. \qquad (9.6\text{-}9)$$

Substituting Eqs. 9.6-6 and 9.6-7 into Eqs. 9.6-9, we have, after canceling the $\Delta p R^2 / 4\mu_p L$ factor,

$$h_l = h_\kappa \frac{\int_0^\kappa [(1/\eta_{rc})(\kappa^2 - \xi^2) + (1 - \kappa^2)] \xi \, d\xi}{\int_0^\kappa [(1/\eta_{rc})(\kappa^2 - \xi^2) + (1 - \kappa^2)] \xi \, d\xi + \int_\kappa^1 (1 - \xi^2) \xi \, d\xi}$$

$$= h_\kappa \kappa^2 \frac{(\kappa^2/\eta_{rc}) + 2(1 - \kappa^2)}{1 - \kappa^4 [1 - (1/\eta_{rc})]}. \qquad (9.6\text{-}10)$$

Using Eqs. 9.6-8 and 9.6-10, we find that the hematocrit reduction as defined in Eq. 9.6-3 is

$$h_R = (1 - \kappa^2) \left[1 - \frac{1}{2 + [\kappa^2/\eta_{rc}(1 - \kappa^2)]} \right]. \qquad (9.6\text{-}11)$$

If we use the interaction model, Eq. 9.2-6, we can substitute

$$\eta_{rc} = \frac{1}{1 - Bh_\kappa} \qquad (9.6\text{-}12)$$

or*

$$\frac{1}{\eta_{rc}} = 1 - B\frac{h_c}{\kappa^2} \qquad (9.6\text{-}13)$$

into Eq. 9.6-11 to obtain

$$\frac{h_l - h_c}{h_l} = (1 - \kappa^2)\left[1 - \frac{1}{2 + (\kappa^2 - Bh_c)/(1 - \kappa^2)} \right]$$

Extracting h_c, we have

$$\left(\frac{B}{h_l}\right)h_c^2 - \left(B\kappa^2 + \frac{2 - \kappa^2}{h_l}\right)h_c + 1 = 0$$

or

$$h_c = \frac{h_l}{2B}\left[B\kappa^2 + \frac{2 - \kappa^2}{h_l} \pm \sqrt{\left(B\kappa^2 + \frac{2 - \kappa^2}{h_l}\right)^2 - \frac{4B}{h_l}} \right]. \qquad (9.6\text{-}14)$$

That is, knowing the hematocrit of the feeding arteries, we should be able to compute h_c and subsequently h_R, since other parameters such as B and κ^2 are also known. Of the two solutions for h_c, only one is a reasonable answer, as will be shown in the following illustration.

Thomas [34] has shown, using the data of Maude and Whitmore [16], that the dimensionless thickness of the plasmatic layer is related to the relative particle size by

$$1 - \kappa = 0.76\frac{\delta}{R}, \qquad (9.6\text{-}15)$$

where δ is the radius of the particles. Using the values

$$\frac{\delta}{R} = 0.1316,$$

$$B = 2.5,$$

$$h_l = 25\%,$$

*With the substitution of Eq. 9.6-8.

h_c can be calculated to be either 21.64% or 46.21%, of which the former is obviously the correct one. In fact, it can be shown that in Eq. 9.6-14 the negative sign in the plus-or-minus sign should always be chosen.

Exercise. Repeat the calculation above, but use a physiologically more realistic large-vessel hematocrit of 40%. Explain the result.

Example 9.6-2   Power-Function Cell Distribution. Noting that the family of curves shown in Figure 9.6-4 resembles the cell distribution observed in the various studies under different conditions, Lih [23] proposed the use of a power function, Eq. 9.5-1, to represent the cell distribution such as shown by curve *A* in Figure 9.5-1. When this is substituted into the Einstein model, Eq. 9.2-1, we have

$$\mu(r) = \mu_0 \left\{ 1 + Bh_m \left[1 - \left(\frac{r}{R} \right)^n \right] \right\}, \tag{9.6-16}$$

which can be used in the pseudo-Newtonian formula*

$$\tau_{rz} = -\mu(r) \frac{dv_z}{dr} \tag{9.6-17}$$

to yield

$$\tau_{rz} = -\mu_0 \left\{ 1 + Bh_m \left[1 - \left(\frac{r}{R} \right)^n \right] \right\} \frac{dv_z}{dr}. \tag{9.6-18}$$

The equation of motion, Eq. C of Table 4.12-2(b), when reduced, gives

$$\frac{1}{r} \frac{d}{dr} (r\tau_{rz}) = \frac{\Delta p}{L}, \tag{9.6-19}$$

which can be integrated with the boundary condition that at $r = 0$, $\tau_{rz} = 0$, so as to render

$$\tau_{rz} = \frac{\Delta p r}{2L}. \tag{9.6-20}$$

Combining Eqs. 9.6-18 and 9.6-20 and rearranging into dimensionless form, we have

$$\frac{d\phi}{d\xi} = -\frac{2\xi}{k(K - \xi^n)} \tag{9.6-21}$$

*That is, Newtonian in form but with a variable viscosity.

where

$$\xi = \frac{r}{R} \quad \text{(dimensionless radial distance)}$$

$$\phi = \frac{v_z}{(\Delta p R^2 / 4\mu_0 L)} \quad \text{(dimensionless velocity)}$$

$$k = Bh_m$$

$$K = \frac{k+1}{k}.$$

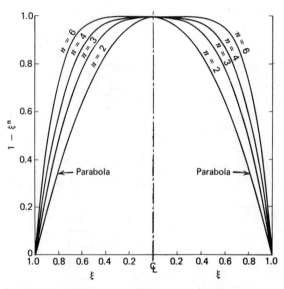

Figure 9.6-4. Empirical model for cell distribution in blood flow.

Noting that both B and h_m are always positive, K must be greater than unity. Since ξ is always between 0 and 1, so should ξ^n/K be (since n is positive). Thus, we can expand the right-hand side of Eq. 9.6-21 into infinite series form. The subsequent integration of this series with the no-slip boundary condition that at $\xi = 1$, $\phi = 0$ gives, as already shown in Example 2.5-4,

$$v_z = \frac{\Delta p R^2}{4\mu_0 L} \cdot \frac{2}{k} \sum_{j=0}^{\infty} \frac{1 - \xi^{jn+2}}{(jn+2)K^{j+1}}. \tag{9.6-22}$$

This solution has also been shown graphically in Figure 2.5-1. The volumetric flow rate and average velocity can thus be obtained as

$$Q = \int_0^{2\pi} \int_0^R v_z(r) r \, dr \, d\theta$$

$$= 2\pi R^2 \int_0^1 \phi(\xi) \xi \, d\xi$$

$$= \frac{\pi \Delta p R^4}{8 \mu_0 L} \cdot \frac{4}{k} \sum_{j=0}^{\infty} \frac{1}{(jn+4) K^{j+1}}, \tag{9.6-23}$$

$$\langle v_z \rangle = \frac{Q}{\pi R^2} = \frac{\Delta p R^2}{8 \mu_0 L} \cdot \frac{4}{k} \sum_{j=0}^{\infty} \frac{1}{(jn+4) K^{j+1}}. \tag{9.6-24}$$

The capillary hematocrit can be easily obtained by averaging $h(r)$ from Eq. 9.5-1 over the circular cross-sectional area, using Eq. 9.6-1:

$$\frac{h_c}{h_m} = 2 \int_0^1 (1 - \xi^n) \xi \, d\xi$$

$$= 2 \left[\frac{\xi^2}{2} - \frac{\xi^{n+2}}{n+2} \right]_0^1$$

$$= \frac{n}{n+2}. \tag{9.6-25}$$

For the large-vessel hematocrit, if we used Eqs. 9.6-22 and 9.6-24 in Eq. 9.6-2, we would find two complicated infinite series in quotient form, an almost hopeless situation as far as obtaining a simple solution is concerned. However, we can get around this by using Eq. 2.5-10 directly, first, in calculating the average velocity:

$$\langle v_z \rangle = 2 \int_0^1 v_z(\xi) \xi \, d\xi$$

$$= \frac{\Delta p R^2}{4 \mu_0 L} \cdot \frac{4}{k} \int_0^1 \left(\int_\xi^1 \frac{\bar{\xi} d\bar{\xi}}{K - \bar{\xi}^n} \right) \xi \, d\xi, \tag{9.6-26}$$

where the overbar on ξ is used to avoid confusing the two "generations" of dummy variables of integration.

The formality of obtaining Eq. 2.5-10 from Eq. 9.6-21 is

$$\phi = -\frac{2}{k} \int \frac{\xi d\xi}{K - \xi^n} + C, \qquad (9.6\text{-}27)$$

where C is the integration constant. Using the boundary condition that at $\xi = 1$, $\phi = 0$, we have

$$0 = -\frac{2}{k} \int \frac{\xi d\xi}{K - \xi^n}\bigg|_{\xi = 1} + C. \qquad (9.6\text{-}28)$$

Realizing that the integral on the right-hand side of Eq. 9.6-27 is actually for $\xi = \xi$, subtracting Eq. 9.6-28 from Eq. 9.6-27 gives us

$$\phi = \frac{2}{k} \int_{\xi}^{1} \frac{\xi d\xi}{K - \xi^n} \qquad (9.6\text{-}29)$$

or

$$v_z = \frac{\Delta p R^2}{4\mu_0 L} \cdot \frac{2}{k} \int_{\xi}^{1} \frac{\xi d\xi}{K - \xi^n}. \qquad (2.5\text{-}10)$$

Equation 9.6-26, just like Eq. 2.5-10, cannot readily be integrated analytically. However, by interchanging the order of the two integrations, we can at least "peel off" one of them. Thus, it becomes*

$$\langle v_z \rangle = \frac{\Delta p R^2}{4\mu_0 L} \cdot \frac{4}{k} \int_{0}^{1} \left(\int_{0}^{\bar{\xi}} \xi d\xi \right) \frac{\bar{\xi} d\bar{\xi}}{K - \bar{\xi}^n}$$

$$= \frac{\Delta p R^2}{4\mu_0 L} \cdot \frac{2}{k} \int_{0}^{1} \frac{\bar{\xi}^3 d\bar{\xi}}{K - \bar{\xi}^n}.$$

At this point the overbar can be dropped without causing confusion since ξ

*With re-expression of the integration limits

is now the only variable of integration; thus,*

$$\langle v_z \rangle = \frac{\Delta p R^2}{4\mu_0 L} \cdot \frac{2}{k} \int_0^1 \frac{\xi^3 d\xi}{K - \xi^n} . \tag{9.6-30}$$

Although this expression cannot be integrated further, later on we will be able to use it the way it is, especially if we first rewrite it as

$$\int_0^1 \frac{\xi^3 d\xi}{K - \xi^n} = \frac{k\langle \phi \rangle}{2} = \frac{8k}{\text{Po}} , \tag{9.6-31}$$

where $\text{Po} = \Delta p D^2 / \mu_0 L \langle v_z \rangle$, where $D = 2R$, the Poiseuille number† for the suspending fluid (plasma) without cells.

We now proceed to obtain the large-vessel hematocrit by using Eqs. 9.5-1, 9.6-2, and 2.5-10:

$$h_l = \frac{2h_m}{\langle v_z \rangle} \cdot \frac{\Delta p R^2}{4\mu_0 L} \cdot \frac{2}{k} \int_0^1 (1 - \xi^n) \left[\int_\xi^1 \frac{\bar{\xi} d\bar{\xi}}{K - \bar{\xi}^n} \right] \xi d\xi, \tag{9.6-32}$$

*Note here that if v_z were not our objective, we could bypass all of the steps between Eqs. 9.6-22 and 9.6-30 and obtain $\langle v_z \rangle$ directly from Eq. 9.6-21 through Q in Eq. 9.1-17 as

$$\langle v_z \rangle = \frac{Q}{\pi R^2} = - \int_0^1 \xi^2 \left(\frac{dv_z}{d\xi} \right) d\xi$$

$$= \frac{\Delta p R^2}{4\mu_0 L} \cdot \frac{2}{k} \int_0^1 \frac{\xi^3 d\xi}{K - \xi^n} .$$

†The Poiseuille number, having a value of 32 for H–P flow, is also the ratio of the Kármán number to the Reynolds number:

$$\text{Po} = \frac{\Delta p D^2}{\mu_0 L \langle v_z \rangle} = \frac{\Delta p D^3 \rho}{\mu_0^2 L} \cdot \frac{\mu_0}{D \langle v_z \rangle \rho} = \text{Ka}/\text{Re}.$$

Furthermore, since the Kármán number can be expanded as

$$\text{Ka} = \frac{\Delta p D^3 \rho}{\mu_0^2 L} = 2 \left(\frac{D \langle v_z \rangle \rho}{\mu_0} \right)^2 \left(\frac{1}{4} \cdot \frac{D}{L} \cdot \frac{\Delta p}{\frac{1}{2}\rho \langle v_z \rangle^2} \right) = 2\text{Re}^2 f,$$

where f is the Fanning friction factor, it follows that

$$\text{Po} = 2\text{Re} \cdot f.$$

where we again use the overbar to distinguish between the two generations of the dummy variable ξ. Again, we reverse the order of integration
so as to peel off one layer:

$$h_l = \frac{4h_m}{\langle\phi\rangle k} \int_0^1 \left[\int_0^{\bar\xi} (\xi - \xi^{n+1}) d\xi \right] \frac{\bar\xi d\bar\xi}{K - \bar\xi^n}$$

$$= \frac{4h_m}{\langle\phi\rangle k} \left\{ \frac{1}{2} \int_0^1 \frac{\bar\xi^3 d\bar\xi}{K - \bar\xi^n} - \frac{1}{n+2} \int_0^1 \frac{\bar\xi^{n+3} d\bar\xi}{K - \bar\xi^n} \right\}. \qquad (9.6\text{-}33)$$

Although neither of the two integrations here can be analytically
evaluated, we can see that the first one is the same as the expression in Eq.
9.6-31, and a little insight will show that the second one can be made a
function of the same after some manipulations, since

$$\int_0^1 \frac{\xi^{n+3} d\xi}{K - \xi^n} = -\int_0^1 \xi^3 d\xi + K \int_0^1 \frac{\xi^3 d\xi}{K - \xi^n}$$

$$= -\frac{1}{4} + K \frac{k\langle\phi\rangle}{2}. \qquad (9.6\text{-}34)$$

Substituting Eqs. 9.6-31 and 9.6-34 into Eq. 9.6-33 yields

$$h_l = \frac{4h_m}{\langle\phi\rangle k} \left[\frac{1}{2} \cdot \frac{k\langle\phi\rangle}{2} - \frac{1}{n+2} \left(-\frac{1}{4} + K \frac{k\langle\phi\rangle}{2} \right) \right]$$

or

$$\frac{h_l}{h_m} = 1 + \frac{1}{(n+2)k\langle\phi\rangle} - \frac{2K}{n+2}. \qquad (9.6\text{-}35)$$

Using Eqs. 9.6-3, 9.6-25, and 9.6-35, the hematocrit reduction can finally
be found:

$$h_R = \frac{2 + (1/k\langle\phi\rangle) - 2K}{n + 2 + (1/k\langle\phi\rangle) - 2K},$$

which, in view of Eq. 9.6-31 and that $K=(k+1)/k$, can be transformed into

$$h_R = \frac{h_l - h_c}{h_l} = \frac{Po-32}{Po-32+16nk}. \qquad (9.6\text{-}36)$$

Using the data taken by Seshadri and Sutera [35] and an estimated value of $Ka = 10{,}000$, Lih [23] has shown that the shape of the cell distribution as represented by the parameter n is indeed a function of the flow parameters in the form of the Reynolds number; that is,

$$n = \frac{C_1}{Re^p}, \qquad (9.6\text{-}37)$$

while k is a function of the particle (cell) diameter d, where $p = 1.15$ and $C_1 k(d)$ has the following combined values:

d	$C_1 k(d)$
920μ	4602
545μ	4830
300μ	6421
127μ	18070

As k is a measure of the magnitude of the suspension (blood) viscosity above that of the pure fluid (plasma), its increase with increasing fineness (decreasing d) of the particles (cells) is entirely reasonable in view of the increasing solid–fluid interfacial area per unit mass that causes the additional friction. However, on increasing the particle diameter, the k value does *not* fall indefinitely, because as it is increased to a certain extent—namely, of the same order of magnitude as the tube size—the assumption of the suspension as a continuum, which forms the basis of our analysis, no longer holds.

Exercise. What would happen to Eqs. 9.6-30, 9.6-32, and 9.6-36 for H–P flow (i.e., when there were no particles present)?

PROBLEMS

9.A   Alternate Procedure for Obtaining Volumetric Flow Rate of a Casson Fluid

Instead of using Eq. 9.1-20, substitute Eqs. 9.1-11 and 9.1-12 into Eq. 9.1-15 to obtain Eq. 9.1-22.

9B.   Wall Slippage of a Casson Fluid [36]

The slippage phenomenon of a Newtonian fluid has been described mathematically in Problem 6A. However, some biological fluids exhibiting this phenomenon (e.g., blood) are inherently non-Newtonian.

(a)   Show that the modified version of the general expression Eq. 9.1-17 for this case is

$$Q = \frac{\pi R^3}{\tau_R^3} \int_0^{\tau_R} \tau_{rz}^2 \dot\gamma(\tau_{rz})\, d\tau_{rz} + \pi R^2 v_R, \qquad (9.B-1)$$

where

$$\dot\gamma(\tau_{rz}) = -\frac{dv_z}{dr} \qquad \text{(a function of } \tau_{rz}\text{)}$$

$$\tau_R = \frac{\Delta p R}{2L} \qquad \text{(shear stress at the wall)}$$

$$v_R = v_R(\tau_R) \qquad \text{(velocity of the fluid at the wall, } r = R, \text{ function of } \tau_R\text{).}$$

(b)   Show that Eq. 9.B-1 gives Eq. 6.A-2 for a Newtonian fluid.
(c)   Show that for the Casson fluid (Eqs. 9.1-1 and 9.1-3), the volumetric flow rate is

$$Q = \frac{\pi R^4 \Delta P}{8\eta L} F(\psi) + \pi R^2 v_R \qquad (9.B-2)$$

where

$$F(\psi) = 1 - \tfrac{16}{7}\psi^{1/2} + \tfrac{4}{3}\psi - \tfrac{1}{21}\psi^4$$

$$\psi = \frac{\tau_y}{\tau_R},$$

so that the apparent *fluidity*, $1/\eta_a$, is $1/\eta_a = (1/\eta)F(\psi) + [4\zeta(\tau_R)]/R$, where $\zeta(\tau_R) = v_R(\tau_R)/\tau_R$.

9.C Expression for Obtaining Rheological Information from Flow-vs-Pressure Data.

Verify that Eq. 9.1-22 is consistent with the general expression Eq. 9.1-24.

9.D The Interaction Model of Blood Suspensions [37]

Assuming that blood viscosity obeys the interaction model [12] (Eq. 9.2-5 or 9.2-6) and that erythrocyte distribution obeys the power-law model [23] (Eq. 9.5-1), show that the velocity distribution of its laminar flow in a circular tube is

$$v_z = \frac{\Delta P R^2}{4\mu_0 L}\left[1 - \frac{nk}{n+2} - (1-k)\xi^2 - \frac{2k}{n+2}\xi^{n+2}\right], \qquad (9.D\text{-}1)$$

where $\xi = r/R$ and $k = Bh_m$. Plot the profiles for various values of h_m and discuss the result. Then obtain the expressions for large-vessel and capillary hematocrits and for hematocrit reduction.

REFERENCES

1. Copley, A. L., and Stainsby, G., Eds. *Flow Properties of Blood.* Pergamon: New York, 1960.
2. Whitmore, R. L. *Rheology of the Circulation.* Pergamon: New York, 1968.
3. Goldsmith, H. L., and Mason, S. G. In F. R. Eirich, Ed., *Rheology*, Vol. IV. Academic: New York, 1967.
4. Brenner, H. In T. B. Drew and J. W. Hoopes, Jr., Eds., *Advances in Chemical Engineering*, Vol. VI. Academic: New York, 1966, pp. 287–438.
5. Casson, N. In C. C. Mill, Ed., *Rheology of Disperse Systems.* Pergamon: New York, 1959, p. 84.
6. Merrill, E. W., and Pelletier, G. A. *J. Appl. Physiol.*, **23** (2), 178 (1967).
7. Merrill, E. W.; Benis, A. M.; Gilliland, E. R.; Sherwood, T. K.; and Salzman, E. W. *J. Appl. Physiol.*, **20** (5), 954 (1965).
8. Einstein, A. *Ann. Physik.*, **19**, 289 (1906); with a correction in *Ann. Physik.*, **34**, 591 (1911). English trans. by Marion Liang Lih is available through the present author.
9. Trevan, J. W. *Biochem. J.*, **12**, 60 (1918).
10. Hatschek, E. *Kolloid. Z.*, **8**, 34 (1911); **11**, 280 (1912).
11. Vand, V., *J. Phys. Coll. Chem.*, **52**, 277 (1948).
12. Oliver, D. R., and Ward, S. G., Nature, :**4348**, 396 (Feb. 28, 1953).

13. Cokelet, G. R., Merrill, E. W., Gilliland, E. R., Shin, H., Britten, A. and Wells, R. E., Jr., *Trans. Soc. Rheol.*, 7, 303 (1963)

14. Fåhraeus, R., *Physiol. Rev.*, 9, 241 (1929)

15. Fåhraeus, R. and Lindqvist, T., *Am. J. Physiol.*, 96, 562 (1931)

16. Maude, A. D. and Whitmore, R. L., *Brit. J. Appl. Phys.*, 7, 98 (1956)

17. Maude, A. D. and Whitmore, R. L., *J. Appl. Physiol.*, 12 (1) 105 (1958)

18. Dix, F. J., and Scott Blair, G. W. *J. Appl. Phys.*, 11, 574 (1940).

19. *CRC Handbook of Tables for Mathematics*, 4th ed., Chemical Rubber Co.: Cleveland, 1970, p. 37.

20. Haynes, R. H., and Burton, A. C. *Am. J. Physiol.*, 197 (5), 943 (1959).

21. Haynes, R. H., *Am. J. Physiol.*, 198 (6), 1193 (1960)

22. Gordon, W., *Biorheol.*, 7, 125 (1970)

23. Lih, M. M. *Bull. Math. Biophys.*, 31 (1), 143 (1969).

24. Segré, G., and Silberberg, A. *Nature*, 189, 209 (1961).

25. Segré, G., and Silberberg, A. *J. Fluid Mech.*, 14, 115, 136 (1962).

26. Goldsmith, H. L. In H. Harder, Ed., *4th Eur. Conf. Microcir.* Karger: Basel–New York, 1966.

27. Anczurowski, E., and Mason, S. G. *J. Coll. Interf. Sci.*, 23 (4), 522, 533, 547 (1967).

28. Goldsmith, H. L., and Mason, S. G. In F. R. Eirich, *Rheology: Theory and Applications*, Vol. IV. Academic: New York, 1967.

29. Karnis, A., and Mason, S. G. *J. Coll. Interf. Sci.*, 23, 120 (1967).

30. Takano, M.; Goldsmith, H. L.; and Mason, S. G. *J. Coll. Interf. Sci.*, 23, 248 (1967).

31. Deakin, M. A. B. *Bull. Math. Biophys.*, 29, 549, 565, 649 (1967).

32. Deakin, M. A. B. *Bull. Math. Biophys.*, 30, 27 (1968).

33. Jeffery, G. B. *Proc. Roy. Soc. (London)*, 102A, 161 (1922).

34. Thomas, H. W. *Biorheol.*, 1, 41 (1962).

35. Seshadri, V., and Sutera, S. P. *J. Coll. Interf. Sci.*, 27 (1), 101 (1968).

36. Oka, S. In A. L. Copley, Ed., *Hemorheology*, Pergamon: New York, 1968, p. 55.

37. Lih, M. M. *Proc. 8th Int. Conf. Med. Biol. Eng.*, Paper 11-7, 1969.

38. Tesfagaber, A.; and Lih, M. M. *Bull. Math. Biol.*, 35, 577–589 (1973).

CHAPTER TEN

Oxygen Transport

Oxygen transport is one of the most vital processes in the living body. It is also one for which transport principles and equations discussed in previous chapters find ample applications. In this concluding chapter of the book, we shall demonstrate some of these applications.

Before we present the details of these examples, we shall first take an overall look at the entire process so as to identify the location of each event to be discussed later and to give some perspectives on how one step or example will be related to another. This will be followed by a brief description of the kinetics of the chemical combination* between oxygen and hemoglobin in the red cell.

In systemic capillaries, where internal respiration occurs, oxygen is transported from the red cells, through plasma, across the capillary wall, and finally into the tissue. In Section 10.3 we will show some examples of the diffusion in the tissue phase, in both radial and axial directions, under various kinetic conditions.

In pulmonary capillaries, where "external respiration" occurs, the red cells obtain oxygen from the alveolar sac, again through plasma, across walls, and membranes. We will show an example of the oxygen diffusion in liquid-filled alveolar space in Section 10.4.

In both of the above, the simultaneous diffusion and chemical reaction of oxygen with hemoglobin are of prime importance. This we shall discuss in Section 10.5, particularly the classical works of Roughton and his co-workers. Of special interest is the explanation of the so-called *facilitated transport* based on fundamental principles.

As in previous chapters, we will arrange the examples in each of the categories in Sections 10.3–10.5 according to the transport type involved rather than by their physiological significance. For a physiologically more coherent treatment of this subject including membrane transport and thermodynamics, two excellent, recently published texts are highly recommended [1, 15]. The present chapter is not meant to be a complete and organized discussion of the subject. Rather, it shows how previously derived equations have been applied in analyzing several aspects of the oxygen transport process.

---

*Note that we use the word *combination* instead of *reaction* because this is different from a chemical reaction in the ordinary sense as will be discussed later.

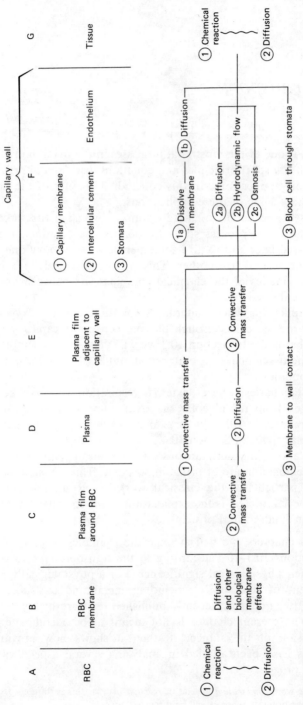

Figure 10.1-1. Breakdown of oxygen-transport steps in the systemic capillary.

10.1 TRANSPORT STEPS

Figure 10.1-1 is a schematic diagram showing all conceivable constituent steps of oxygen transport in the systemic capillary. They are properly placed either in series or in parallel with one another. The numbers in this diagram correspond to those in Figure 10.1-2, in which a section of the capillary is enlarged and partially "opened" to show where each step occurs.

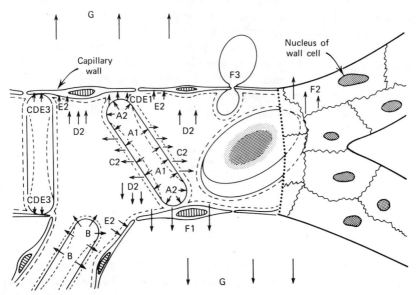

Figure 10.1-2.   Graphical representation of oxygen-transport steps in the systemic capillary.

With the exception of what happens outside the capillary, and perhaps the slightly different structure of the capillary walls, oxygen transport in the pulmonary capillary is similar but in the reverse direction, of course. In the ultimate combination of these steps into the overall process, a well-known transport principle governs. That is, the resistances of steps in series and the conductances (or reciprocals of resistances) of steps in parallel are additive. This also means that when we talk about the rate-limiting step in a series of steps, we refer to the slowest one, while that in parallel is the fastest step.

10.2 OXYGEN–HEMOGLOBIN EQUILIBRIUM AND KINETICS

It is a well-known fact that the physical solubility of oxygen in plasma is extremely small, comparable to that in water. Were it not for the existence of the red corpuscles as an oxygen carrier, no creature could ever live. With the erythrocytes the oxygen-carrying capacity of blood is increased 30–40 times, to approximately 20 ml $O_2/100$ ml blood.

The compound chiefly responsible for the affinity to oxygen is hemoglobin, denoted by Hb. It does not, however, chemically react with oxygen in the normal sense. Instead, the O_2 molecule is attached to, or associated with, the hemoglobin, forming oxyhemoglobin (oxygenated hemoglobin, denoted by HbO_2, but *not* an oxide). The fraction of oxygenated hemoglobin in the total hemoglobin is the fractional saturation (s) of hemoglobin. It is related to the partial pressure of oxygen (or oxygen tension), p_{O_2}, with which it is in equilibrium as shown in Figure 10.2-1. Experimentally determined by Dill [2], this is the well-known oxygen dissociation (or saturation) curve.

As early as 1910, Hill [3] proposed the use of the following empirical relationship*

$$s = \frac{[HbO_2]}{[HbO_2] + [Hb]}$$

$$= \frac{K(p_{O_2})^n}{1 + K(p_{O_2})^n}, \tag{10.2-1}$$

where K is a constant (but depending on the pH and temperature of the hemoglobin solution) and n is an empirical constant between 2.5 and 2.6.

However, each hemoglobin molecule actually contains four heme groups each of which can attract an oxygen molecule. Noting this, Adair [4] proposed the celebrated *intermediate compound hypothesis*, in which the four O_2 molecules are considered to be attached to the four heme groups in a stepwise fashion:

$$Hb_4 + O_2 \underset{k_1}{\overset{k_1'}{\rightleftharpoons}} Hb_4O_2 \tag{10.2-2a}$$

$$Hb_4O_2 + O_2 \underset{k_2}{\overset{k_2'}{\rightleftharpoons}} Hb_4O_4 \tag{10.2-2b}$$

---

*The sigmoidal type of oxygen dissociation curve had already been known at Hill's time except that the data obtained later by Dill are more reliable, more universally accepted, and, thus, used here.

$$Hb_4O_4 + O_2 \underset{k_3}{\overset{k_3'}{\rightleftharpoons}} Hb_4O_6 \qquad (10.2\text{-}2c)$$

$$Hb_4O_6 + O_2 \underset{k_4}{\overset{k_4'}{\rightleftharpoons}} Hb_4O_8, \qquad (10.2\text{-}2d)$$

where k_i and k_i' are the dissociation and association rate constants,

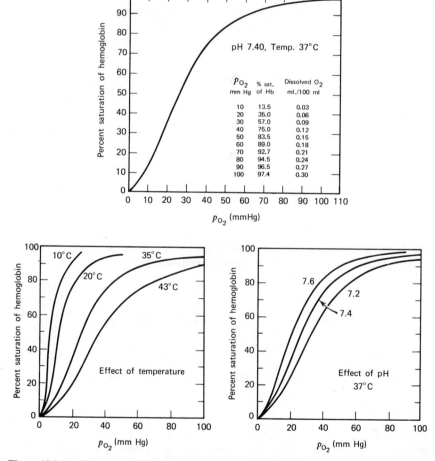

Figure 10.2-1. The oxygen dissociation curve [2]. It shifts with p_{CO_2} value as well.

respectively,* and $K_i = k'_i/k_i$ is the equilibrium constant. Based on this mechanism, the following expression can be derived

$$
s = \frac{[Hb_4O_8]}{[Hb_4O_8]+[Hb]}
$$

$$
= \frac{K_1 p_{O_2} + 2K_1 K_2 p_{O_2}^2 + 3K_1 K_2 K_3 p_{O_2}^3 + 4K_1 K_2 K_3 K_4 p_{O_2}^4}{4\left(1 + K_1 p_{O_2} + K_1 K_2 p_{O_2}^2 + K_1 K_2 K_3 p_{O_2}^3 + K_1 K_2 K_3 K_4 p_{O_2}^4\right)}. \quad (10.2\text{-}3)
$$

Using numerical procedures, DeHaven and DeLand [5] have found that data at pH 7.4 and 37°C are best fitted by the following set of dissociation constants:

$$
K_1 = \frac{k'_1}{k_1} = 0.04
$$

$$
K_2 = \frac{k'_2}{k_2} = 0.034
$$

$$
K_3 = \frac{k'_3}{k_3} = 0.003
$$

$$
K_4 = \frac{k'_4}{k_4} = 0.8.
$$

However, Forbes and Roughton [6], using the assumption that the rates of the first three association steps are proportional to the number of unattached iron atoms in the heme groups, predicted that

$$
k'_1 : k'_2 : k'_3 = 4 : 3 : 2. \quad (10.2\text{-}4)
$$

If the same mass action law applies to the dissociation steps, we have

$$
k_1 : k_2 : k_3 = 1 : 2 : 3, \quad (10.2\text{-}5)
$$

so that the ratios between the three equilibrium constants are

$$
K_1 : K_2 : K_3 = 4 : \tfrac{3}{2} : \tfrac{2}{3}, \quad (10.2\text{-}6)
$$

*Normally, k_i pertains to a forward reaction, while k'_i refers to the reverse; however, here we are looking from the dissociation point of view. Also, this model can be adapted to other gases, such as CO, by simply replacing the O_2.

which is quite different from that from DeHaven and DeLand's values. In fact, the latter authors reported that another set of dissociation constants that also approximate this ratio is $K_1 = 0.04$, $K_2 = 0.016$, $K_3 = 0.0097$, and $K_4 = 0.375$ [5].

Although it would seem that, following the same trend as in Eq. 10.2-6, K_4 ought to be $1/4$, the actually obtained value exceeds this proportion. In fact, it is larger even than K_1. This can be explained using the assumption of Pauling [7] that the association with the fourth heme group is greatly enhanced by the occupation of the previous three groups. In any case, the empirical equation (Eq. 10.2-1) is used more frequently because of its mathematical simplicity.

As far as the kinetics (reaction rate) is concerned, experimentation with the intermediate steps are not practical and a rate equation for the simplified overall model

$$ Hb + O_2 \underset{k}{\overset{k'}{\rightleftharpoons}} HbO_2 \qquad (10.2\text{-}7) $$

can be written as

$$ \frac{d}{dt}[HbO_2] = k'[Hb][O_2] - k[HbO_2]. \qquad (10.2\text{-}8) $$

From initial-reaction rate* data, Gibson et al. [8] found an average value of $k' = 3.5 \times 10^9$ ml/mole-sec. It must be stressed that this k' value is for the simplified overall model only and may vary with the prevailing percent of saturation.

10.3 OXYGEN TRANSPORT BETWEEN CAPILLARY BLOOD AND TISSUE

In Example 6.3-1 we presented the simplest and most basic case of oxygen transport in the tissue around capillaries [9] upon which almost all later works have been built. Although the analysis was not as sophisticated as present-day work, Krogh's cylinder itself has proven to be a useful model on which more accurate analyses have been performed. One of the earliest was that of Bloch [10], who applied the mass-transfer coefficient (or permeability) concept at the capillary wall to link the oxygen tension on the tissue side of the capillary boundary to that in the blood stream. (Krogh's original analysis, as presented in Example 6.3-1, assumes that the oxygen tension on the tissue side of the capillary wall is known, which is not very realistic.) He also introduced the more realistic oxygen consump-

---

*That is, a reaction with only a very small conversion so that the product concentration is low and the reverse reaction can be neglected.

tion model, Eq. 10.2-1. Unfortunately, he did not proceed to solve the simultaneous partial differential equations resulted probably because of mathematical complexity and the lack of computational technology at the time.

Blum [11] made several assumptions and was able to solve the capillary and tissue equations together for three slightly simplified cases, two of which we will present as Examples 10.3-1 and 10.3-2. Others have also made further advances. However, the work of Reneau et al. [12] on oxygen diffusion in the gray matter of the brain, while extremely thorough and enlightening, involves a great deal of numerical computation that would be difficult to report here. The convective oxygen transport, particularly that in the plasmatic gap between two red cells in the bolus flow, is also very interesting but requires the use of the so-called *random-walk* mathematical technique[13], which we are not prepared to deal with here.

Before presenting Blum's work, let us first write down the pertinent steady-state equations of continuity and the accompanying boundary conditions. Referring to Figure 6.3-1, for the tissue region, we have, from Eq. E of Table 5.5-1,*

$$\frac{\mathcal{D}_r}{r} \frac{\partial}{\partial r}\left(r \frac{\partial c_T}{\partial r}\right) + \mathcal{D}_z \frac{\partial^2 c_T}{\partial z^2} = \frac{A c_T}{B + c_T}, \tag{10.3-1}$$

where c_T is the oxygen concentration in the tissue and we distinguish between the oxygen diffusivities in the radial ($\mathcal{D}_r$) and axial ($\mathcal{D}_z$) directions. We have also used the Michaelis–Menten kinetics for the metabolic oxygen consumption,* with A and B being the maximum rate and the Michaelis–Menten constant, respectively. Within the capillary, we assume that its size is so small and the flow condition is such that the radial concentration and velocity gradients can be neglected and that we can instead use their area averaged values $\langle c \rangle$ and $\langle v_z \rangle$, respectively. Using a differential macroscopic mass balance over a length dz, similar to Eq. 8.8-25, we have

$$-\pi R_1^2 \langle v_z \rangle d\langle c \rangle = h \cdot 2\pi R_1 \, dz \left(\langle c \rangle - c_{T_1}\right), \tag{10.3-2}$$

where R_1 is the radius of the capillary and c_{T_1} is the oxygen concentration at the capillary wall on the tissue side.[†] The negative sign is there because $\langle c \rangle$ decreases with increasing z. Equation 10.3-2 can be rearranged to

$$\frac{d\langle c \rangle}{dz} = -\frac{2h}{R_1 \langle v_z \rangle}\left(\langle c \rangle - c_{T_1}\right). \tag{10.3-3}$$

*$R_A = -A c_T/(B + c_T)$ for consumption.

[†] Here we have "borrowed" the symbol "h" for mass transfer coefficient to avoid confusion with the reaction rate constants k_0 and k_1 later.

The boundary conditions are

B.C. 1: at $r = R_1$, $-\mathcal{D}_r \left(\dfrac{\partial c_T}{\partial r} \right)_{r = R_1} = h(\langle c \rangle - c_{T_1})$

B.C. 2: at $r = R_2$, $\dfrac{\partial c_T}{\partial r} = 0$

B.C. 3: at $z = 0$, $\dfrac{\partial c_T}{\partial z} = 0$, $c_T = \langle c \rangle = c_0$

B.C. 4: at $z = L$, $\dfrac{\partial c_T}{\partial z} = 0$,

where R_2 is the outer radius of the tissue cylinder and c_0 is the arterial oxygen concentration.

Example 10.3-1   Hypoxia, Longitudinal Diffusion in Tissue Negligible. When the oxygen concentration in the tissue is low, $c_T \ll B$ and the reaction term becomes $k_1 c_T$ (i.e., approaching first order), where $k_1 = A/B$. This, plus the condition that $\mathcal{D}_z = 0$, simplifies Eq. 10.3-1 to

$$\frac{1}{r} \frac{d}{dr} \left(r \frac{dc_T}{dr} \right) = \gamma^2 c_T, \tag{10.3-4}$$

where $\gamma = \sqrt{k_1 / \mathcal{D}_r}$. From Section 2.9 the solution of this equation should include the modified Bessel functions. While leaving the details to Appendix G, we write below the solution of Eq. (10.3-4) and that of its companion, Eq. 10.3-3,

$$\langle c \rangle = c_0 \exp \left(-\frac{2}{1 + \beta} \cdot \frac{h}{\langle v_z \rangle} \cdot \frac{z}{R_1} \right) \tag{10.3-5}$$

$$c_T = \langle c \rangle \frac{\beta}{1 + \beta} \cdot \frac{K_1(\gamma R_2) I_0(\gamma r) + I_1(\gamma R_2) K_0(\gamma r)}{K_1(\gamma R_2) I_0(\gamma R_1) + I_1(\gamma R_2) K_0(\gamma R_1)}, \tag{10.3-6}$$

where

$$\beta = \frac{h}{\gamma B \mathcal{D}_r} \tag{10.3-7}$$

and

$$B = \frac{-K_1(\gamma R_2) I_1(\gamma R_1) + I_1(\gamma R_2) K_1(\gamma R_1)}{K_1(\gamma R_2) I_0(\gamma R_1) + I_1(\gamma R_2) K_0(\gamma R_1)}. \tag{10.3-8}$$

Equation 10.3-6 indicates that the tissue oxygen concentration is proportional to that available from the capillary $\langle c \rangle$ and its radial variation is expressed by the Bessel-function group. Through Eq. 10.3-5 the tissue oxygen concentration is also very sensitive to the flow rate. A small drop in $\langle v_z \rangle$ would result in an exponential decrease in the oxygen concentration. Thus, under hypoxia condition the oxygen concentration is flow-limited. Furthermore, from Eq. 10.3-5, it is also an exponentially decaying function of the axial distance z. This is because when the tissue is starved for oxygen, the section close to the entrance will take up a large portion of the supply, leaving relatively little for the tissue surrounding the downstream section.

Example 10.3-2 Hyperbaric Oxygen Content with Longitudinal Diffusion in Tissue. The reason for using this example is that when the tissue is oxygen-rich, it exhibits a different behavior from that shown in the previous example. Also, there are available some numerical computations showing oxygen profiles, both radial and longitudinal, in the cerebral gray matter, as will be seen later. With a high oxygen content, the oxygen consumption rate becomes concentration-independent, since $c_T \gg B$ and A becomes the zero-order rate constant k_0. With a longitudinal diffusional coefficient $\mathcal{D}_z$, which could be different from $\mathcal{D}_r$, we have

$$\frac{\mathcal{D}_r}{r} \frac{\partial}{\partial r}\left(r \frac{\partial c_T}{\partial r} \right) + \mathcal{D}_z \frac{\partial^2 c_T}{\partial z^2} = k_0. \tag{10.3-9}$$

Equation 10.3-3 and the boundary conditions shown on p. 423 still apply. Using a separation-of-variable technique (see Appendix H for details), the solutions have been found to be

$$\langle c \rangle = c_0 - \frac{k_0}{\langle v_z \rangle}(n^2 - 1)\left\{ z - \frac{4\mathcal{D}_z}{R_1^2 \langle v_z \rangle} \sum_{i=1}^{\infty} \frac{W_i}{\alpha_i^2} \right.$$

$$\left. \cdot \left[(1 - \cosh\lambda_i L)\frac{\sinh\lambda_i z}{\sinh\lambda_i L} + \cosh\lambda_i z - 1 \right] \right\}, \tag{10.3-10}$$

$$c_T = c_0 - \frac{k_0 R_1^2}{4\mathcal{D}_r}(1 - \xi^2 + 2n^2 \ln \xi) - \frac{k_0}{\langle v_z \rangle}(n^2 - 1)$$

$$\cdot \left\{ z + \frac{R_1 \langle v_z \rangle}{2h} + \frac{2}{R_1} \sum_{i=1}^{\infty} \frac{W_i}{\alpha_i^2} \left[\frac{2\mathcal{D}_z}{\langle v_z \rangle R_1} + \frac{\sqrt{\mathcal{D}_z}}{\sqrt{\mathcal{D}_r}} \frac{Z_0(\alpha_i r)}{Z_1(\alpha_i R)} \right. \right.$$

$$\left. \left. \times \left(\frac{1 - \cosh \lambda_i L}{\sinh \lambda_i L} \cosh \lambda_i z + \sinh \lambda_i z \right) \right] \right\}, \qquad (10.3\text{-}11)$$

where

$$n = \frac{R_2}{R_1}, \qquad \xi = \frac{r}{R_1}, \qquad (10.3\text{-}12)$$

$$\lambda_i = \frac{\alpha_i}{\sqrt{\frac{\mathcal{D}_r}{\mathcal{D}_z}}} \qquad (10.3\text{-}13)$$

$$Z_k(\alpha_i r) = Y_0(\alpha_i R_1) J_k(\alpha_i r) - J_0(\alpha_i R_1) Y_k(\alpha_i r), \qquad (10.3\text{-}14)$$

$$W_i = \frac{[R_1 Z_1(\alpha_i R_1)]^2}{[R_2 Z_0(\alpha_i R_z)]^2 - [R_1 Z_1(\alpha_i R_1)]^2}, \qquad (10.3\text{-}15)$$

and α_i are the roots of $Z_1(\alpha_i R_2) = 0$; that is,

$$Y_0(\alpha_i R_1) J_1(\alpha_i R_2) - J_0(\alpha_i R_1) Y_1(\alpha_i R_2) = 0. \qquad (10.3\text{-}16)$$

The salient feature of this set of solutions can best be gleaned from a special case where longitudinal diffusion is again made negligible ($\mathcal{D}_z = 0$) and the capillary wall perfectly permeable ($h = \infty$). Then Eqs. 10.3-10 and 10.3-11 are simplified to

$$\langle c \rangle = c_0 - \frac{k_0}{\langle v_z \rangle}(n^2 - 1)z, \qquad (10.3\text{-}17)$$

$$c_T = c_0 - \frac{k_0 R_1^2}{4\mathcal{D}_r}(1 - \xi^2 + 2n^2 \ln \xi) - \frac{k_0}{\langle v_z \rangle}(n^2 - 1)z$$

$$= \langle c \rangle - \frac{k_0 R_1^2}{4\mathcal{D}_r}(1 - \xi^2 + 2n^2 \ln \xi). \qquad (10.3\text{-}18)$$

We can see that Eq. 10.3-18 is essentially the same as Eq. 6.3-3, with $\langle c \rangle$ corresponding to the p_{A_i} in the earlier case. This should come as no surprise, since the simplification here ($\mathcal{D}_z = 0$ and zero-order reaction) reduces the present case to that of Example 6.3-1. The only difference is that the $\langle c \rangle$ here is a function of z instead of being constant as p_{A_i} was in the previous case. Equation 10.3-17 also shows that $\langle c \rangle$, and hence c_T, again decrease with decreasing flow rate but not as drastically as the exponential decay function in Eq. 10.3-5 in the previous example. This is somewhat expected because in the oxygen-rich tissue there is more "reserve" so to speak, and it does not depend on fresh supply as heavily.

Equations 10.3-17 and 10.3-18 show that both $\langle c \rangle$ and c_T decrease linearly in the longitudinal direction, except that c_T is further lowered by the radially dependent function. This can be verified by plotting them in a three-dimensional coordinate system. Figure 10.3-1 is such a plot based on

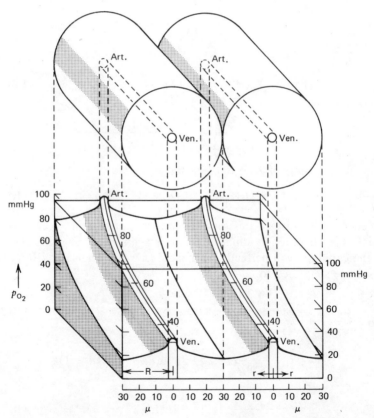

Figure 10.3-1. Radial and longitudinal oxygen tension in and around capillaries [14].

the work of Opitz and Schneider [14] on brain tissue. Here, the oxygen tension (partial pressure) is used, and because of the nonlinear behavior of the oxygen dissociation curve, the longitudinal decrease in oxygen tension decline is also nonlinear.

10.4 PULMONARY GAS EXCHANGE

One obvious difference between oxygen transports in pulmonary and systemic capillaries is that the former involves exchange between blood and alveolar space instead of between blood and tissue. In the alveolar space there is only diffusion but no metabolic reaction as in the tissue. Furthermore, the transport in the lung has to be more efficient on a per-unit-area basis, because sufficient oxygen has to be picked up in this comparatively small volume to be distributed to all capillaries of the entire body.

Depending on the particular application, the lung can be considered as a lumped-parameter unit in a multicompartment model of the respiratory system [16] or a distributed-parameter system with nonuniform profiles of oxygen and carbon dioxide. Even in parts of the lung there is an uneven distribution of these gases. In the only example we present in this section, we will examine the concentration profiles in one basic lung unit in the interesting experiment of liquid ventilation, to see how accurately such a physiological event can be described by physical principles of diffusion in a sphere.

Example 10.4-1 Liquid Ventilation in a Spherical Lobule [17]. Various mammals have been known to survive, without apparent adverse effects, breathing with such oxygen-rich liquids as oxygenated fluorocarbon emulsion and Ringer solution in their lungs. In one such experiment dogs were used and each of their primary lobules was considered a spherical exchange unit of radius R as shown in Figure 10.4-1. Ringer solution with known initial O_2 and CO_2 concentrations were introduced into the lungs by gravity at frequencies between 6 to 21 breaths per minute. Their subsequent concentration changes were monitored at various times and radial positions in the lobule.

To analyze the situation, we start with the equation of continuity, Eq. F from Table 5.5-1, without bulk flow or chemical reaction in the liquid:

$$\frac{\partial c_A}{\partial t} = \mathcal{D}_{AB} \frac{1}{r^2} \frac{\partial}{\partial r}\left(r^2 \frac{\partial c_A}{\partial r}\right), \qquad (10.4\text{-}1)$$

where subscript A represents either oxygen or carbon dioxide in the two cases discussed below. They are considered separately because of the different boundary conditions used for these two cases. $\mathcal{D}_{AB}$ is the diffusivity of A in the oxygenated liquid, B.

Case 1: Oxygen, $A = O_2$. In this case, the oxygen flux, F_{O_2}, from the liquid to surrounding capillaries at the boundary of the sphere is taken to be constant, so that we have

$$\text{I.C.:} \quad \text{at } t = 0, c_{O_2} = c_{O_2}^0$$

$$\text{B.C. 1: at } r = 0, \frac{\partial c_{O_2}}{\partial r} = 0 \quad \text{or} \quad c_{O_2} = \text{finite}$$

$$\text{B.C. 2: at } r = R, \; -\mathcal{D}_{O_2} \frac{\partial c_{O_2}}{\partial r} = F_{O_2} \quad \text{(constant)}.$$

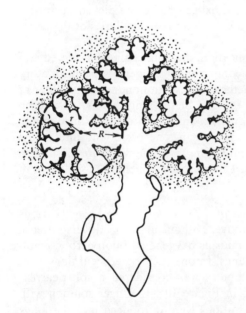

Figure 10.4-1. The primary lobule as a hypothetical gas-exchange unit (from [17]).

Defining the dimensionless quantities

$$\Gamma(\xi, \tau) = \frac{c_{O_2}^0 - c_{O_2}(r, t)}{F_{O_2} R / \mathcal{D}_{O_2}} \qquad \text{(dimensionless concentration)}$$

$$\xi = \frac{r}{R} \qquad \text{(dimensionless radial coordinate)}$$

$$\tau = \frac{\mathcal{D}_{O_2} t}{R^2}, \qquad \text{(dimensionless time)}$$

we have

$$\frac{\partial \Gamma}{\partial \tau} = \frac{1}{\xi^2} \frac{\partial}{\partial \xi} \left(\xi^2 \frac{\partial \Gamma}{\partial \xi} \right) \qquad (10.4\text{-}2)$$

I.C.:   at $\tau = 0$, $\Gamma = 0$

B.C. 1: at $\xi = 0$, $\dfrac{\partial \Gamma}{\partial \xi} = 0$ or $\Gamma = \text{finite}$

B.C. 2: at $\xi = 1$, $\dfrac{\partial \Gamma}{\partial \xi} = 1$.

Following a similar reasoning to that of Example 6.2-6, we can see that starting initially at a uniform concentration c_{O_2}, a nonuniform oxygen profile gradually develops until a long time[‡] later when the shape of the profile no longer changes but the level of the concentration keeps on decreasing linearly with time because of the combination of constant wall flux and lack of fresh oxygen supply.[‡] Symbolically, we can write

$$\Gamma(\xi,\tau) = \Gamma_\infty(\xi,\tau) - \Gamma_t(\xi,\tau) \qquad (10.4\text{-}3)$$

and

$$\Gamma_\infty(\xi,\tau) = a_0\tau + f(\xi), \qquad (10.4\text{-}4)$$

where a_0 is a constant, Γ_∞ is the expression for Γ at long times[*] and Γ_t is the transient solution in between. Using Eq. 10.4-4 in Eq. 10.4-2, we have

$$a_0 = \frac{1}{\xi^2} \frac{d}{d\xi} \left(\xi^2 \frac{df}{d\xi} \right), \qquad (10.4\text{-}5)$$

for which the solution is, with the application of B.C. 1 and B.C. 2,[†]

$$f(\xi) = \tfrac{1}{2}\xi^2 + C_2, \qquad (10.4\text{-}6)$$

so that

$$\Gamma_\infty = 3\tau + \tfrac{1}{2}\xi^2 + C_2 \qquad (10.4\text{-}7)$$

The second integration constant, C_2, has to be evaluated from the "newly created" boundary condition that the rate of decrease of total oxygen content in the lobule must be equal to the rate of oxygen diffusion at the

---

[‡] Here the time reference is one breathing cycle. Fresh oxygen is supplied with each intake of the liquid, but within each cycle there is no fresh supply; thus the oxygen level gradually diminishes.

[*] We cannot call this "steady state," because no matter how long we wait, it still changes with time, unless, of course, when O_2 is entirely depleted or the cycle terminates, whichever occurs first.

[†] Which now becomes $df/d\xi = 0$ and 1, respectively. Note that the use of B.C. 2 only yields the value of $a_0 (= 3)$, not the second integration constant C_2.

spherical walls, or

$$-\frac{d}{dt}\int_V c_{O_2}\,dV = F_{O_2}\cdot 4\pi R^2,\qquad(10.4\text{-}8)$$

which can be rewritten as*

$$\frac{d}{dt}\int_0^R\left[c_{O_2}^0 - c_{O_2}(r,t)\right]4\pi r^2 dr = F_{O_2}\cdot 4\pi R^2$$

and made dimensionless:

$$\frac{d}{d\tau}\int_0^1\Gamma(\xi,\tau)\xi^2 d\xi = 1.\qquad(10.4\text{-}9)$$

Integrating Eq. 10.4-9 between $\tau=0$ and $\tau=\tau_\infty$ (a long time) and since $\Gamma(\xi,0)=0$ according to B.C. 1 and $\Gamma(\xi,\tau_\infty)=\Gamma_\infty(\xi,\tau)$ by definition, we have

$$\int_0^1\Gamma_\infty(\xi,\tau)\xi^2 d\xi = \tau_\infty.\qquad(10.4\text{-}10)$$

Using Eq. 10.4-7 in Eq. 10.4-10 and performing the integration, we obtain

$$C_2 = -\frac{3}{10}$$

and hence

$$\Gamma_\infty(\xi,\tau) = 3\tau + \tfrac{1}{2}\xi^2 - \frac{3}{10}.\qquad(10.4\text{-}11)$$

Now, using Eq. 10.4-3 in Eq. 10.4-2 and the boundary and initial conditions, we have

$$\frac{\partial\Gamma_t}{\partial\tau} = \frac{1}{\xi^2}\frac{\partial}{\partial\xi}\left(\xi^2\frac{\partial\Gamma_t}{\partial\xi}\right)\qquad(10.4\text{-}12)$$

I.C.':   at $\tau=0$, $\Gamma_t=\Gamma_\infty=\tfrac{1}{2}\xi^2-\dfrac{3}{10}$

B.C. 1': at $\xi=0$, $\dfrac{\partial\Gamma_t}{\partial\xi}=0$ or $\Gamma_t=$ finite

B.C. 2': at $\xi=1$, $\dfrac{\partial\Gamma_t}{\partial\xi}=\dfrac{\partial\Gamma_\infty}{\partial\xi}-\dfrac{\partial\Gamma}{\partial\xi}=1-1=0.$

---

*This involves the use of Figure 2.7-3(c) and integration over θ and ϕ. Note that the addition of $c_{O_2}^0$ in the integrand does not affect the equality, since $c_{O_2}^0$ is a constant so that $dc_{O_2}^0/dt=0$.

Assuming a solution of the form

$$\Gamma_t(\xi,\tau) = G(\xi)\cdot H(\tau) \qquad (10.4\text{-}13)$$

and following the discussion between Eqs. 2.10-4 and 2.10-5 on spherical geometry, the general solutions of Eq. 10.4-12 after separation can be found to be

$$H = C_3 e^{-\beta^2 \tau} \qquad (10.4\text{-}14)$$

$$G = \frac{C_4}{\xi}\sin\beta\xi + \frac{C_5}{\xi}\cos\beta\xi, \qquad (10.4\text{-}15)$$

where $-\beta^2$ is the "separation constant" and C_3, C_4, and C_5 are the integration constants. Using B.C. 1' eliminates C_5, and using B.C. 2' specifies that β must be the roots of

$$\tan\beta = \beta, \qquad (10.4\text{-}16)$$

which means that β must assume values $\beta_k (k=1,2,\dots,\infty),$* as shown in Figure 10.4-2. Therefore, our combined general solution is

$$\Gamma_t = \sum_{k=1}^{\infty} \frac{C_k'}{\xi}(\sin\beta_k\xi)\exp(-\beta_k^2\tau). \qquad (10.4\text{-}17)$$

where the remaining constants C_3 and C_4 have been lumped into C_k'. Applying the initial condition to Eq. 10.4-17, multiplying both sides by $\xi\sin\beta_l\xi d\xi$, and integrating from $\xi=0$ to $\xi=1$ gives us, with the help of the orthogonality relationship Eq. D of Table 2.11-1 (with $N=1$) and Eqs. 11 and 19 of Table 2.4-1 and Eq. 10.4-16, we obtain

$$C_k' = \frac{2}{\beta_k^2\sin\beta_k},$$

so that the complete solution now is

$$\Gamma(\xi,\tau) = 3\tau + \tfrac{1}{2}\xi^2 - \frac{3}{10}$$

$$-\frac{2}{\xi}\sum_{k=1}^{\infty}\frac{\sin\beta_k\xi}{\beta_k^2\sin\beta_k}\exp(-\beta_k^2\tau), \qquad (10.4\text{-}18)$$

---

*It should actually be $k=0, \pm 1, \pm 2,\dots,\pm\infty$ but because of the symmetrical nature of β_k (see Fig. 10.4-2), $\beta_k = -\beta_k$. Thus the negative side contributes nothing new to the series which can then be "folded" at $k=0$ where $\beta_0=0$ (see also pp. 466–467).

where, it should be remembered, β_k is defined by Eq. 10.4-16. In dimensional form, we have

$$\frac{c_{O_2}^0 - c_{O_2}(r,t)}{F_{O_2}R/\mathcal{D}_{O_2}} = 3\frac{\mathcal{D}_{O_2}t}{R^2} + \frac{1}{2}\left(\frac{r}{R}\right)^2 - \frac{3}{10}$$

$$-\frac{2R}{r}\sum_{k=1}^{\infty}\frac{\sin(\beta_k r/R)}{\beta_k^2 \sin\beta_k}\exp\left(\frac{-\beta_k^2 \mathcal{D}_{O_2}t}{R^2}\right). \qquad (10.4\text{-}19)$$

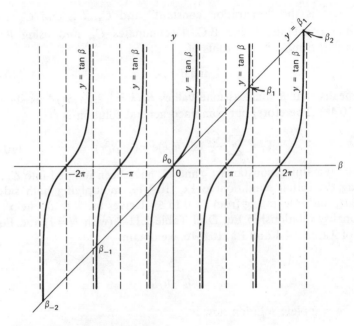

Figure 10.4-2. Roots of $\tan\beta_k = \beta_k$.

This solution is plotted in Figure 10.4-3, in which we have put the Γ scale upside down because $\Gamma = 0$ is the highest concentration and the concentration actually decreases as Γ value increases. Also note that by using dimensionless quantities, a seemingly complex function such as Eq. 10.4-19 can be plotted in a rather simple and compact manner, without specifying individual values of $\mathcal{D}_{O_2}$, F_{O_2}, R, and so on.

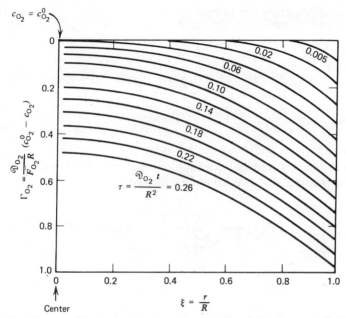

Figure 10.4-3. Theoretically calculated oxygen-concentration profiles in a spherical lobule. Note that depending on the relative values of $c_{O_2}^0$ and $F_{O_2}R/\mathcal{D}_{O_2}$, the Γ scale may end before or beyond 1.

Case 2: Carbon Dioxide, $A = CO_2$. For the absorption of CO_2 by the oxygenated Ringer solution in the spherical lobule, we consider that there is an ample supply of CO_2 at constant concentration $c_{CO_2}^R$ from the pulmonary capillaries. The initial CO_2 concentration in the Ringer solution is $c_{CO_2}^0$ (which could be zero if one does not wish to put it in the starting liquid). Equation 10.4-1 still applies, but the dimensionless concentration should be defined differently in order to use the method of separation of variables:

$$\Gamma(\xi, \tau) = \frac{c_{CO_2}^R - c_{CO_2}(r,t)}{c_{CO_2}^R - c_{CO_2}^0} \qquad \text{(unaccomplished concentration ratio).}$$

With other dimensionless quantities still the same, except that subscript O_2 should be changed to CO_2, we still have Eq. 10.4-2 but with the following initial and boundary conditions:

$$\text{I.C.:} \quad \text{at } \tau = 0, \Gamma = 1$$

$$\text{B.C. 1: at } \xi = 0, \frac{\partial \Gamma}{\partial \xi} = 0 \quad \text{or} \quad \Gamma = \text{finite}$$

$$\text{B.C. 2: at } \xi = 1, \Gamma = 0.$$

Assuming $\Gamma(\xi, \tau) = G(\xi) H(\tau)$ and following the "usual" procedure of applying the boundary and initial conditions and the orthogonal relationship, Eq. A in Table 2.11-1, we have the solution*

$$\Gamma(\xi, \tau) = \frac{2}{\pi \xi} \sum_{k=1}^{\infty} \frac{(-1)^{k+1}}{k} \sin k\pi\xi \cdot e^{-k^2\pi^2\tau} \qquad (10.4\text{-}20)$$

or

$$\frac{c_{CO_2}^R - c_{CO_2}(r,t)}{c_{CO_2}^R - c_{CO_2}^0} = \frac{2R}{\pi r} \sum_{k=1}^{\infty} \frac{(-1)^{k+1}}{k} \sin \frac{k\pi r}{R}$$

$$\cdot \exp\left(\frac{-k^2\pi^2 \mathcal{D}_{CO_2} t}{R^2}\right). \qquad (10.4\text{-}21)$$

The solution is plotted in Figure 10.4-4, where the temperature actually increases as the Γ scale goes down from 1 to 0 because it is an unaccomplished ratio. The corresponding accomplished ratio runs in the opposite direction on the right-hand scale.

The accuracy of this spherical model for both cases can be seen from Figure 10.4-5, where the experimentally measured oxygen and carbon dioxide tensions (partial pressures)† for a typical dog are found to lie very neatly on or close to the theoretically calculated curve. The diffusivities used at 37°C were $\mathcal{D}_{O_2} = 3.22 \times 10^{-5}$ cm^2/sec, $\mathcal{D}_{CO_2} = 2.55 \times 10^{-5}$ cm^2/sec (the same as those in water). Of the seven dogs tested, it was determined that the average one had approximately half a million exchange units each 420 μ in radius and the total gas transfer area for the entire lung was about 1 m^2.

---

*The reader may wish to derive this himself. After Case 1, anything is simpler.

†Related to the concentrations via the solubility coefficient α.

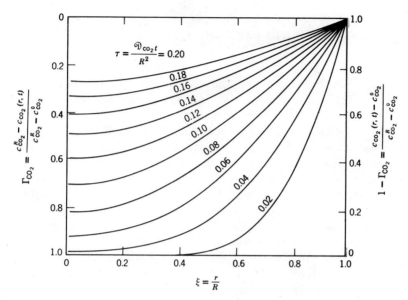

Figure 10.4-4. Theoretically calculated CO_2 concentration profiles in a spherical lobule. Note that unlike Figure 10.4-3, the Γ scale here is bounded by 0 and 1.

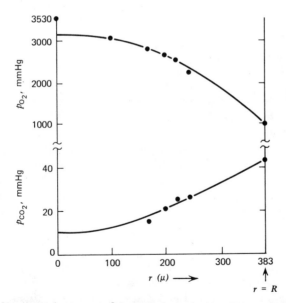

Figure 10.4-5. Agreement between experimental data (dots) and theoretically calculated oxygen and carbon dioxide tension profiles (curves) in a spherical lobule [17].

435

10.5 SIMULTANEOUS DIFFUSION AND CHEMICAL REACTION IN HE-MOGLOBIN SOLUTION AND THE RED CELL

Of the many investigators who have studied the chemical and physical processes involving hemoglobin and such physiologically related gases as O_2, CO, CO_2, and N_2, F. J. W. Roughton unquestionably holds the deanship. He has devoted his distinguished career, which spanned approximately half a century, almost entirely to the theoretical formulation, mathematical solution, and experimental verification of problems in this area. Some of his work on the equilibrium and kinetics of the O_2–Hb combination have been cited in Section 10.2. Much of the rest of his work and that of his co-workers and students have dealt with the particular problem of simultaneous diffusion-chemical reactions between these gases and hemoglobin, both in "loose" solution and in red cells suspended in plasma. This aspect of their work has been summarized in a review article [18] that, among other things, explains on physical grounds why oxygen uptake is slowed down by having the hemoglobin contained in red cells instead of being freely mixed in plasma as a solution. It also concludes that the red-cell membrane offers significant resistance to the uptake process. We cannot hope to cover this monumental career in the limited space here, but in Example 10.5-1 we will work out a simplified case to show the nature of this simultaneous physical (diffusion) and chemical process.

In Example 10.5-2 we shall see how Fick's simple law of diffusion, coupled with the oxygen-dissociation curve, can be used to formally explain the process of *facilitated transport*, which seemed to puzzle many researchers for some time. This type of transport is so named because it is "facilitated" by the diffusion of chemically combined O_2—oxyhemoglobin.

Again, we do not pretend that these two examples represent the entire problem area of simultaneous diffusion and chemical reaction involving O_2 and Hb. Rather, they are merely two simplified cases through which to demonstrate how the basic equations of change (continuity) and state (Fick's law) can be applied and solved in order to draw some conclusions that, although quantitatively not exact, are quite enlightening and qualitatively useful. Most of the recently developed numerical solutions are too lengthy to be reported here.

Example 10.5-1   Initial Rate of Simultaneous Diffusion and Chemical Reaction in the Erythrocyte. If we approximate the biconcave red corpuscle by a thin disk of which the combined area of the two circular surfaces is much larger than the rim area, such as shown in Figure 10.5-1, then we can view it as a finite slab of thickness $2b$ and containing hemoglobin at initial oxygen concentration c_i subject to diffusion from both sides with oxygen concentration at both surfaces known to be c_b.

Since the diffusing oxygen molecules also combine chemically with the hemoglobin in the slab, we can write, from Eq. D of Table 5.5-1,

$$\frac{\partial c_{O_2}}{\partial t} = \mathcal{D}_{O_2-Hb} \frac{\partial^2 c_{O_2}}{\partial x^2} - k' c_{O_2} c_{Hb} + k c_{HbO_2}, \qquad (10.5\text{-}1)$$

where HbO_2 denotes oxyhemoglobin and $\mathcal{D}_{O_2-Hb}$ the pure diffusivity of O_2 in hemoglobin (i.e., *not* an effective one including chemical reaction). Here we use the one-step oxygenation model for the sake of simplicity.

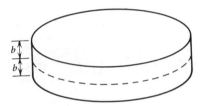

Figure 10.5-1. The thin-disk model of the red blood cell.

Suppose we are only interested in the solution for short contact times during which not enough oxyhemoglobin has built up to make the reverse reaction (dissociation) significant; then the last term in Eq. 10.5-1 can be dropped. Moreover, during this initial stage, only a small change in hemoglobin concentration takes place so that c_{Hb} can be considered virtually constant. Thus, Eq. 10.5-1 becomes

$$\frac{\partial c_{O_2}}{\partial t} = \mathcal{D}_{O_2-Hb} \frac{\partial^2 c_{O_2}}{\partial x^2} - k'' c_{O_2}, \qquad (10.5\text{-}2)$$

where $k'' = k' c_{Hb}$ with the following initial and boundary conditions:

I.C.: at $t = 0$, $c_{O_2} = c_i$

B.C.1: at $x = 0$, $\dfrac{\partial c_{O_2}}{\partial x} = 0$

B.C.2: at $x = \pm b$, $c_{O_2} = c_b$.

In so doing, we have neglected for the time being resistance offered by the red cell membrane.

Now three different cases can be examined:

Case 1: Diffusion Only, No Chemical Reaction $(k''=0)$. In this case the simplified partial differential equation

$$\frac{\partial c_{O_2}}{\partial t} = \mathcal{D}_{O_2-Hb} \frac{\partial^2 c_{O_2}}{\partial x^2} \tag{10.5-3}$$

can be solved by either the separation of variables or the Laplace transform method (see analogous Example 11.1-2 in Ref. 19 or the "slab" part of Eq. 6.2-65, Example 6.2-5), with the following result [20]:

$$\Gamma(\xi,\tau) = \frac{c_b - c_{O_2}(r,t)}{c_b - c_i}$$

$$= \frac{2}{\pi} \sum_{n=0}^{\infty} \frac{(-1)^n}{n+\frac{1}{2}} \exp\left[-(n+\tfrac{1}{2})^2 \pi^2 \tau\right] \cos\left(n+\tfrac{1}{2}\right)\pi\xi, \tag{10.5-4}$$

where

$$\xi = \frac{x}{b}$$

$$\tau = \frac{\mathcal{D}_{O_2-Hb} t}{b^2}$$

and the average oxygen concentration in the disk shape corpuscle is

$$\langle\Gamma\rangle(\tau) = \frac{c_b - \langle c_{O_2}\rangle(t)}{c_b - c_i}$$

$$= \frac{\int_{-1}^{+1}\Gamma(\xi,\tau)d\xi}{\int_{-1}^{+1}d\xi}$$

$$= \frac{2\int_{0}^{1}\Gamma(\xi,\tau)d\xi*}{2}$$

$$= \frac{2}{\pi^2}\sum_{n=0}^{\infty}\frac{1}{(n+\frac{1}{2})^2}\exp\left[-(n+\tfrac{1}{2})^2\pi^2\tau\right]. \tag{10.5-5}$$

*This is true because Γ is symmetrical with respect to ξ (see the cosine function). It may no longer be true if otherwise.

For $\tau > 0.2$,[†] neglecting all but the first term in the series in Eq. 10.5-5 introduces no more than 0.2% error. For $\tau < 0.2$, the following approximation

$$\langle \Gamma \rangle (\tau) = 1 - 2 \sqrt{\frac{\tau}{\pi}} \qquad (10.5\text{-}6)$$

or

$$\langle c_{O_2} \rangle (t) = c_i + \frac{2}{b}(c_b - c_i) \sqrt{\frac{\mathfrak{D}_{O_2 - Hb}\, t}{\pi}} \qquad (10.5\text{-}7)$$

gives rise to a maximum error of only 0.4%.

Case 2: Chemical Reaction Only, No Diffusional Resistance.[*]   Equation 10.5-2 becomes

$$\frac{dc_{O_2}}{dt} = -k'' c_{O_2} \qquad (10.5\text{-}8)$$

for which the solution is simply

$$c_{O_2} = c_i e^{-k'' t} \qquad (10.5\text{-}9)$$

or independent of position. Therefore, its average is also independent of the thickness, $2b$:

$$\langle c_{O_2} \rangle (t) = \int_{-b}^{+b} c_{O_2}(t)\, dx \Big/ \int_{-b}^{+b} dx$$

$$= c_i e^{-k'' t}. \qquad (10.5\text{-}10)$$

Case 3: Simultaneous Diffusion and Chemical Reaction.   For the combined effect, the full Eq. 10.5-2 has to be solved. This can be accomplished by taking the Laplace transform with respect to t and subsequently solving the resulting ordinary differential equation. After inverting the Laplace

†From the generally accepted values of $\mathfrak{D}_{O_2 - Hb} = 7.0 \times 10^{-6}$ cm^2/sec and $2b = 2.2$ μ, this would correspond to $t > 0.035$ sec.

*Note that we are not saying "no diffusion," in which case nothing would be supplied to continue the chemical reaction. Rather, this is a case in which the reactants are immediately available everywhere in the red cell, or infinite diffusion.

transform, the following expression can be obtained [20]:*

$$\Lambda(\xi, \tau) = \frac{c_{O_2}(r,t) - c_i}{c_b - c_i}$$

$$= \frac{\cosh \beta \xi}{\cosh \beta} - 2 \sum_{n=0}^{\infty} \frac{(-1)^n (n+\frac{1}{2})\pi}{\gamma_n^2} \cos(n+\tfrac{1}{2})\pi\xi \cdot e^{-\gamma_n^2 \tau}, \quad (10.5\text{-}11)$$

where

$$\beta = b\sqrt{k''/\mathcal{D}_{O_2-Hb}} \ , \text{ forward reaction parameter}$$

$$\gamma_n^2 = (n+\tfrac{1}{2})^2 \pi^2 + \beta^2$$

with the average concentration being

$$\langle \Lambda \rangle(\tau) = \frac{\langle c_{O_2} \rangle(t) - c_i}{c_b - c_i}$$

$$= \frac{\displaystyle\int_{-1}^{+1} \Lambda(\xi,\tau)\, d\xi}{\displaystyle\int_{-1}^{+1} d\xi}$$

$$= \frac{2\displaystyle\int_0^1 \Lambda(\xi,\tau)\, d\xi}{2}$$

$$= \frac{1}{\beta} \tanh \beta - 2 \sum_{n=0}^{\infty} \frac{1}{\gamma_n^2} e^{-\gamma_n^2 \tau}. \quad (10.5\text{-}12)$$

With the average concentration of available oxygen known, the average rate of chemical formation of oxyhemoglobin can be expressed as

$$\frac{d\langle c_{HbO_2} \rangle}{dt} = k''\langle c_{O_2} \rangle(t), \quad (10.5\text{-}13)$$

since only the forward reaction is important. The average oxyhemoglobin

---

*Note that here we use the symbol Λ for the accomplished concentration ratio as distinguished from the previously defined unaccomplished ratio Γ.

concentration at time t is, if $c_i = 0$ for a simple illustration,

$$\langle c_{HbO_2} \rangle = \int_0^t k'' \langle c_{O_2} \rangle (t) \, dt$$

$$= c_b \left[\beta\tau \tanh\beta - 2\beta^2 \sum_{n=0}^{\infty} \frac{1 - e^{-\gamma_n^2 \tau}}{\gamma_n^4} \right]. \qquad (10.5\text{-}14)$$

From the definition of γ_n it is quite obvious that regardless of the value of β (i.e., b, k'', and $\mathcal{D}_{O_2Hb}$), γ_n^2 increases quite rapidly as index n increases starting at zero. This makes all subsequent terms in the series of Eq. 10.5-14 negligible compared with the first term. But, since τ is usually quite small for initial stages, even this first term is near zero, which leaves only the first term in the bracket for consideration. From the definition

$$\tanh\beta = \frac{e^\beta - e^{-\beta}}{e^\beta + e^{-\beta}},$$

it can easily be shown that if $\beta > 3$, the maximum possible error by writing $\tanh\beta = 1$ is only 0.5%. Thus, we can confidently write

$$\langle c_{HbO_2} \rangle = c_b \beta\tau$$

or

$$t = \frac{b \langle c_{HbO_2} \rangle}{c_b \sqrt{k'' \mathcal{D}_{O_2-Hb}}}. \qquad (10.5\text{-}15)$$

Remember that for Case 1 (diffusion only), we can obtain from Eq. 10.5-7, similarly with $c_i = 0$,

$$t = \frac{\pi b^2 \langle c_{O_2} \rangle^2}{4 c_b^2 \mathcal{D}_{O_2-Hb}}, \qquad (10.5\text{-}16)$$

while for Case 2 (chemical reaction only), we can readily see from Eq. 10.5-10 that the time to reach a certain $\langle c_{O_2} \rangle$, and hence $\langle c_{HbO_2} \rangle$, will not be a function of the half-thickness b. Now for simultaneous diffusion-chemical reaction it is proportional to b. We can thus conclude that the combined action is some sort of geometric mean between the two individual actions.* This also lends credence to the argument that, although physically occurring simultaneously in a common environment, procedurally these two actions are steps in series rather than in parallel.[†] That is, diffusion feeds, and carries away the product of, the chemical reaction.

Actually, we have only scratched the surface of Roughton's work in this area. In Figure 10.5-2 we show an oxygen-saturation time diagram in which our complex Eq. 10.5-14 or the simpler version, Eq. 10.5-15, contributes to only a small section (*AB*) of one curve. For longer times Roughton first estimated the solution between a maximum and a minimum [20] and the problem was later solved numerically [21,22] for both cases with (curve 3) [23] and without (curve 2) red cell membrane. Compared with curve 1 for hemoglobin in free solution form, one can readily conclude that both the erythrocyte and its membrane offer significant resistances of comparable magnitude. Since our objective here is only to show some simple examples of using transport equations, the reader is referred to an excellent summary that appeared in one of this current Biomedical Engineering Series [1].

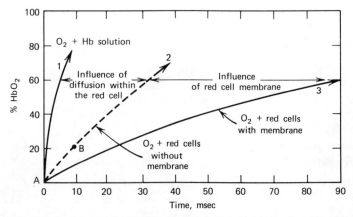

Figure 10.5-2. Oxygenation rates of hemoglobin solution and red cells with and without membrane [18].

---

*In fact this geometric-mean concept applies to dependencies on c_b, $\mathcal{D}_{O_2-Hb}$ and k'' as well.

†Because if in parallel, the two rates would complement each other with the result that the time for the combined action would be a mean value between the two. Rather, it would be shorter than both.

Example 10.5-2   Facilitated Diffusion. In the transport of gases through a flat layer of hemoglobin solution, if the species were chemically inert or if the rate of diffusion were not influenced by chemical reaction, then, according to Fick's law of diffusion, the rate or flux of transfer would only be dependent upon the partial-pressure difference of the species across the layer, not the partial pressure itself, as shown by the broken lines for nitrogen in Figure 10.5-3. However, this is not the case for oxygen. As shown by the solids lines, at sufficiently low overall oxygen tensions (partial pressures), the rate of oxygen transport suddenly rises even though the tension difference across this layer is kept constant. This phenomenon puzzled many investigators for some time but can be readily explained by suitably combining Fick's law with the saturation curve.

Referring to Figure 10.5-4, since the total oxygen flux through the hemoglobin layer comes from the diffusion of physically dissolved O_2 as well as that of chemically combined O_2 as oxyhemoglobin, HbO_2, we can write

$$J_{O_2} = - \mathcal{D}_{O_2} \frac{dc_{O_2}}{dx} - \mathcal{D}_{HbO_2} \frac{dc_{HbO_2}}{dx}. \qquad (10.5\text{-}17)$$

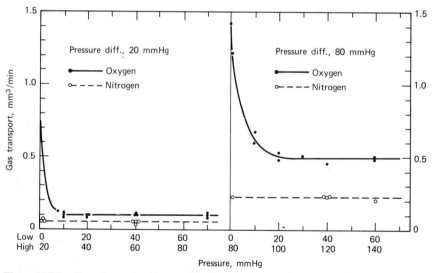

Figure 10.5-3. Experimental evidence of facilitated transport [24]. Note that throughout the tension change on the abscissa the same tension differences of 20 and 80 mmHg are maintained. Also the rate over the flat portion for an 80 mmHg tension difference is approximately four times that for a 20 mmHg tension difference. Note also that there is no facilitation on nitrogen transport.

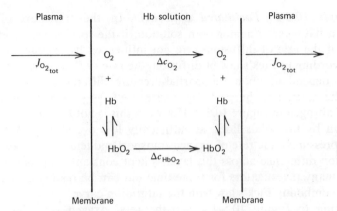

Figure 10.5-4. Schematic illustration for facilitated transport.

From the definition for fractional saturation

$$s = \frac{c_{HbO_2}}{c_{HbO_2} + c_{Hb}} \qquad (10.5\text{-}18)$$

and observing that the total hemoglobin and oxyhemoglobin concentration is constant (since these two molecules are of approximately equal size and molecular weight, diffusion of one in one direction is always compensated by that of the other in the opposite direction, thus maintaining the same combined concentration), we can write

$$\frac{dc_{HbO_2}}{dx} = (c_{HbO_2} + c_{Hb})\frac{ds}{dx}. \qquad (10.5\text{-}19)$$

Further observing that

$$\frac{ds}{dx} = \frac{ds}{dc_{O_2}} \cdot \frac{dc_{O_2}}{dx}, \qquad (10.5\text{-}20)$$

where

$$\frac{ds}{dc_{O_2}} = \frac{1}{\alpha} \cdot \frac{ds}{dp_{O_2}}, \qquad (10.5\text{-}21)$$

in which α is the solubility coefficient and ds/dp_{O_2} is the slope ot the oxygen dissociation curve (Figure 10.2-1), we can combine Eq. 10.5-17, and Eqs. 10.5-19 through 10.5-21 to form

$$J_{O_2} = - \mathcal{D}_{eff} \frac{dc_{O_2}}{dx}, \qquad (10.5\text{-}22)$$

where the effective diffusivity is

$$\mathcal{D}_{eff} = \mathcal{D}_{O_2} + \frac{\mathcal{D}_{HbO_2}}{\alpha}(c_{Hb} + c_{HbO_2})\frac{ds}{dp_{O_2}}. \qquad (10.5\text{-}23)$$

We can see from Figure 10.2-1 that at high oxygen tension the slope ds/dp_{O_2} is near zero, so that $\mathcal{D}_{eff} \doteq \mathcal{D}_{O_2}$, practically no facilitation. However, where p_{O_2} is sufficiently low, ds/dp_{O_2} becomes quite large and adds a significant magnitude to $\mathcal{D}_{O_2}$. Thus,

$$\frac{\mathcal{D}_{eff}}{\mathcal{D}_{O_2}} - 1 = \frac{\mathcal{D}_{HbO_2}}{\alpha \mathcal{D}_{O_2}}(c_{Hb} + c_{HbO_2})\frac{ds}{dp_{O_2}} \qquad (10.5\text{-}24)$$

is a measure of the degree of facilitation. However, since the concentration changes throughout the layer, we use the overall or average value

$$F = \frac{\overline{\mathcal{D}}_{eff}}{\mathcal{D}_{O_2}} - 1 \qquad (10.5\text{-}25)$$

corresponding to the definition

$$J_{O_2} = \overline{\mathcal{D}}_{eff}\frac{\Delta c_{O_2}}{b}. \qquad (10.5\text{-}26)$$

Physically, this process involves in effect O_2 being combined with Hb on one side of the hemoglobin solution, carried across it as part of the HbO_2 molecule and released at the other end, as shown in Figure 10.5-4. This is sometimes referred to as the *bucket-brigade transport*.

PROBLEMS

10.A Steady-State Oxygen-Tension Profile in the Cornea [25]

The cornea is one of the few mammalian tissues in which diffusion, convective water flow and consumption (or production) of the diffusing substances are operating simultaneously. Let us consider oxygen transport and assess the importance of the convective process.

(a)   Assuming the dissolved oxygen obeys Henry's law, write the equation of continuity in terms of the oxygen tensions in the stroma $(0 \leqslant x \leqslant a$, where x is the distance from the aqueous) and the epithelium $(a \leqslant x \leqslant L)$ as follows:

$$\frac{d^2 P_1}{dx^2} - \frac{v}{\mathcal{D}_1} \cdot \frac{dP_1}{dx} - \frac{Q_1}{\mathcal{D}_1 k_1} = 0, \qquad (10.\text{A-1})$$

$$\frac{d^2 P_2}{dx^2} - \frac{v}{\mathcal{D}_2} \cdot \frac{dP_2}{dx} - \frac{Q_2}{\mathcal{D}_2 k_2} = 0, \qquad (10.\text{A-2})$$

where v is convective velocity of the solution, Q stands for the oxygen consumption rate, $\mathcal{D}$ is the diffusivity, k is the solubility constant of oxygen in water, and subscripts 1 and 2 refer to the stroma and epithelium regions, respectively. What are the assumptions made?

(b)   These two equations are *not* simultaneous ones in the sense that they can be solved separately in a straightforward manner (see Recipe D2 of Table 2.8-1). However, they are coupled through the following boundary conditions:

B.C.1: at $x = 0$, $P_1 = P_i$

B.C.2: at $x = a$, $P_1 = P_2$

B.C.3: at $x = a$, $-\mathcal{D}_1 k_1 \dfrac{dP_1}{dx} + vk_1 P_1 = -\mathcal{D}_2 k_2 \dfrac{dP_2}{dx} + vk_2 P_2$

B.C.4: at $x = L$, $P_2 = P_0$,

where $P_i = 55$ mmHg and $P_0 = 155$ mmHg for the normal open eye. What is the meaning of B.C.3.?

(c)   Show that the solutions to Eqs. 10.A-1 and 10.A-2 with the boundary conditions are

$$P_1 = M e^{vx/\mathcal{D}_1} - \frac{Q_1 x}{vk_1} + N, \qquad (10.\text{A-3})$$

$$P_2 = R e^{vx/\mathcal{D}_2} - \frac{Q_2 x}{vk_2} + S, \qquad (10.\text{A-4})$$

where M, N, R, and S are constants consisting of the parameters given. These solutions have been calculated numerically based on the following data:

	(1) *Stroma*	(2) *Epithelium*
Thickness (cm)	4.6×10^{-2}	0.4×10^{-2}
Oxygen consumption, Q (ml O_2/sec-ml)	1.1×10^{-5}	3.8×10^{-4}
Diffusional coefficient of oxygen, $\mathcal{D}$ (cm^2/sec)	1.3×10^{-5}	0.85×10^{-5}
Solubility coefficient of oxygen, k(ml O_2/ml-mmHg)	2.3×10^{-5}	2.09×10^{-4}
Convective flow rate v(cm/sec)	10^{-6}	10^{-6}

and are graphically represented in Figure 10.A.

(d)   To assess the importance of convection, re-solve Eqs. 10.A-1 and 10.A-2 *without* the convection term and with the same set of boundary conditions. Plot your result on Figure 10.A and compare. What is your conclusion?

(e)   For the closed eye, Mishima and Hedbys [26] showed that convective flow is indeed negligible. This would give us solutions in the same mathematical form as the results of (d), but now, since the eye is closed, $P_0 = P_i = 55$ mmHg. Plot your solution and compare with Figure 10.A.

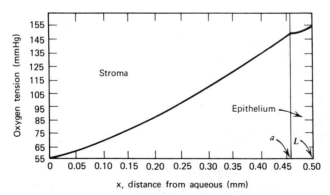

Figure 10.A.   Oxygen-tension profile in the cornea.

10.B Low-Order Approximation for Membrane Blood Oxygenators of Cylindrical Geometry [27]

Although many realistic solutions of oxygenation problems require the use of numerical methods, and hence computers, others can be analytically solved by use of simplfying assumptions. Although some of these assumptions may be quite far from the truth, they yield results that are at least qualitatively valid and that give us some physical insight into the nature of the process being studied.

Let us consider a membrane tube oxygenator in which the blood in each thin tube is separated into two regions—a shell layer in which oxygenation (chemical combination) is considered to have been completed so that the free hemoglobin concentration c_{Hb} is zero and so that oxygen is transferred radially only by diffusion, and a central unoxygenated core in which the blood is essentially the same as venous blood; that is, $c_{O_2} = c_0, c_{Hb} = c_{Hb_0}$ (both constants). Naturally the shell layer thickens and the central core shrinks as the blood flows from the oxygenator inlet ($z = 0$) toward the exit. Thus, the point $r = \lambda R$ at which one region ends and the other starts is a function of the axial distance; that is, $0 \leqslant \lambda(\zeta) \leqslant 1$, where $\zeta = z/Z_0$ with $Z_0 = \text{Re} \cdot \text{Sc} \cdot [c_{Hb_0}/c_w - c_0)]D$ (Re = Reynolds number, Sc = Schmidt number, D = diameter of the tube). The oxygen concentration at the tube wall is known to be c_w.

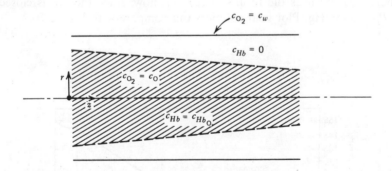

Figure 10.B. The shrinking unoxygenated venous core in blood flowing through a membrane tube oxygenator.

(a) Show that from the standpoint of dimension, ζ is a correct and reasonable choice of dimensionless axial distance. What is the physical meaning of this quantity?

(b)   By use of the appropriate equation of continuity (Table 5.5-1) and Fick's law, show that for the oxygenated layer $(\lambda \leqslant \xi \leqslant 1, \xi = r/R)$

$$N_{O_{2r}} = J_{O_{2r}} = \mathscr{D}_{O_{2m}} \frac{c_{O_2} - c_w}{-r \ln \dfrac{r}{R}}, \qquad (10.\text{B-}1)$$

where $D_{O_{2m}}$ is the effective diffusivity of dissolved O_2 through blood.

(c)   Show that for the unoxygenated core $(0 \leqslant \xi \leqslant \lambda)$ the local flow rate of unoxygenated hemoglobin down the tube is

$$Q(z) = 2\pi \int_0^{\lambda R} c_{Hb_0} v_z(r) r \, dr, \qquad (10.\text{B-}2)$$

since λ is a function of z.

(d)   Using a fully developed parabolic velocity profile for the H–P flow, show that

$$Q(\zeta) = \pi R^2 v_{z_{max}} c_{Hb_0} \left(\lambda^2 - \frac{\lambda^4}{2} \right). \qquad (10.\text{B-}3)$$

Does this seem to be a reasonable expression?

(e)   Via a hemoglobin balance over a differential tube length dz, show that

$$dQ = 2\pi \lambda R N_{O_2}\big|_{r = \lambda R} dz \qquad (10.\text{B-}4)$$

(f)   Using Eqs. 10.B-1 and 10.B-3, show that

$$\frac{d}{d\zeta} \left(\frac{\lambda^2}{4} - \frac{\lambda^4}{8} \right) = \frac{1}{\ln \lambda} \qquad (10.\text{B-}5)$$

What is the end condition at $\zeta = 0$?

(g)   Integrate Eq. 10.B-5 with the end condition to yield

$$\zeta = \frac{1}{8} \left[(2\xi^2 - \xi^4) \ln \xi + \frac{3}{4} - \xi^2 + \frac{\xi^4}{4} \right]. \qquad (10.\text{B-}6)$$

This is the trajectory of the boundary between the oxygenated layer and the unoxygenated core.

(h)   If we were to design an oxygenator based on Eq. 10.B-6, what would be the *useful length* of the tube—that is, the length beyond which no additional mass transfer occurs because the entire blood is oxygenated?

What is the value of Q at this length. Show mathematically and explain physically.

(i)   Plot the trajectory represented by Eq. 10.B-6 in a dimensionless plot.

(j)   If we define the fractional oxygenation as

$$f(\zeta) = \frac{Q(\zeta=0) - Q(\zeta)}{Q(\zeta=0)}$$

show that, based on Eq. 10.B-3,

$$f(\zeta) = (1-\lambda^2)^2. \tag{10.B-7}$$

Plot this fraction on the same chart as the trajectory.

(k)   Using the data used by Weissman and Mockros [28],

$$c_{Hb_0} = 2.73 \times 10^{-3} \text{ g-moles}/1*$$

$$c_w = 0.942 \times 10^{-3} \text{ g-moles}/1$$

$$c_0 = 0.0943 \times 10^{-3} \text{ g-moles}/1,$$

compare their numerical results with the present approximation. Note that this is a severe (i.e., pessimistic) test of the model.

10.C   Turbulent Flow in an Oxygenator

The velocity profile of turbulent flow in a circular tube is

$$v_z = v_{z_{max}} \left(1 - \frac{r}{R}\right)^{1/7}, \tag{10.C-1}$$

while the average velocity is

$$\langle v_z \rangle = \frac{49}{60} v_{z_{max}}. \tag{10.C-2}$$

Following a procedure similar to that of Problem 10.B, find the profiles for the axial decline of hemoglobin and of fractional oxygen in a tubular membrance oxygenator under turbulent flow condition. Is there an increase of the efficiency of oxygen transport? If so, does the increase justify the risk of hemolysis due to the turbulence?

*This is an *effective* value of c_{Hb_0}, after having allowed for physical oxygen capacity of the system.

10.D Approximation for a Rectangular Oxygenator

Noting the velocity profiles and average velocity as given in Example 6.1-1, find the hemoglobin decline and fractional oxygenator profiles in the slit of a rectangular oxygenator.

REFERENCES

1. Middleman, S. *Transport Phenomena in the Cardiovascular System.* Wiley-Interscience: New York, 1972.
2. Dill, D. B. In *Handbook of Respiratory Data in Aviation.* National Research Council: Washington, D. C., 1958.
3. Hill, A. V. *J. Physiol.,* **40**, 4 (1910).
4. Adair, G. S. *J. Biol. Chem.,* **63**, 529 (1925).
5. DeHaven, J. C., and DeLand, E. C. *Rand Memorandum RM-3212-PR.* Rand Corp.: Santa Monica, Calif., 1962.
6. Forbes, W. H., and Roughton, F. J. W. *J. Physiol.,* **71**, 229 (1931).
7. Pauling, L. *Proc. Nat. Acad. Sci.,* **21**, 186 (1935).
8. Gibson, Q. H.; Kreuzer, F.; Meda, E.; and Roughton, F. J. W. *J. Physiol.,* **129**, 65 (1955).
9. Krogh, A. *J. Physiol.,* **52**, 391, 409 (1919).
10. Bloch, I. *Bull. Math. Biophys.,* **5**, 1 (1943).
11. Blum, J. J. *Am. J. Physiol.,* **198**, 991 (1960).
12. Reneau, D. D., Jr.; Bruley, D. F.; and Knisely, M. H. *A. I.Ch. E. J.* **15**, 916 (1969).
13. Bugliarello, G.; and Jackson, E. D., III. *Proc. Am. Soc. Civ. Engrs., J. Eng. Mech. Div.,* **90** (EM4), 49–77 (1964).
14. Opitz, E., and Schneider, M. *Ergebn. Physiol.,* **46**, 126 (1950).
15. Lightfoot, E. N., Jr. *Transport Phenomena in Living Systems.* Wiley-Interscience: New York, 1974.
16. Warner, H. R., and Seagrave, R. C. In *Mass Transfer in Biological Systems*, A. L. Shrier and T. G. Kaufman, Eds. American Institute of Chemical Engineers: New York, 1970, p. 12.
17. Kylstra, J. A.; Paganelli, C. V.; and Lanphier, E. H. *J. Appl. Physiol.,* **21**, 177 (1966).
18. Roughton, F. J. W. *Prog. Biophys. Biophys. Chem.,* **9**, 55 (1959).
19. Bird, R. B.; Stewart, W. E.; and Lightfoot, E. N., Jr. *Transport Phenomena.* Wiley: New York, 1960.
20. Roughton, F. J. W. *Proc. Roy. Soc.* (London), **B111**, 1 (1932).
21. Moll, W. *Resp. Physiol.,* **6**, 1 (1968).

22. Kutchai, H. *Resp. Physiol.*, **10**, 273 (1970).
23. Nicolson, P., and Roughton, F. J. W. *Proc. Roy. Soc.* (London) **B138**, 241 (1951).
24. Hemmingsen, E., and Scholander, P. F., *Science*, **132**, 1379 (1960).
25. Bert, J. L., and Fatt, I. *Bull. Math. Biophys.*, **31** (3), 569 (1969).
26. Mishima, S., and Hedbys, B. O. *Exptl. Eye Res.*, **6**, 10 (1967).
27. Lightfoot, E. N., Jr. *A. I. Ch. E. J.*, **14** (4), 669 (1968).
28. Weissman, M. H., and Mockros, L. F. *J. Eng. Mech. Div., Am. Soc. Civ. Engrs.*, *93, 225 (1967);* **94**, 857 (1968).

APPENDIX A

Determination of C_m'' from Equation 6.1-26

The left-hand side can be split into two parts:

$$(I) = \int_0^1 \xi J_0(k_m \xi) \, d\xi. \tag{A.1}$$

It should be noted that the integral relationships in Table 2.9-2(C) cannot be applied directly because they are for $J_p(r)$, while in the present case there is a coefficient in front of the variable ξ. We can get around this simply by making the substitution $r = k_m \xi$ so that

$$(I) = \frac{1}{k_m^2} \int_0^{k_m} r J_0(r) \, dr$$

$$= \frac{1}{k_m^2} [r J_1(r)]_0^{k_m} \quad \text{[by setting } p = 1 \text{ in Eq. 1 of Table 2.9-2(C)]}$$

$$= \frac{1}{k_m} J_1(k_m) \quad [J_1(0) = 0, \text{ Fig. 2.9-1(a)}] \tag{A.2}$$

$$(II) = \int_0^1 \xi^3 J_0(k_m \xi) \, d\xi. \tag{A.3}$$

Even with $r = k_m \xi$, none of the integral relationships in Table 2.9-2(C) can be applied directly without using one of the algebraic relationships in Table 2.9-2(A) to change the Bessel function into higher-order ones.

Setting $p=1$ in Eq. 1 of Table 2.9-2(A), we have

$$J_0(r) = \frac{2}{r}J_1(r) - J_2(r), \qquad (A.4)$$

so that

$$(II) = \frac{1}{k_m^4}\int_0^{k_m} 2r^2 J_1(r)\,dr - \frac{1}{k_m^4}\int_0^{k_m} r^3 J_2(r)\,dr. \qquad (A.5)$$

Then, using Eq. 1 of Table 2.9-2(C) and setting $p=2$ and $p=3$ in turn, we obtain

$$(II) = \frac{2}{k_m^4}\left[r^2 J_2(r)\right]_0^{k_m} - \frac{1}{k_m^4}\left[r^3 J_3(r)\right]_0^{k_m}$$

$$= \frac{2}{k_m^2}J_2(k_m) - \frac{1}{k_m}J_3(k_m), \qquad (A.6)$$

since $J_2(0) = J_3(0) = 0$, and $J_2(k_m)$ and $J_3(k_m)$ are nonzero but finite. Using Eq. 1 of Table 2.9-2(A) with $p=3$, we have

$$J_1(r) + J_3(r) = \frac{4}{r}J_2(r),$$

which can be used to reduce the order in the Bessel functions in Eq. A.6. That is,

$$(II) = \frac{1}{k_m^2}\left[k_m J_1(k_m) - 2J_2(k_m)\right]. \qquad (A.7)$$

Using Eq. A.4 again and remembering that $J_0(k_m) = 0$, Eq. A.7 becomes

$$(II) = \left(\frac{1}{k_m} - \frac{4}{k_m^3}\right)J_1(k_m). \qquad (A.8)$$

Combining Eqs. A.2 and A.8, the left-hand side of Eq. 6.1-26 becomes

$$\text{LHS} = (I) - (II) = \frac{4}{k_m^3}J_1(k_m). \qquad (A.9)$$

On the right-hand side of Eq. 6.1-26, $J_0'(k_m)$ can be transformed by using a proper differential relationship, namely, Eq. 1 of Table 2.9-2(B), with

$p = 0$, which gives

$$J_0'(r) = -J_1(r), \qquad (A.10)$$

upon application of which we obtain

$$\text{RHS} = C_m'' \cdot \tfrac{1}{2} [J_1(k_m)]^2 \qquad (A.11)$$

Equating the expressions in Eqs. A.9 and A.11 and canceling common factors, we finally have

$$C_m'' = \frac{8}{k_m^3 J_1(k_m)}. \qquad (6.1\text{-}27)$$

APPENDIX B

Inverse Laplace Transform of Equation 6.1-41

Consider

$$\bar{g}(p) = \frac{N(p)}{D(p)}, \tag{B.1}$$

where

$$N(p) = I_0\left(\sqrt{\frac{p}{\nu}}\,R\right) - I_0\left(\sqrt{\frac{p}{\nu}}\,r\right), \tag{B.2}$$

$$D(p) = pI_0\left(\sqrt{\frac{p}{\nu}}\,R\right). \tag{B.3}$$

But we will encounter a problem if we attempt to obtain the roots of $D(p) = 0$. That is, we have

$$p = 0 \tag{B.4}$$

$$I_0\left(\sqrt{\frac{p}{\nu}}\,R\right) = 0, \tag{B.5}$$

and we know from Figure 2.9-1(c) that $I_0(x)$ never intersects the x-axis. However, this does *not* mean that roots do not exist at all; in fact, we can see that the roots are in the imaginary space if we remember Eq. 2.9-19, which in this case gives us

$$J_0\left(i\sqrt{\frac{p}{\nu}}\,R\right) = 0 \tag{B.6}$$

so that

$$p = a_k = -\frac{\alpha_k^2 \nu}{R^2},$$
(B.7)

where α_k are the roots of $J_0(\alpha_k) = 0$, as we recall from Figure 2.9-1(a) that the $J_0(x)$ function fluctuates in a diminishing manner between positive and negative values, thus intersecting the x-axis at an infinite number of points, $\alpha_1 = 2.40483 \cdots$, $\alpha_2 = 5.52009 \cdots$, $\alpha_3 = 8.65373 \cdots$, $\cdots$ which are real values. This is also to say that if we insist on using the roots of $I_0(\sqrt{p/\nu} \, R) = 0$, we would end up with imaginary values. Thus, we now have

$$N(p) = J_0\left(i\sqrt{\frac{p}{\nu}} \, R\right) - J_0\left(i\sqrt{\frac{p}{\nu}} \, r\right)$$
(B.8)

$$D(p) = pJ_0\left(i\sqrt{\frac{p}{\nu}} \, R\right),$$
(B.9)

so that, using Eq. 1 of Table 2.9-2(B), and Eqs. (c) and (e) of Table 2.2-1 (with the footnote)

$$D'(p) = J_0\left(i\sqrt{\frac{p}{\nu}} \, R\right) + p\frac{d}{dp}\left[J_0\left(i\sqrt{\frac{p}{\nu}} \, R\right)\right]$$

$$= J_0\left(i\sqrt{\frac{p}{\nu}} \, R\right) + p\left[-J_1\left(i\sqrt{\frac{p}{\nu}} \, R\right) \cdot \frac{1}{2} \cdot \frac{i}{\sqrt{p\nu}} R\right]$$

$$= J_0\left(i\sqrt{\frac{p}{\nu}} \, R\right) - \frac{1}{2}\left(i\sqrt{\frac{p}{\nu}} \, R\right)J_1\left(i\sqrt{\frac{p}{\nu}} \, R\right).$$

Then

$$D'(a_k) = -\frac{1}{2}\alpha_k J_1(\alpha_k),$$
(B.10)

since by definition $J_0(i\sqrt{p/\nu R}) = J_0(\alpha_k) = 0$. For the same reason, for Eq. B.8, we have

$$N(a_k) = -J_0\left(\frac{\alpha_k r}{R}\right).$$
(B.11)

Using Eqs. B.10 and B.11 with the Heaviside partial fraction expansion formula (Eq. 2.12-4), we have

$$g(t) = \mathcal{L}^{-1}\{\bar{g}(p)\} = \mathcal{L}^{-1}\left\{\frac{N(p)}{D(p)}\right\}$$

$$= 2 \sum_{k=1}^{\infty} \frac{J_0\left(\dfrac{\alpha_k r}{R}\right)}{\alpha_k J_1(\alpha_k)} \exp\left(-\frac{\alpha_k^2 \nu t}{R^2}\right). \tag{6.1-43}$$

APPENDIX C

Obtaining $p(r,z)-p(0,0)$ from $\partial p/\partial r$ and $\partial p/\partial z$

Known:

$$\frac{\partial p}{\partial r} = -\frac{8\mu r}{R^3}(a_0 + a_1 z) \qquad \text{(6.1-83)}$$

$$\frac{\partial p}{\partial z} = -\frac{4a_1\mu}{R}\left[\left(\frac{r}{R}\right)^2 + \frac{2Q(z)}{a_1\pi R^3} + \frac{1}{3}\right]. \qquad \text{(6.1-84)}$$

The most straightforward, but a lengthy way is to recognize

$$p = \int \frac{\partial p}{\partial r}\, dr + M(z), \qquad \text{(C.1)}$$

where $M(z)$ is a function of z only, to be evaluated later. Substitution for $\partial p/\partial r$ from Eq. 6.1-83 yields

$$p = -\frac{4\mu r^2}{R^3}(a_0 + a_1 z) + M(z). \qquad \text{(C.2)}$$

Differentiating p partially with respect to z in Eq. C.2 and equating that to the one in Eq. 6.1-84 give, after canceling equal terms on both sides,

$$\frac{dM}{dz} = -\frac{4a_1\mu}{R}\left[\frac{2Q(z)}{a_1\pi R^3} + \frac{1}{3}\right], \qquad \text{(C.3)}$$

459

which upon integration results in

$$M = -\frac{8\mu}{\pi R^4}\int Q(z)\,dz - \frac{4a_1\mu}{3R}z + k, \qquad (C.4)$$

where k is the integrating constant and $\int Q(z)\,dz$ can be rewritten, using an average flow rate $\overline{Q}(z)$ over tube length z:

$$\overline{Q}(z) = \frac{1}{z}\int_0^z Q(z)\,dz, \qquad (C.5)$$

so that we have, remembering Eq. C.2,

$$p(r,z) = -\frac{4\mu r^2}{R^3}(a_0 + a_1 z) - \left(\frac{4a_1\mu}{3R} + \frac{8\mu\overline{Q}}{\pi R^4}\right)z + k. \qquad (C.6)$$

Noting that $p = p(0,0)$ at $r = 0$ and $z = 0$, we can evaluate the integration constant to be

$$k = p(0,0). \qquad (C.7)$$

Thus, finally, we have

$$p(r,z) - p(0,0) = -\frac{4\mu}{R}(a_0 + a_1 z)\left(\frac{r}{R}\right)^2 - \mu\left(\frac{4a_1}{3R} + \frac{8\overline{Q}}{\pi R^4}\right)z. \qquad (6.1\text{-}85)$$

It is of interest to note that the same Eq. 6.1-85 can be obtained by integrating over z first—that is,

$$p = \int\frac{\partial p}{\partial z}\,dz + N(r), \qquad (C.8)$$

and then evaluating $N(r)$ in a process similar to the above for obtaining $M(z)$.

An alternate procedure is to use the differential expansion

$$dp = \frac{\partial p}{\partial r}\,dr + \frac{\partial p}{\partial z}\,dz. \qquad (C.9)$$

However, in integrating dp to obtain p, one should realize that if we follow the path in Figure C-1, the first integration along r is performed at $z = 0$;

that is,

$$p(r,z) - p(0,0) = \int_{(0,0)}^{(r,z)} dp$$

$$= \int_0^r \frac{\partial p}{\partial r}(r,0)\,dr + \int_0^z \frac{\partial p}{\partial z}(r,z)\,dz. \qquad (C.10)$$

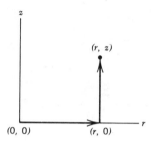

Figure C-1.

Substituting for $\partial p/\partial r$ and $\partial p/\partial z$ from Eqs. 6.1-83 and 6.1-84 but re-membering that $z=0$ in $\partial p/\partial r$, we have

$$p(r,z) - p(0,0) = \int_0^r -\frac{8\mu a_0 r}{R^3}\,dr + \int_0^z -\frac{4a_1\mu}{R}\left[\left(\frac{r}{R}\right)^2 + \frac{2Q(z)}{a_1\pi R^3} + \frac{1}{3}\right]dz$$

$$= -\frac{4\mu a_0 r^2}{R^3} - \frac{4a_1\mu}{R}\left(\frac{r}{R}\right)^2 z - \frac{8\mu}{\pi R^4}\int_0^z Q\,dz - \frac{4a_1\mu}{3R}$$

$$= -\frac{4\mu}{R}(a_0 + a_1 z)\left(\frac{r}{R}\right)^2 - \mu\left(\frac{4a_1}{3R} + \frac{8\overline{Q}}{\pi R^4}\right)z,$$

which is the same as Eq. 6.1-85.

As p is an exact function,* it does not depend on path. That is, one can

*One can prove this by differentiating Eq. 6.1-83 with respect to z and Eq. 6.1-84 with respect to r, finding

$$\frac{\partial^2 p}{\partial r \partial z} = \frac{\partial^2 p}{\partial z \partial r}.$$

also obtain Eq. 6.1-85 by following the path depicted in Figure C-2:

$$p(r,z) - p(0,0) = \int \frac{\partial p}{\partial z}(0,z)\,dz + \int \frac{\partial p}{\partial r}(r,z)\,dr. \qquad (C.11)$$

It should be noted that we cannot use Eq. C.9 "directly" by saying

$$p = \int dp = \int \frac{\partial p}{\partial r}\,dr + \int \frac{\partial p}{\partial z}\,dz + k,$$

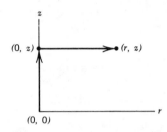

Figure C-2.

because an extra term would appear. Referring to a similar differential equation involving an exact differential, Formula A(2) of Table 2.8-1, we note that the correct formula to use is

$$p = \int dp$$

$$= \int \frac{\partial p}{\partial r}\,dr + \int \left(\frac{\partial p}{\partial z} - \frac{\partial}{\partial z} \int \frac{\partial p}{\partial r}\,dr \right) dz. \qquad (C.12)$$

The reader can immediately verify that evaluating

$$\frac{\partial p}{\partial z} - \frac{\partial}{\partial z} \int \frac{\partial p}{\partial r}\,dr$$

is the same as obtaining dM/dz above (Eq. C.3).

APPENDIX D

Solutions to Equations 6.2-42 and 6.2-43

The parallel* partial differential equations to be solved are

$$\frac{1}{\xi}\frac{\partial}{\partial \xi}\left(\xi\frac{\partial \Theta^{I}}{\partial \xi}\right) + \frac{\partial^{2}\Theta^{I}}{\partial \zeta^{2}} + \phi^{I} = 0 \qquad (6.2\text{-}42)$$

$$\frac{1}{\xi}\frac{\partial}{\partial \xi}\left(\xi\frac{\partial \Theta^{II}}{\partial \xi}\right) + \frac{\partial^{2}\Theta^{II}}{\partial \zeta^{2}} + \phi^{II} = 0 \qquad (6.2\text{-}43)$$

with the following boundary conditions:

B.C.1: at $\zeta = 0$, $\Theta^{I} = \Theta^{II} = 1$ for all $0 \leqslant \xi \leqslant 1$

B.C.2: at $\zeta = \nu$, $\dfrac{\partial \Theta^{I}}{\partial \zeta} = \dfrac{\partial \Theta^{II}}{\partial \zeta} = 0$ for all $0 \leqslant \xi \leqslant 1$

B.C.3: at $\xi = 0$, $\dfrac{\partial \Theta^{I}}{\partial \xi} = 0$ for all $0 \leqslant \zeta \leqslant \nu$

B.C.4: at $\xi = \kappa$, $\Theta^{I} = \Theta^{II}$, $k^{I}\dfrac{\partial \Theta^{I}}{\partial \xi} = k^{II}\dfrac{\partial \Theta^{II}}{\partial \xi}$ for all $0 \leqslant \zeta \leqslant \nu$

B.C.5: at $\xi = 1$, $-\dfrac{\partial \Theta^{II}}{\partial \xi} = \text{Bi}\,\Theta^{II} + \Lambda$ for all $0 \leqslant \zeta \leqslant \nu$

Of the three alternative methods described in Example 6.2-4 (p. 238), the third one is preferred in view of the parallel nature of the boundary conditions on ζ. This means that we first find the solution Θ_{∞}^{I} and Θ_{∞}^{II},

---

*These are *not* simultaneous partial differential equations in that neither Θ^{I} nor Θ^{II} appears in *both* equations simultaneously. Rather, Θ^{I} and Θ^{II} are linked through B.C.4.

463

assuming the system to be of infinite extent in the radial (r or ξ) dimension. This, in effect, gives us an infinite slab of finite thickness onto which we superimpose the infinite cylinder and "carve out" the finite cylinder (see Figure 6.2-12). Thus from Eqs. 6.2-42 and 6.2-43 we first ignore the effect of the radial coordinate and write the infinite-slab heat transfer equations as

$$\frac{d^2\Theta_\infty^I}{d\zeta^2} + \phi^I = 0 \tag{D.1}$$

$$\frac{d^2\Theta_\infty^{II}}{d\zeta^2} + \phi^{II} = 0 \tag{D.2}$$

Since the infinite slab can be considered as an extreme case of the finite cylinder (when the radius is equal to infinity), B.C.1 and B.C.2 should hold for Θ_∞^I and Θ_∞^{II} as well. Applying them to the general solutions of Eqs. D.1 and D.2, we have

$$\Theta_\infty^I = 1 + \frac{\phi^I \zeta}{2}(2\nu - \zeta) \tag{D.3}$$

and

$$\Theta_\infty^{II} = 1 + \frac{\phi^{II} \zeta}{2}(2\nu - \zeta) \tag{D.4}$$

Now we define the "transient" solutions Θ_t^I and Θ_t^{II} as the difference between those of the finite cylinder and of the infinite slab*

$$\Theta_t^I(\xi,\zeta) = 1 + \frac{\phi^I \zeta}{2}(2\nu - \zeta) - \Theta^I(\xi,\zeta) \tag{D.5}$$

$$\Theta_t^{II}(\xi,\zeta) = 1 + \frac{\phi^{II} \zeta}{2}(2\nu - \zeta) - \Theta^{II}(\xi,\zeta) \tag{D.6}$$

---

*We now omit the quotation marks on t to make the symbol look simpler. See p. 238 for their meaning.

with the boundary conditions now becoming

B.C.1: at $\zeta=0$, $\left.\begin{array}{l}\Theta_t^I=\Theta_\infty^I-\Theta^I=1-1=0\\\Theta_t^{II}=\Theta_\infty^{II}-\Theta^{II}=1-1=0\end{array}\right\}$ for all $0\leqslant\xi\leqslant1$

B.C.2: at $\zeta=\nu$, $\dfrac{\partial\Theta_t^I}{\partial\zeta}=\dfrac{\partial\Theta_t^{II}}{\partial\zeta}=0$ for all $0\leqslant\xi\leqslant1$

B.C.3: at $\xi=0$, $\dfrac{\partial\Theta_t^I}{\partial\xi}=0$ for all $0\leqslant\zeta\leqslant\nu$

B.C.4: at $\xi=\kappa$, $\left.\begin{array}{l}\Theta_t^I-\dfrac{\phi^I\zeta}{2}(2\nu-\zeta)=\Theta_t^{II}-\dfrac{\phi^{II}\zeta}{2}(2\nu-\zeta)\\[2mm]k^I\dfrac{\partial\Theta_t^I}{\partial\xi}=k^{II}\dfrac{\partial\Theta_t^{II}}{\partial\xi}\end{array}\right\}$ for all $0\leqslant\zeta\leqslant\nu$

B.C.5: at $\xi=1$, $\dfrac{\partial\Theta_t^{II}}{\partial\xi}=\text{Bi}\left[1+\dfrac{\phi^{II}\zeta}{2}(2\nu-\zeta)-\Theta_t^{II}\right]+\Lambda.$ for all $0\leqslant\zeta\leqslant\nu$

Substituting Eqs. D.5 and D.6 into Eqs. 6.2-42 and 6.2-43, we have

$$\frac{1}{\xi}\frac{\partial}{\partial\xi}\left(\xi\frac{\partial\Theta_t^I}{\partial\xi}\right)+\frac{\partial^2\Theta_t^I}{\partial\zeta^2}=0 \tag{D.7}$$

$$\frac{1}{\xi}\frac{\partial}{\partial\xi}\left(\xi\frac{\partial\Theta_t^{II}}{\partial\xi}\right)+\frac{\partial^2\Theta_t^{II}}{\partial\zeta^2}=0 \tag{D.8}$$

Postulating

$$\Theta_t^I(\xi,\zeta)=f(\xi)g(\zeta) \tag{D.9}$$

$$\Theta_t^{II}(\xi,\zeta)=p(\xi)q(\zeta) \tag{D.10}$$

and using the principle that a pure function of ξ can be identically equal to one of ζ only when both are equal to the same constant (see Example 6.1-2), we have

$$\frac{1}{f\xi}\frac{d}{d\xi}\left(\xi\frac{df}{d\xi}\right)=-\frac{1}{g}\frac{d^2g}{d\zeta^2}=\beta^2 \tag{D.11}$$

$$\frac{1}{p\xi}\frac{d}{d\xi}\left(\xi\frac{dp}{d\xi}\right)=-\frac{1}{q}\frac{d^2q}{d\zeta^2}=\gamma^2 \tag{D.12}$$

for which we have the solutions

$$f(\xi) = D_1 I_0(\beta\xi) + D_2 K_0(\beta\xi) \tag{D.13}$$

$$g(\zeta) = D_3 \sin\beta\zeta + D_4 \cos\beta\zeta \tag{D.14}$$

$$p(\xi) = E_1 I_0(\gamma\xi) + E_2 K_0(\gamma\xi) \tag{D.15}$$

$$q(\zeta) = E_3 \sin\gamma\zeta + E_4 \cos\gamma\zeta \tag{D.16}$$

with the first three boundary conditions splitting into

B.C.1: at $\zeta = 0$, $g = 0$ and $q = 0$

B.C.2: at $\zeta = \nu$, $\dfrac{dg}{d\zeta} = \dfrac{dq}{d\zeta} = 0$

B.C.3: at $\xi = 0$, $\dfrac{df}{d\xi} = 0$

while B.C.4 and B.C.5 remain the same.

The application of B.C.1 and B.C.2 to Eqs. D.14 and D.16 yields $D_4 = E_4 = 0$ and

$$\beta = \gamma = \beta_n = (n + \tfrac{1}{2})\frac{\pi}{\nu} \tag{D.17}$$

where $n = 0, \pm 1, \pm 2, \dots, \pm\infty$. Although the application of B.C.3 to Eq. D.13 gives $D_2 = 0$ (since $K_1(0) = \infty$), a similar result for E_2 cannot be deduced from B.C.4 or B.C.5. Rather, the separate solutions are linearly recombined via the principle of superimposition (see Eqs. 2.8-7 and 6.1-23) to yield

$$\Theta_t^I = \sum_{n=-\infty}^{+\infty} C_n' I_0(\beta_n\xi)\sin\beta_n\zeta \tag{D.18}$$

$$\Theta_t^{II} = \sum_{n=-\infty}^{+\infty} \{B_{1n}' I_0(\beta_n\xi) + B_{2n}' K_0(\beta_n\xi)\}\sin\beta_n\zeta \tag{D.19}$$

where $C_n' = D_{1n}D_{3n}$, $B_{1n}' = E_{1n}E_{3n}$, and $B_{2n}' = E_{2n}E_{3n}$. Notice here that we can shorten the series by "folding" them into halves at between $n = -1$ and $n = 0$. That is, in view of such relationships as $\sin(-x) = -\sin x$, $I_0(x) = I_0(-x)$, and so on, terms with n and $-(n+1)$ can be paired off with their coefficients combined to form new ones. Thus we can rewrite Eqs.

D.18 and D.19 into

$$\Theta_t^I = \sum_{n=0}^{\infty} C_n I_0(\beta_n\xi)\sin\beta_n\zeta \tag{D.18'}$$

$$\Theta_t^{II} = \sum_{n=0}^{\infty} \{ B_{1n}I_0(\beta_n\xi) + B_{2n}K_0(\beta_n\xi) \}\sin\beta_n\zeta \tag{D.19'}$$

where $C_n = C_n' - C_{-(n+1)}'$, $B_{1n} = B_{1n}' - B_{1[-(n+1)]}'$, and so on.
In order to use B.C.4, we obtain from Eqs. D.18' and D.19'

$$\Theta_t^I(\kappa,\zeta) = \sum_{n=0}^{\infty} C_n I_0(\beta_n\kappa)\sin\beta_n\zeta \tag{D.20}$$

$$\Theta_t^{II}(\kappa,\zeta) = \sum_{n=0}^{\infty} \{ B_{1n}I_0(\beta_n\kappa) + B_{2n}K_0(\beta_n\kappa) \}\sin\beta_n\zeta \tag{D.21}$$

$$\frac{\partial\Theta_t^I}{\partial\xi}\bigg|_{\xi=\kappa} = \sum_{n=0}^{\infty} C_n\beta_n I_1(\beta_n\kappa)\sin\beta_n\zeta \tag{D.22}$$

$$\frac{\partial\Theta_t^{II}}{\partial\xi}\bigg|_{\xi=\kappa} = \sum_{n=0}^{\infty} \beta_n \{ B_{1n}I_1(\beta_n\kappa) - B_{2n}K_1(\beta_n\kappa) \}\sin\beta_n\zeta \tag{D.23}$$

the substitution of which into B.C.4 gives

$$\frac{\phi^I\zeta}{2}(2\nu-\zeta) - \sum_{n=0}^{\infty} C_n I_0(\beta_n\kappa)\sin\beta_n\zeta$$

$$= \frac{\phi^{II}\zeta}{2}(2\nu-\zeta) - \sum_{n=0}^{\infty} \{ B_{1n}I_0(\beta_n\kappa) + B_{2n}K_0(\beta_n\kappa) \}\sin\beta_n\zeta \tag{D.24}$$

$$k^I\sum_{n=0}^{\infty} C_n\beta_n I_1(\beta_n\kappa)\sin\beta_n\zeta$$

$$= k^{II}\sum_{n=0}^{\infty} \beta_n \{ B_{1n}I_1(\beta_n\kappa) - B_{2n}K_1(\beta_n\kappa) \}\sin\beta_n\zeta \tag{D.25}$$

While Eq. (D.24) can be condensed to

$$\sum_{n=0}^{\infty} \{ (B_{1n} - C_n) I_0(\beta_n \kappa) + B_{2n} K_0(\beta_n \kappa) \} \sin \beta_n \zeta = \frac{\phi^{II} - \phi^{I}}{2} \zeta(2\nu - \zeta)$$

(D.26)

Equation D.25 dictates that

$$C_n = \frac{k^{II}}{k^{I}} \left\{ B_{1n} - B_{2n} \frac{K_1(\beta_n \kappa)}{I_1(\beta_n \kappa)} \right\}$$

(D.27)

since for the equality of the two series to hold for any value of ζ, the corresponding terms must be equal individually.

The coefficients C_n, B_{1n}, and B_{2n} can be "released" from the series in Eq. D.26 via the use of the orthogonality relationship of the sine function, that is, multiplying both sides by $\sin \beta_m \zeta \, d\zeta$ and integrating from 0 to ν. Since* (see footnote while remembering Eq. D.17)

$$\int_0^{\nu} \sin\left[(n+\tfrac{1}{2}) \frac{\pi \zeta}{\nu} \right] \sin\left[(m+\tfrac{1}{2}) \frac{\pi \zeta}{\nu} \right] d\zeta = \begin{cases} 0 & \text{for all} \quad m \neq n \\ \nu/2 & \text{for} \quad m = n \end{cases}$$

(D.28)

Equation D.26 becomes

$$\{ (B_{1n} - C_n) I_0(\beta_n \kappa) + B_{2n} K_0(\beta_n \kappa) \} \frac{\nu}{2}$$

$$= \frac{\phi^{II} - \phi^{I}}{2} \int_0^{\nu} \zeta(2\nu - \zeta) \sin\left[(n+\tfrac{1}{2}) \frac{\pi \zeta}{\nu} \right] d\zeta$$

or

$$(B_{1n} - C_n) I_0(\beta_n \kappa) + B_{2n} K_0(\beta_n \kappa) = \frac{2(\phi^{II} - \phi^{I})}{\nu \beta_n^3}$$

(D.29)

*From Eqs. 13 and 14 of Table 2.4-1, two additional orthogonality relationships can be obtained:

$$\int_0^1 \sin(m+\tfrac{1}{2})\pi\xi \sin(n+\tfrac{1}{2})\pi\xi \, d\xi = \int_0^1 \cos(m+\tfrac{1}{2})\pi\xi \cos(n+\tfrac{1}{2})\pi\xi \, d\xi = \tfrac{1}{2}\delta_{mn}$$

where m and n are integers and δ_{mn} is the Kronecker delta (see Eq. 2.16-5, p. 97).

Now applying Eq. D.19′ to B.C.5, we have

$$\sum_{n=0}^{\infty} \beta_n \{ B_{1n} I_1(\beta_n) - B_{2n} K_1(\beta_n) \} \sin \beta_n \zeta$$

$$= \mathrm{Bi} \left[1 + \frac{\phi^{\mathrm{II}} \zeta}{2} (2\nu - \zeta) - \sum_{n=0}^{\infty} \{ B_{1n} I_0(\beta_n) + B_{2n} K_0(\beta_n) \} \sin \beta_n \zeta \right] + \Lambda$$

or

$$\sum_{n=0}^{\infty} (B_{1n} F_n + B_{2n} G_n) \sin \beta_n \zeta = \mathrm{Bi} \left[1 + \frac{\phi^{\mathrm{II}} \zeta}{2} (2\nu - \zeta) \right] + \Lambda \qquad (D.30)$$

where

$$F_n = \mathrm{Bi} I_0(\beta_n) + \beta_n I_1(\beta_n) \qquad (D.31)$$

$$G_n = \mathrm{Bi} K_0(\beta_n) - \beta_n K_1(\beta_n) \qquad (D.32)$$

The coefficients B_{1n} and B_{2n} can again be released from the series as before by multiplying both sides by $\sin \beta_m \zeta \, d\zeta$ and integrating from 0 to ν. Thus, we obtain

$$\frac{\nu}{2} (B_{1n} F_n + B_{2n} G_n) = (\mathrm{Bi} + \Lambda) \int_0^{\nu} \sin \beta_n \zeta \, d\zeta$$

$$+ \frac{\mathrm{Bi} \phi^{\mathrm{II}}}{2} \int_0^{\nu} \zeta (2\nu - \zeta) \sin \beta_n \zeta \, d\zeta$$

$$= \frac{1}{\beta_n} (\mathrm{Bi} + \Lambda) + \frac{\mathrm{Bi} \phi^{\mathrm{II}}}{2} \cdot \frac{2}{\beta_n^3}$$

or

$$B_{1n} F_n + B_{2n} G_n = \frac{2}{\nu \beta_n} (\mathrm{Bi} + \Lambda) + \frac{2 \mathrm{Bi} \phi^{\mathrm{II}}}{\nu \beta_n^3} \qquad (D.33)$$

In summary, the solutions to Eqs. 6.2-42 and 6.2-43 are

$$\Theta^{\mathrm{I}} = 1 + \frac{\phi^{\mathrm{I}} \zeta}{2} (2\nu - \zeta) - \sum_{n=0}^{\infty} C_n I_0(\beta_n \xi) \sin \beta_n \zeta \qquad (6.2\text{-}46)$$

$$\Theta^{II} = 1 + \frac{\phi^{II}\zeta}{2}(2\nu - \zeta) - \sum_{n=0}^{\infty} \{ B_{1n}I_0(\beta_n\xi) + B_{2n}K_0(\beta_n\xi) \} \sin \beta_n \zeta \quad (6.2\text{-}47)$$

where

$$\beta_n = \frac{(n+\frac{1}{2})\pi}{\nu} \qquad (n=0,1,2,\ldots,\infty)$$

and the coefficients C_n, B_{1n}, and B_{2n} are obtained by solving the simultaneous equations D.27, D.29, and D.33:

$$\frac{k^I}{k^{II}} C_n - B_{1n} + \frac{K_1(\beta_n\kappa)}{I_1(\beta_n\kappa)} B_{2n} = 0$$

$$-I_0(\beta_n\kappa)C_n + I_0(\beta_n\kappa)B_{1n} + K_0(\beta_n\kappa)B_{2n} = \frac{2(\phi^{II}-\phi^I)}{\nu\beta_n^3}$$

$$F_nB_{1n} + G_nB_{2n} = \frac{2(\text{Bi}+\Lambda)}{\nu\beta_n} + \frac{2\text{Bi}\phi^{II}}{\nu\beta_n^3}$$

where

$$F_n = \text{Bi}\,I_0(\beta_n) + \beta_n I_1(\beta_n)$$

$$G_n = \text{Bi}\,K_0(\beta_n) - \beta_n K_1(\beta_n)$$

APPENDIX E

Evaluating A_{mn} from Equation 6.2-63

In order to utilize the orthogonality relationships, we multiply Eq. 6.2-63 by $\xi J_0(\alpha_k \xi)\cos(l-\frac{1}{2})\pi\zeta\,d\xi\,d\zeta$ and integrate both sides over ξ and ζ from 0 to 1. Now all terms with $k \neq n$, $l \neq m$ on the right-hand side disappear except the one with $k = n$, $l = m$. According to Eq. D.1 of Table 2.9-2 and the footnote to Eq. D.28,

$$\text{RHS} = A_{lk}\tfrac{1}{2}\big[J_0'(\alpha_k)\big]^2 \cdot \tfrac{1}{2}. \tag{E.1}$$

Observing Eq. B.1 of Table 2.9-2, Eq. E.1 becomes

$$\text{RHS} = A_{lk} \cdot \tfrac{1}{4}\big[J_1(\alpha_k)\big]^2. \tag{E.2}$$

The left-hand side is

$$\text{LHS} = \int_0^1 \int_0^1 \xi J_0(\alpha_k \xi)\cos(l-\tfrac{1}{2})\pi\zeta\,d\xi\,d\zeta$$

$$= \int_0^1 \xi J_0(\alpha_k \xi)\,d\xi \int_0^1 \cos(l-\tfrac{1}{2})\pi\zeta\,d\zeta. \tag{E.3}$$

The first part is the same as Eq. A.1 and the second part is

$$\int_0^1 \cos(l-\tfrac{1}{2})\pi\zeta\,d\zeta = \left[\frac{1}{(l-\tfrac{1}{2})\pi}\sin(l-\tfrac{1}{2})\pi\zeta\right]_0^1$$

$$= \frac{1}{(l-\tfrac{1}{2})\pi}\sin(l-\tfrac{1}{2})\pi. \tag{E.4}$$

Since l is an integer ranging from 1 to ∞, we can see that $\sin(l-\tfrac{1}{2})\pi$ alternate between 1 and -1 as l changes in steps of 1. Equation E.4 can be rewritten as

$$\int_0^1 \cos\left(l-\tfrac{1}{2}\right)\pi\zeta\, d\zeta = \frac{(-1)^{l-1}}{(l-\tfrac{1}{2})\pi}. \tag{E.5}$$

Substituting Eqs. A.2 and E.5 into Eq. E.3, we have

$$\text{LHS} = \frac{(-1)^{l-1}}{(l-\tfrac{1}{2})\pi} \cdot \frac{1}{\alpha_k} J_1(\alpha_k). \tag{E.6}$$

Putting Eqs. E.2 and E.6 together, we have

$$A_{lk} \cdot \tfrac{1}{4}\left[J_1(\alpha_k)\right]^2 = \frac{(-1)^{l-1}}{(l-\tfrac{1}{2})\pi} \cdot \frac{1}{\alpha_k} J_1(\alpha_k)$$

or

$$A_{mn} = 4 \cdot \frac{(-1)^{m-1}}{(m-\tfrac{1}{2})\pi} \cdot \frac{1}{\alpha_n J_1(\alpha_n)}. \tag{E.7}$$

APPENDIX F

Solutions for the Maternal and Fetal Transport Fractions, Example 8.8-5

In Example 8.8-5, for the countercurrent model Eqs. 8.8-25 and 8.8-26 can be rewritten as

$$\frac{dc_M}{d\zeta} = \alpha_1(c_F - c_M) \tag{F.1}$$

$$\frac{dc_F}{d\zeta} = \alpha_2(c_F - c_M), \tag{F.2}$$

where $\zeta = z/L$, the dimensionless longitudinal coordinate, and $\alpha_1 = P/Q_M$, $\alpha_2 = P/Q_F$. The boundary conditions are

B.C.1: $\zeta = 0$, $c_M = c_{MA}$, $c_F = c_{FV}$

B.C.2: $\zeta = 1$, $c_M = c_{MV}$, $c_F = c_{FA}$.

Taking the Laplace transforms of both Eqs. F.1 and F.2 with B.C.1, we have

$$p\bar{c}_M - c_{MA} = \alpha_1 \bar{c}_F - \alpha_1 \bar{c}_M$$

$$p\bar{c}_F - c_{FV} = \alpha_2 \bar{c}_F - \alpha_2 \bar{c}_M,$$

which can be rearranged to give

$$(p + \alpha_1)\bar{c}_M - \alpha_1 \bar{c}_F = c_{MA} \tag{F.3}$$

$$\alpha_2 \bar{c}_M + (p - \alpha_2)\bar{c}_F = c_{FV}. \tag{F.4}$$

Solving these two equations simultaneously, we have

$$\bar{c}_M = \frac{\begin{vmatrix} c_{MA} & -\alpha_1 \\ c_{FV} & p - \alpha_2 \end{vmatrix}}{\begin{vmatrix} p + \alpha_1 & -\alpha_1 \\ \alpha_2 & p - \alpha_2 \end{vmatrix}} = \frac{(p - \alpha_2) c_{MA} + \alpha_1 c_{FV}}{p(p + \alpha_1 - \alpha_2)} \tag{F.5}$$

$$\bar{c}_F = \frac{\begin{vmatrix} p + \alpha_1 & c_{MA} \\ \alpha_2 & c_{FV} \end{vmatrix}}{\begin{vmatrix} p + \alpha_1 & -\alpha_1 \\ \alpha_2 & p - \alpha_2 \end{vmatrix}} = \frac{(p + \alpha_1) c_{FV} - \alpha_2 c_{MA}}{p(p + \alpha_1 - \alpha_2)}. \tag{F.6}$$

Inverting the Laplace transform, we note that in Eq. F.5, following the notations in Eq. 2.12-4,

$$\bar{f}(p) = (p - \alpha_2) c_{MA} + \alpha_1 c_{FV},$$

$$\bar{g}(p) = p(p + \alpha_1 - \alpha_2).$$

Thus, we have

$$\bar{g}'(p) = 2p + \alpha_1 - \alpha_2$$

and for $g(p) = 0$, the roots are

$$a_1 = 0, \qquad a_2 = \alpha_2 - \alpha_1.$$

According to the HPFET (Eq. 2.12-4), we have

$$c_M(\zeta) = \mathcal{L}^{-1}\{\bar{c}_M(p)\}$$

$$= \frac{\bar{f}(0)}{\bar{g}'(0)} e^{0 \cdot \zeta} + \frac{\bar{f}(\alpha_2 - \alpha_1)}{\bar{g}'(\alpha_2 - \alpha_1)} e^{(\alpha_2 - \alpha_1)\zeta}$$

$$= \frac{-\alpha_2 c_{MA} + \alpha_1 c_{FV}}{\alpha_1 - \alpha_2} + \frac{-\alpha_1 c_{MA} + \alpha_1 c_{FV}}{\alpha_2 - \alpha_1} e^{(\alpha_2 - \alpha_1)\zeta}$$

$$= \frac{R}{1 - R} \left\{ \left(\frac{c_{MA}}{R} - c_{FV} \right) + (c_{FV} - c_{MA}) \exp\left[\alpha_1 \left(\frac{1}{R} - 1 \right) \zeta \right] \right\}, \tag{8.8-27}$$

where $R = \alpha_1/\alpha_2 = Q_F/Q_M$, the fetal-to-maternal flow ratio. Following the same procedure, Eq. F.6 can be inverted to yield

$$c_F(\zeta) = \frac{R}{1-R}\left\{\left(\frac{c_{MA}}{R} - c_{FV}\right) + \left(\frac{c_{FV} - c_{MA}}{R}\right)\exp\left[\alpha_1\left(\frac{1}{R} - 1\right)\zeta\right]\right\}.$$

(8.8-28)

After Eqs. 8.8-35 and 8.8-36 are obtained, we can substitute the latter into Eq. 8.8-29 to yield

$$T_F = \frac{c_{FV} - \dfrac{R}{1-R}\left[\dfrac{c_{MA}}{R} - c_{FV} + \dfrac{c_{FV} - c_{MA}}{R}e^{-\alpha_1(1-1/R)}\right]}{c_{MA} - \dfrac{R}{1-R}\left[\dfrac{c_{MA}}{R} - c_{FV} + \dfrac{c_{FV} - c_{MA}}{R}e^{-\alpha_1(1-1/R)}\right]}$$

$$= \frac{c_{FV}[(1-R)/R] - c_{MA}/R + c_{FV} - (c_{FV}/R - c_{MA}/R)e^{-P/Q_2}}{c_{MA}[(1-R)/R] - c_{MA}/R + c_{FV} - (c_{FV}/R - c_{MA}/R)e^{-P/Q_2}}$$

$$= \frac{(c_{MA}/R - c_{FV}/R)(e^{-P/Q_2} - 1)}{(c_{MA} - c_{FV})[(1/R)e^{-P/Q_2} - 1]}$$

$$= \frac{1 - e^{-P/Q_2}}{R - e^{-P/Q_2}}.$$

(8.8-39)

For the maternal-pool model, we use Eq. 8.8-42, which can now be rewritten as

$$\frac{dc_F}{d\zeta} = \alpha_2(c_{MV} - c_F).$$

(F.7)

The solution of this equation with the boundary condition that $c_F = c_{FA}$ at $\zeta = 0$, is simply

$$\ln\frac{c_F - c_{MV}}{c_{FA} - c_{MV}} = -\alpha_2\zeta$$

or

$$\frac{c_F - c_{MV}}{c_{FA} - c_{MV}} = e^{-\alpha_2\zeta},$$

(F.8)

which readily yields Eq. 8.8-43.

For the fetal transport fraction, we note that $c_F = c_{FV}$ at $\zeta = 1$, so that

$$\frac{c_{FV} - c_{MV}}{c_{FA} - c_{MV}} = e^{-\alpha_2}$$

or

$$\frac{c_{FV} - c_{FA}}{c_{FA} - c_{MV}} = e^{-\alpha_2} - 1. \tag{F.9}$$

Also,

$$\frac{c_{MA} - c_{FA}}{c_{FV} - c_{FA}} = \frac{c_{MA} - c_{MV}}{c_{FV} - c_{FA}} - \frac{c_{FA} - c_{MV}}{c_{FV} - c_{FA}}$$

$$= R - \frac{1}{e^{-\alpha_2} - 1} \tag{F.10}$$

the latter being due to Eqs. 8.8-34 and F.9.

We note that Eq. F.10 is the reciprocal of the definition for the fetal transport fraction, T_F (Eq. 8.8-29). It follows that

$$T_F = \frac{1}{R - [1/(e^{-\alpha_2} - 1)]}$$

$$= \frac{e^{-\alpha_2} - 1}{Re^{-\alpha_2} - R - 1}.$$

Reversing the signs in both numerator and denominator, we finally have

$$T_F = \frac{1 - e^{-\alpha_2}}{1 + R(1 - e^{-\alpha_2})}. \tag{8.8-45}$$

APPENDIX G

Solutions for Oxygen Distribution in Hypoxia

The equation for the longitudinal oxygen variation in the capillary

$$\frac{d\langle c\rangle}{dz} = -\frac{2h}{R_1\langle v_z\rangle}(\langle c\rangle - c_{T_1}) \tag{10.3-3}$$

and that for the radial distribution in tissue (longitudinal diffusion negligible)

$$\frac{1}{r}\frac{d}{dr}\left(r\frac{dc_T}{dr}\right) = \gamma^2 c_T \tag{10.3-4}$$

are parallel, linked through B.C.1 of the following boundary conditions*:

B.C.1: at $r = R_1$, $\quad -\mathcal{D}_r\left(\dfrac{dc_T}{dr}\right)_{r=R_1} = h(\langle c\rangle - c_{T_1})$

B.C.2: at $r = R_2$, $\quad \dfrac{dc_T}{dr} = 0$

B.C.3: at $z = 0$, $\quad \langle c\rangle = c_0$

*Here B.C.'s 3 and 4 for c_T at the two ends of z disappear because c_T ceased to be a function of z. Note also that the derivatives of c_T with respect to r are no longer partial ones.

477

The general solution to Eq. 10.3-4 is (see Section 2.9)

$$c_T = C_1 I_0(\gamma r) + C_2 K_0(\gamma r) \tag{G.1}$$

differentiating which with respect to r, we obtain

$$\frac{dc_T}{dr} = \gamma [C_1 I_1(\gamma r) - C_2 K_1(\gamma r)] \tag{G.2}$$

Applying B.C.2, we have

$$C_1 = \frac{C_2 K_1(\gamma R_2)}{I_1(\gamma R_2)} \tag{G.3}$$

Using Eq. G.3 and realizing that at $r = R_1$, $c_T = c_{T_1}$, Eq. (G.1) becomes

$$c_T = c_{T_1} Z_0(\gamma r) \tag{G.4}$$

where

$$Z_0(\gamma r) = \frac{K_1(\gamma R_2) I_0(\gamma r) + I_1(\gamma R_2) K_0(\gamma r)}{K_1(\gamma R_2) I_0(\gamma R_1) + I_1(\gamma R_2) K_0(\gamma R_1)} \tag{G.5}$$

To eliminate c_{T_1}, we differentiate Eq. G.4 with respect to r to obtain

$$\frac{dc_T}{dr} = c_{T_1} \gamma \frac{K_1(\gamma R_2) I_1(\gamma r) - I_1(\gamma R_2) K_1(\gamma r)}{K_1(\gamma R_2) I_0(\gamma R_1) + I_1(\gamma R_2) K_0(\gamma R_1)} \tag{G.6}$$

and, for $r = R_1$

$$\left(\frac{dc_T}{dr} \right)_{r=R_1} = -c_{T_1} \gamma B \tag{G.7}$$

where

$$\frac{-K_1(\gamma R_2) I_1(\gamma R_1) + I_1(\gamma R_2) K_1(\gamma R_1)}{K_1(\gamma R_2) I_0(\gamma R_1) + I_1(\gamma R_2) K_0(\gamma R_1)} \tag{10.3-8}$$

Substituting Eq. G.7 into B.C.1, we have

$$h(\langle c \rangle - c_{T_1}) = c_{T_1} \gamma B \mathcal{D}_r$$

or

$$c_{T_1} = \frac{h \langle c \rangle}{h + \gamma B \mathcal{D}_r} = \frac{\beta \langle c \rangle}{1 + \beta} \tag{G.8}$$

where

$$\beta = \frac{h}{\gamma B \, \mathcal{D}_r} \tag{10.3-7}$$

Incorporation of Eq. G.8 into Eq.10.3-3 yields

$$\frac{d\langle c\rangle}{dz} = -\frac{2h}{R_1\langle v_z\rangle} \cdot \frac{\langle c\rangle}{1+\beta} \tag{G.9}$$

the solution of which, with B.C.3, is

$$\langle c\rangle = c_0 \exp\left[-\frac{2h}{R_1\langle v_z\rangle}\left(\frac{z}{1+\beta}\right)\right] \tag{10.3-5}$$

Substitution of Eq. G.8 into Eq. G.4 gives

$$c_T = \langle c\rangle \frac{\beta}{1+\beta} Z_0(\gamma r) \tag{10.3-6}$$

where $Z_0(\gamma r)$ is given by Eq. G.5.

APPENDIX H

Solutions for Hyperbaric Oxygen Distributions in Capillary and in Tissue, Equations 10.3-3 and 10.3-9

To make the operations look neater, we first convert the differential equations and boundary conditions into dimensionless forms:

$$\frac{d\theta}{d\zeta} = -H(\theta - \phi_1) \tag{10.3-3$'$}$$

$$\frac{1}{\xi}\frac{\partial}{\partial\xi}\left(\xi\frac{\partial\phi}{\partial\xi}\right) + \chi^2\frac{\partial^2\phi}{\partial\zeta^2} = \gamma_0^2 \tag{10.3-9$'$}$$

B.C.1: at $\quad \xi = 1,\quad$ (a) $\phi = \phi_1(\zeta)\quad$ (b) $-\dfrac{\partial\phi}{\partial\xi} = \dfrac{Sh}{2}(\theta - \phi_1)$

B.C.2: at $\quad \xi = n,\quad \dfrac{\partial\phi}{\partial\xi} = 0$

BC.3: at $\quad \zeta = 0,\quad$ (a) $\dfrac{\partial\phi}{\partial\zeta} = 0\quad$ (b) $\theta = 1$

B.C.4: at $\quad \zeta = 1,\quad \dfrac{\partial\phi}{\partial\zeta} = 0$

where

$$\theta(\zeta) = \frac{\langle c \rangle}{c_0}$$

$$\phi(\xi,\zeta) = \frac{c_T}{c_0}, \qquad \phi_1 = \frac{c_{T_1}}{c_0}$$

$$\xi = \frac{r}{R_1}, \qquad n = \frac{R_2}{R_1}$$

$$\zeta = \frac{z}{L}$$

$$\chi = \sqrt{\frac{\mathcal{D}_z}{\mathcal{D}_r}} \cdot \frac{R_1}{L}$$

$$\gamma_0 = R_1 \sqrt{\frac{k_0}{\mathcal{D}_r c_0}}$$

$$H = \frac{2hL}{R_1 \langle v_z \rangle}$$

$$\mathrm{Sh} = \frac{2hR_1}{\mathcal{D}_r} \quad \text{(Sherwood number)}$$

To solve Eq. 10.3-9' we use the technique used in Appendix D by viewing the finite annular cylinder as the orthogonal intersection between an infinite one and a slab. The ordinary differential equation corresponding to Eq. 10.3-9' for an infinite cylinder is

$$\frac{1}{\xi}\frac{d}{d\xi}\left(\xi\frac{d\phi_\infty}{d\xi}\right) = \gamma_0^2 \qquad (\text{H.1})$$

for which the solution with B.C.2 and $\phi_\infty(\xi=1)=\phi_1$ (since ϕ_∞ is but an extreme case of ϕ) is

$$\phi_\infty = \phi_1 - \frac{\gamma_0^2}{4}(1 + 2n^2\ln\xi - \xi^2) \qquad (\text{H.2})$$

Now defining

$$\phi_t(\xi,\zeta)=\phi_\infty(\xi)-\phi(\xi,\zeta)$$

$$=\phi_1-\frac{\gamma_0^2}{4}(1+2n^2\ln\xi-\xi^2)-\phi \tag{H.3}$$

but realizing that in the finite-cylinder case, ϕ_1 is a function of $\zeta,$* instead of being a constant. Inserting Eq. H.3 into Eq. 10.3-9' yields

$$\frac{1}{\xi}\frac{\partial}{\partial\xi}\left(\xi\frac{\partial\phi_t}{\partial\xi}\right)+\chi^2\frac{\partial^2\phi_t}{\partial\zeta^2}=\chi^2\frac{d^2\phi_1}{d\zeta^2} \tag{H.4}$$

Postulating

$$\phi_t(\xi,\zeta)=f(\xi)\cdot g(\zeta)$$

Eq. H.4 becomes

$$\frac{g}{\xi}\frac{d}{d\xi}\left(\xi\frac{df}{d\xi}\right)+\chi^2 f\frac{d^2g}{d\zeta^2}=\chi^2\frac{d^2\phi_1}{d\zeta^2} \tag{H.5}$$

The only way that this equation can be separated and satisfied is first, for

$$\frac{d^2\phi_1}{d\zeta^2}=0 \tag{H.6}$$

so that the remainder of Eq. H.5 can be rearranged to yield two individual functions of ξ and ζ separately:

$$\frac{1}{f\xi}\frac{d}{d\xi}\left(\xi\frac{df}{d\xi}\right)=-\frac{\chi^2}{g}\frac{d^2g}{d\zeta^2} \tag{H.7}$$

Again, for a pure function of ξ to be equal to that of ζ only at all times, both of them must be equal to the same constant, $-\alpha^2$, so that

$$\frac{1}{\xi}\frac{d}{d\xi}\left(\xi\frac{df}{d\xi}\right)+\alpha^2 f=0 \tag{H.8}$$

$$\frac{d^2g}{d\zeta^2}-\left(\frac{\alpha}{\chi}\right)^2 g=0 \tag{H.9}$$

*This may seem contradictory to Eq. H.2 wherein the same ϕ_1 cannot be a function of ζ (since ϕ_∞ is a function of ξ only). However, we may regard $\phi(\xi,\zeta)$ in Eq. H.3 as actually consisting of two parts, $\phi_2(\zeta)$ and $\phi_3(\xi,\zeta)$, and that $\phi_2(\zeta)$ is combined with the constant ϕ_1 to form a new $\phi_1(\zeta)$. The important point here is that at $\xi=1$, ϕ must still be a function of ζ [see B.C.1 (a)] and the present arrangement with $\phi_1=\phi_1(\zeta)$ insures that.

The general solutions for these two equations are, then (see Table 2.10-1),

$$f = C_1 J_0(\alpha\xi) + C_2 Y_0(\alpha\xi) \tag{H.10}$$

$$g = C_3 \cosh \frac{\alpha\zeta}{\chi} + C_4 \sinh \frac{\alpha\zeta}{\chi} \tag{H.11}$$

while that for Eq. H.6 is

$$\phi_1 = B_1\zeta + B_2 \tag{H.12}$$

so that the entire solution is

$$\phi = B_1\zeta + B_2 - \frac{\gamma_0^2}{4}(1 + 2n^2 \ln\xi - \xi^2) - [C_1 J_0(\alpha\xi) + C_2 Y_0(\alpha\xi)]$$

$$\times \left(C_3 \cosh \frac{\alpha\zeta}{\chi} + C_4 \sinh \frac{\alpha\zeta}{\chi} \right) \tag{H.13}$$

Remembering that at $\xi = 1, \phi = \phi_1 = B_1\zeta + B_2$, Eq. H.13 becomes

$$0 = [C_1 J_0(\alpha) + C_2 Y_0(\alpha)]\left(C_3 \cosh \frac{\alpha\zeta}{\chi} + C_4 \sinh \frac{\alpha\zeta}{\chi} \right) \tag{H.14}$$

so that

$$C_1 = \frac{-C_2 Y_0(\alpha)}{J_0(\alpha)} \tag{H.15}$$

Substituting Eq. H.15 back into Eq. H.13, we obtain

$$\phi = B_1\zeta + B_2 - \frac{\gamma_0^2}{4}(1 + 2n^2 \ln\xi - \xi^2) + [Y_0(\alpha)J_0(\alpha\xi) - J_0(\alpha)Y_0(\alpha\xi)]$$

$$\times \left(C_3' \cosh \frac{\alpha\zeta}{\chi} + C_4' \sinh \frac{\alpha\zeta}{\chi} \right) \tag{H.16}$$

where

$$C_3' = \frac{C_2 C_3}{J_0(\alpha)} \quad \text{and} \quad C_4' = \frac{C_2 C_4}{J_0(\alpha)}$$

If we use the shorthand notation

$$Z_k(\alpha\xi) = Y_0(\alpha)J_k(\alpha\xi) - J_0(\alpha)Y_k(\alpha\xi) \tag{H.17}$$

we may write

$$Z_0(\alpha\xi) = Y_0(\alpha)J_0(\alpha\xi) - J_0(\alpha)Y_0(\alpha\xi) \tag{H.18}$$

so that Eq. H.16 becomes

$$\phi = B_1 \zeta + B_2 - \frac{\gamma_0^2}{4}(1 + 2n^2 \ln \xi - \xi^2) + Z_0(\alpha\xi)\left(C_3' \cosh \frac{\alpha\zeta}{\chi} + C_4' \sinh \frac{\alpha\zeta}{\chi}\right)$$

$$(\text{H.19})$$

In order to use B.C.2, we first differentiate Eq. H.19 partially with respect to ξ:

$$\frac{\partial\phi}{\partial\xi} = -\frac{\gamma_0^2}{2}\left(\frac{n^2}{\xi} - \xi\right) - \alpha Z_1(\alpha\xi)\left(C_3' \cosh \frac{\alpha\zeta}{\chi} + C_4' \sinh \frac{\alpha\zeta}{\chi}\right) \quad (\text{H.20})$$

where according to the definition of Eq. H.17

$$Z_1(\alpha\xi) = Y_0(\alpha)J_1(\alpha\xi) - J_0(\alpha)Y_1(\alpha\xi) \quad (\text{H.21})$$

Now employing B.C.2, we have

$$0 = \alpha Z_1(n\alpha)\left(C_3' \cosh \frac{\alpha\zeta}{\chi} + C_4' \sinh \frac{\alpha\zeta}{\chi}\right) \quad (\text{H.22})$$

Since α cannot be zero and in order for Eq. H.22 to hold for any arbitrary value of ζ, $Z_1(n\alpha)$ must vanish. This means that α can only take on a series of values, $\alpha_i (i = 1, 2, \ldots, \infty)$, such that[*]

$$Z_1(\alpha_i n) = Y_0(\alpha_i)J_1(\alpha_i n) - J_0(\alpha_i)Y_1(\alpha_i n) = 0 \quad (\text{H.23})$$

Then, from the principle of superimposition (see Section 2.8 and Eq. 6.1-23) the general solution Eq. H.19 becomes

$$\phi = B_1 \zeta + B_2 - \frac{\gamma_0^2}{4}(1 + 2n^2 \ln \xi - \xi^2)$$

$$+ \sum_{i=1}^{\infty} Z_0(\alpha_i\xi)\left(C_{3_i}' \cosh \frac{\alpha_i\zeta}{\chi} + C_{4_i}' \sinh \frac{\alpha_i\zeta}{\chi}\right) \quad (\text{H.24})$$

Turning to the capillary side for a moment, in order to fit B.C.1(b), we first evaluate $\partial\phi/\partial\xi$ at $\xi = 1$ according to Eq. H.20 (with the superimposi-

---

[*]Note that the α_i here differs from that in Example 10.3-2 by a factor equal to R_1 (see Eq. 10.3-16).

tion shown in Eq. H.24):

$$\left(\frac{\partial \phi}{\partial \xi}\right)_{\xi=1} = -\frac{\gamma_0^2}{2}(n^2-1) - \sum_{i=1}^{\infty} \alpha_i Z_1(\alpha_i)\left(C_{3_i}' \cosh \frac{\alpha_i \zeta}{\chi} + C_{4_i}' \sinh \frac{\alpha_i \zeta}{\chi}\right)$$

(H.25)

Eliminating $(\theta - \phi_1)$ from Eq. 10.3-3' and B.C.1(b), we have

$$\left(\frac{\partial \phi}{\partial \xi}\right)_{\xi=1} = \frac{\text{Sh}}{2H} \cdot \frac{d\theta}{d\zeta}$$

(H.26)

Insertion of Eq. H.25 into Eq. H.26 leads to

$$\frac{d\theta}{d\zeta} = -\frac{2H}{\text{Sh}}\left\{\frac{\gamma_0^2}{2}(n^2-1) + \sum_{i=1}^{\infty} \alpha_i Z_1(\alpha_i)\left(C_{3_i}' \cosh \frac{\alpha_i \zeta}{\chi} + C_{4_i}' \sinh \frac{\alpha_i \zeta}{\chi}\right)\right\}$$

(H.27)

which, upon integration, yields

$$\theta = -\frac{2H}{\text{Sh}}\left\{\frac{\gamma_0^2}{2}(n^2-1)\zeta \right.$$

$$\left. + \chi \sum_{i=1}^{\infty} Z_1(\alpha_i)\left(C_{3_i}' \sinh \frac{\alpha_i \zeta}{\chi} + C_{4_i}' \cosh \frac{\alpha_i \zeta}{\chi}\right)\right\} + C_5$$

(H.28)

Using B.C.3(b), we obtain

$$1 = -\frac{2H}{\text{Sh}}\chi \sum_{i=1}^{\infty} C_{4_i}' Z_1(\alpha_i) + C_5$$

or

$$C_5 = 1 + \frac{2H\chi}{\text{Sh}} \sum_{i=1}^{\infty} C_{4_i}' Z_1(\alpha_i)$$

(H.29)

the substitution of which into Eq. H.28 gives

$$\theta = 1 - \frac{2H}{\text{Sh}}\left\{\frac{\gamma_0^2}{2}(n^2-1)\zeta \right.$$

$$\left. + \chi \sum_{i=1}^{\infty} Z_1(\alpha_i)\left[C_{3_i}' \sinh \frac{\alpha_i \zeta}{\chi} + C_{4_i}'\left(\cosh \frac{\alpha_i \zeta}{\chi} - 1\right)\right]\right\}$$

(H.30)

Using Eqs. H.25 and H.30 on B.C.1(b) gives us

$$\frac{\gamma_0^2}{2}(n^2-1)+\sum_{i=1}^{\infty}\alpha_i Z_1(\alpha_i)\left(C_{3_i}'\cosh\frac{\alpha_i\zeta}{\chi}+C_{4_i}'\sinh\frac{\alpha_i\zeta}{\chi}\right)$$

$$-\frac{Sh}{2}\left\{1-\frac{2H}{Sh}\left[\frac{\gamma_0^2}{2}(n^2-1)\zeta\right.\right.$$

$$+\chi\sum_{i=1}^{\infty}Z_1(\alpha_i)\left\{C_{3_i}'\sinh\frac{\alpha_i\zeta}{\chi}+C_{4_i}'\left(\cosh\frac{\alpha_i\zeta}{\chi}-1\right)\right\}\right]$$

$$\left.-B_1\zeta-B_2\right\}=0 \tag{H.31}$$

For this to hold for all values of ζ, the sum of coefficients of all terms with same power or function of ζ must vanish, that is, for constant term (ζ^0):

$$\frac{\gamma_0^2}{2}(n^2-1)=\frac{Sh}{2}\left\{1+\frac{2H\chi}{Sh}\sum_{i=1}^{\infty}Z_1(\alpha_i)C_{4_i}'-B_2\right\}$$

$$\therefore B_2=1-\frac{\gamma_0^2}{Sh}(n^2-1)+\frac{2H\chi}{Sh}\sum_{i=1}^{\infty}Z_1(\alpha_i)C_{4_i}' \tag{H.32}$$

for ζ:

$$0=-\frac{\gamma_0^2}{Sh}(n^2-1)H-B_1$$

$$\therefore B_1=-\frac{\gamma_0^2}{Sh}(n^2-1)H \tag{H.33}$$

Since summing the coefficients on $\sinh(\alpha_i\zeta/\chi)$ and $\cosh(\alpha_i\zeta/\chi)$ would yield no useful information, we turn to B.C.3(a) and B.C.4 by first differentiating Eq. H.24 partially with respect to ζ:

$$\frac{\partial\phi}{\partial\zeta}=B_1+\frac{1}{\chi}\sum_{i=1}^{\infty}\alpha_i Z_0(\alpha_i\xi)\left(C_{3_i}'\sinh\frac{\alpha_i\zeta}{\chi}+C_{4_i}'\cosh\frac{\alpha_i\zeta}{\chi}\right) \tag{H.34}$$

Using B.C.3(a), we have

$$0 = B_1 + \frac{1}{\chi} \sum_{i=1}^{\infty} \alpha_i Z_0(\alpha_i \xi) C'_{4_i} \tag{H.35}$$

or (with the help of Eq. H.33)

$$\sum_{i=1}^{\infty} \alpha_i Z_0(\alpha_i \xi) C'_{4_i} = \frac{\gamma_0^2}{Sh}(n^2 - 1)H\chi \tag{H.36}$$

Using B.C.4, we have

$$0 = B_1 + \frac{1}{\chi} \sum_{i=1}^{\infty} \alpha_i Z_0(\alpha_i \xi)\left(C'_{3_i} \sinh \frac{\alpha_i}{\chi} + C'_{4_i} \cosh \frac{\alpha_i}{\chi}\right) \tag{H.37}$$

Comparing Eqs. H.35 and H.37, we can conclude that

$$C'_{4_i} = C'_{3_i} \sinh \frac{\alpha_i}{\chi} + C'_{4_i} \cosh \frac{\alpha_i}{\chi}$$

or

$$C'_{3_i} = C'_{4_i} \frac{1 - \cosh(\alpha_i/\chi)}{\sinh(\alpha_i/\chi)} \tag{H.38}$$

So now the matter becomes that of evaluating C'_{4_i}.

To do so we must use the orthogonality relationship on $Z_0(\alpha_i \xi)$ by multiplying both sides of Eq. H.36 by $\xi Z_0(\alpha_j \xi) d\xi$ and integrate from 1 to n. Omitting the details, we have

$$C'_{4_i} = \frac{-2\gamma_0^2 H\chi(n^2 - 1)[Z_1(\alpha_i)]^2 / Sh}{\alpha_i^2 \left\{[nZ_0(\alpha_i n)]^2 - [Z_1(\alpha_i)]^2\right\} Z_1(\alpha_i)} \tag{H.39}$$

Substituting Eqs. H.32, H.33, H.38, and H.39 into Eqs. H.24 and H.30 gives the dimensionless forms of the desired results Eqs. 10.3-10 and 10.3-11.

We can thus see that the procedure followed in this case is basically the same as that in Appendix G (the hypoxia case) except that in the present case the partial differential equation (Eq. 10.3-9′) makes the details more complex.

APPENDIX I

Buckingham's Π Theorem and Its Application in Establishing General Functional Relations of Interphase Transport

The Buckingham's Π(Pi) theorem states that "the functional relationship among q quantities, whose units may be given in terms of u fundamental units, may be written as a function of $q - u$ dimensionless groups (the Π's)." These groups are automatically formed when we impose a "balance" on the combined units or dimensions of the various quantities involved. However, it should be emphasized that these quantities must be correctly chosen in the first place, either by intuition or by other reliable methods.

Let us take the heat-transfer coefficient (Section 7.4) as an example. By examining the equation of energy, we might conclude that this coefficient should be a function of density, heat capacity, viscosity, and thermal conductivity of the fluid; diameter and length of the tube; and velocity of the flow, or

$$h = h\left(\rho, \hat{C}_p, \mu, k, D, L, V\right). \tag{I.1}$$

There are four fundamental units or dimensions involved:

$$T = \text{temperature (e.g., °K)}$$

$$\theta = \text{time (e.g., seconds)}$$

$$l = \text{length (e.g., centimeters)}$$

$$M = \text{mass (e.g., grams).}$$

One might wonder what happens to H, the energy unit (e.g., the calorie). Actually, it is a derived, not fundamental, unit. In mechanical form, energy is force times distance and thus has the dimension $Ml^2\theta^{-2}$. In thermal form, it is related to temperature, for example, one calorie is defined as the thermal energy required to raise the temperature of 1 g of water by 1°C (from 14.5° to 15.5°C, to be more precise). Thus, as long as we have specified T, θ, l, and M, we can no longer treat H as an independent unit. This we will prove shortly by pretending that we do not know any better and include H as one of the fundamental units as well. We will see that it does not contribute anything new to the analysis.

Whatever the specific form of Eq. I.1, and however the various parameters are arranged, and no matter to what powers they are to be raised, we do know that their combined, or net, dimensions on the right-hand side must be equal to those on the left-hand side, or

$$h[=]\rho^a \hat{C}_p^b \mu^c k^d D^e L^f V^g, \tag{I.2}$$

where $a, b, \ldots, g$ are the powers to which these parameters are raised. In fact, the entire idea is to let these powers "adjust themselves" to insure that the dimensions will be balanced. The various quantities have the following cgs units and fundamental dimensions:

$$h[=] \quad \frac{\text{cal}}{\text{cm}^2 - \text{sec} - {}^\circ\text{K}} \quad [=]Hl^{-2}\theta^{-1}T^{-1},$$

$$\rho[=] \quad \frac{\text{g}}{\text{cm}^3} \quad [=]Ml^{-3},$$

$$\hat{C}_p[=] \quad \frac{\text{cal}}{\text{g} - {}^\circ\text{K}} \quad [=]HM^{-1}T^{-1},$$

$$\mu[=] \quad \frac{\text{g}}{\text{cm} - \text{sec}} \quad [=]Ml^{-1}\theta^{-1},$$

$$k[=] \quad \frac{\text{cal}}{\text{cm} - \text{sec} - {}^\circ\text{K}} \quad [=]Hl^{-1}\theta^{-1}T^{-1},$$

$$D[=] \quad \text{cm} \quad [=]l,$$

$$L[=] \quad \text{cm} \quad [=]l,$$

$$V[=] \quad \frac{\text{cm}}{\text{sec}} \quad [=]l\theta^{-1}.$$

Substituting these dimensions into Eq. I.2 and collecting the same fundamental ones together, we have

$$Hl^{-2}\theta^{-1}T^{-1} = H^{b+d}M^{a-b+c}l^{-3a-c-d+e+f+g}\theta^{-c-d-g}T^{-b-d}. \quad (I.3)$$

Since the units on both sides must be balanced, we can equate the powers of the same unit on both sides to obtain

$$\text{for } H: \qquad 1 = b + d \qquad\qquad (I.4)$$

$$\text{for } M: \qquad 0 = a - b + c \qquad\qquad (I.5)$$

$$\text{for } l: \qquad -2 = -3a - c - d + e + f + g \qquad (I.6)$$

$$\text{for } \theta: \qquad -1 = -c - d - g \qquad\qquad (I.7)$$

$$\text{for } T: \qquad -1 = -b - d. \qquad\qquad (I.8)$$

We can readily see that Eqs. I.4 and I.8 are redundant, supporting our earlier statement the H and T are actually interdependent units. Thus, we have only four truly independent simultaneous equations but seven unknowns $(a,b,\ldots,g)$. From elementary algebra we know that we are short by three equations and everything will eventually be expressed in terms of these three "residual" unknowns. This is actually the basis of the Π-theorem, because the number of quantities, q, is actually the number of unknowns here plus 1 (corresponding to the known exponent on h, unity) and the number of fundamental units, u, the same as the number of independent simultaneous equations. Thus, in our present case, the number of dimensionless groups is

$$q - u = (7+1) - 4 = 4$$

or one can be expressed as a function of the other three.

In order to express $c,d,e,$ and g in terms of a,b and $f,$* we take the following steps:

$$\text{From Eq. I.4:} \qquad\qquad d = 1 - b \qquad (I.9)$$

$$\text{From Eq. I.5:} \qquad\qquad c = b - a \qquad (I.10)$$

$$\text{Eq. I.9 and Eq. I.10 into Eq. I.7:} \qquad g = a \qquad (I.11)$$

$$\text{Eq. I.9 through Eq. I.11 into Eq. I.6:} \quad e = a - f - 1. \quad (I.12)$$

*With other combinations, the results would be the same.

Substituting Eqs. I.9 through I.12 into Eq. I.2, we have

$$h[\; = \;] \rho^a \hat{C}_p^b \mu^{b-a} k^{1-b} D^{a-f-1} L^f V^a, \tag{I.13}$$

which, upon collection of like-power terms, yields the following groupings:

$$\frac{hD}{k}, \left(\frac{DV\rho}{\mu} \right)^a, \left(\frac{\hat{C}_p \mu}{k} \right)^b, \left(\frac{L}{D} \right)^f.$$

Or, in general, we can write

$$\frac{hD}{k} = f\left(\frac{DV\rho}{\mu}, \frac{\hat{C}_p \mu}{k}, \frac{L}{D} \right), \tag{I.14}$$

which agrees completely with Eq. 7.4-11.

Author Index

"Upright" page numbers denote where authors are cited,- *italicized* ones denote where literature is listed at the end of a chapter. Parenthesis indicate editorship while brackets denote translation.

Abbrecht, P. H., 367, *377*
Abramowitz, M., 26, *116*
Adair, G. S., 418, *451*
Aiba, S., 18, *116*
Allen, R. D., 268, *283*
Ames, W. F., 87, *117*
Anczurowski, E., 397, *414*
Aris, R., 102, *117*
Atabek, H. B., 203, *282*
Aukland, K., 332, 333, *376*

Babb, A. L., 360, 362, 366, *377*
Battaglia, F. C., 343, *376*
Bellavia, J. V., 372, *377*
Benis, A. M., 384, 386, *413*
Bergmann, F., 182, *183*
Berliner, R. W., 332, 333, *376*
Bert, J. L., 445, *452*
Besarab, A., 369, 371, 372, *377*
Bird, R. B., 3, *15*, 65, 79, *116*, 129, *132*, 153, *166*, 173, *183*, 245, *282*, 285, 300, 301, 303-306, 314, *318*, 321, 326, 329, 345, *376*, 438, *451*
Birkhoff, G., 79, *116*
Bischoff, K. B., 6, 7, 15, (*318, 377*)
Bixler, H. J., 369, 371, 372, *377*
Bloch, I., 421, *451*
Blum, J. J., 422, *451*
Bower, B. F., 332, 333, *376*
Breckenridge, J. R., 317, *319*
Brenner, H., 378, 397, *413*
Briscoe, W. A., 5, *15*

Britten, A., 388, *414*
Brown, R. G., 6, *15*
Bruley, D. F., 254, *282*, 422, *451*
Bruns, P. D., 343, *376*
Buckles, R. G., 262, *282*
Buettner, K., 312, 313, *319*
Bugliarello, G., 14, *15*, 422, *451*
Burton, A. C., 391, 392, 396, *414*

Carey, F. G., 318, *319*
Carlsen, E., 5, *15*
Carslaw, H. S., 247
Casson, N., 128, *132*, 379, *413*
Chapman, S., 3, *15*, 129, *132*
Chilton, C. H., (288, 289, *318*, 328, 338, *376*)
Chilton, T. H., 307, *318*
Clifford, J., 315, 316, *319*
Cokelet, G. R., 388, *414*
Colburn, A. P., 299, 307, *318*, 358, *377*
Colin, J., 309, 311-314, 317, *319*
Colquhoun, D., 254, 259, *282*
Colton, C. K., 368, *377*
Comroe, J. H., Jr., 5, *15*
Copley, A. L., (378, *413, 414*)
Cowling, T. G., 3, *15*, 129, *132*
Curtiss, C. F., 3, *15*, 129, *132*, 153, *166*

Darwish, M. A., 5, *15*, 217, 234, 235, 239, 240, *282*
Deakin, M. A. B., 398, *414*
Dedrick, R. L., (*318, 377*)

493

Subject Index